Drug-Loaded Colloidal Systems in Nanomedicine II

Drug-Loaded Colloidal Systems in Nanomedicine II

Editor

Leonard Atanase

MDPI • Basel • Beijing • Wuhan • Barcelona • Belgrade • Manchester • Tokyo • Cluj • Tianjin

Editor
Leonard Atanase
Biomaterials, Faculty of
Medical Dentistry
Apollonia University
Iasi
Romania

Editorial Office
MDPI
St. Alban-Anlage 66
4052 Basel, Switzerland

This is a reprint of articles from the Special Issue published online in the open access journal *Polymers* (ISSN 2073-4360) (available at: www.mdpi.com/journal/polymers/special_issues/drug_loaded_colloid_syst_nanomed_II).

For citation purposes, cite each article independently as indicated on the article page online and as indicated below:

LastName, A.A.; LastName, B.B.; LastName, C.C. Article Title. *Journal Name* **Year**, *Volume Number*, Page Range.

ISBN 978-3-0365-5360-3 (Hbk)
ISBN 978-3-0365-5359-7 (PDF)

© 2022 by the authors. Articles in this book are Open Access and distributed under the Creative Commons Attribution (CC BY) license, which allows users to download, copy and build upon published articles, as long as the author and publisher are properly credited, which ensures maximum dissemination and a wider impact of our publications.

The book as a whole is distributed by MDPI under the terms and conditions of the Creative Commons license CC BY-NC-ND.

Contents

About the Editor vii

Preface to "Drug-Loaded Colloidal Systems in Nanomedicine II" ix

Corina Popovici, Marcel Popa, Valeriu Sunel, Leonard Ionut Atanase and Daniela Luminita Ichim
Drug Delivery Systems Based on Pluronic Micelles with Antimicrobial Activity
Reprinted from: *Polymers* **2022**, *14*, 3007, doi:10.3390/polym14153007 1

Min Lu, Wei Wei, Wenhao Xu, Nikolay E. Polyakov, Alexandr V. Dushkin and Weike Su
Preparation of DNC Solid Dispersion by a Mechanochemical Method with Glycyrrhizic Acid and Polyvinylpyrrolidone to Enhance Bioavailability and Activity
Reprinted from: *Polymers* **2022**, *14*, 2037, doi:10.3390/polym14102037 17

Ziyad Binkhathlan, Abdullah H. Alomrani, Olsi Hoxha, Raisuddin Ali, Mohd Abul Kalam and Aws Alshamsan
Development and Characterization of PEGylated Fatty Acid-*Block*-Poly(-caprolactone) Novel Block Copolymers and Their Self-Assembled Nanostructures for Ocular Delivery of Cyclosporine A
Reprinted from: *Polymers* **2022**, *14*, 1635, doi:10.3390/polym14091635 39

Kheira Zanoune Dellali, Mohammed Dellali, Delia Mihaela Rață, Anca Niculina Cadinoiu, Leonard Ionut Atanase and Marcel Popa et al.
Assessment of Physicochemical and In Vivo Biological Properties of Polymeric Nanocapsules Based on Chitosan and Poly(*N*-vinyl pyrrolidone-*alt*-itaconic anhydride)
Reprinted from: *Polymers* **2022**, *14*, 1811, doi:10.3390/polym14091811 57

Razan Haddad, Nasr Alrabadi, Bashar Altaani and Tonglei Li
Paclitaxel Drug Delivery Systems: Focus on Nanocrystals' Surface Modifications
Reprinted from: *Polymers* **2022**, *14*, 658, doi:10.3390/polym14040658 85

Bijuli Rabha, Kaushik Kumar Bharadwaj, Siddhartha Pati, Bhabesh Kumar Choudhury, Tanmay Sarkar and Zulhisyam Abdul Kari et al.
Development of Polymer-Based Nanoformulations for Glioblastoma Brain Cancer Therapy and Diagnosis: An Update
Reprinted from: *Polymers* **2021**, *13*, 4114, doi:10.3390/polym13234114 111

Rafaela Figueiredo Rodrigues, Juliana Barbosa Nunes, Sandra Barbosa Neder Agostini, Paloma Freitas dos Santos, Juliana Cancino-Bernardi and Rodrigo Vicentino Placido et al.
Preclinical Evaluation of Polymeric Nanocomposite Containing Pregabalin for Sustained Release as Potential Therapy for Neuropathic Pain
Reprinted from: *Polymers* **2021**, *13*, 3837, doi:10.3390/polym13213837 133

Raissa Lohanna Gomes Quintino Corrêa, Renan dos Santos, Lindomar José Calumby Albuquerque, Gabriel Lima Barros de Araujo, Charlotte Jennifer Chante Edwards-Gayle and Fabio Furlan Ferreira et al.
Ciprofibrate-Loaded Nanoparticles Prepared by Nanoprecipitation: Synthesis, Characterization, and Drug Release
Reprinted from: *Polymers* **2021**, *13*, 3158, doi:10.3390/polym13183158 153

Iulia Babutan, Alexandra-Delia Lucaci and Ioan Botiz
Antimicrobial Polymeric Structures Assembled on Surfaces
Reprinted from: *Polymers* **2021**, *13*, 1552, doi:10.3390/polym13101552 169

Ana María Martínez-Relimpio, Marta Benito, Elena Pérez-Izquierdo, César Teijón, Rosa María Olmo and María Dolores Blanco
Paclitaxel-Loaded Folate-Targeted Albumin-Alginate Nanoparticles Crosslinked with Ethylenediamine. Synthesis and In Vitro Characterization
Reprinted from: *Polymers* **2021**, *13*, 2083, doi:10.3390/polym13132083 . . . . **205**

About the Editor

Leonard Atanase

Prof. Dr. Leonard-Ionut Atanase studied macromolecular chemistry at the "Gh. Asachi"Technical University in Iasi and achieved his PhD at University de Haute Alsace, Mulhouse, France, under the supervision of Prof. G. Riess. At the beginning of 2016, he returned in Romania, to the "Apollonia"University of Iasi; the experience accumulated in the almost 10 years spent in France enabled him to earn the title of Full Professor in 2017. Regarding his scientific activity, Prof. Atanase is noted through numerous participation in international conferences, with 2 books and 10 book chapters published in international publishers, and 50 articles published in ISI journals. He is currently the Dean of the Faculty of Dental Medicine at the "Apollonia"University in Iasi and the coordinator of the Biomaterials Laboratory.

Preface to "Drug-Loaded Colloidal Systems in Nanomedicine II"

This reprint represents a platform not only for dissemination, but also for debating new results in the field of nanomedicine, which has a very large practical potential. The chapters of this reprint combine fundamental research with practical applications of colloidal systems in the field of nanomedicine. Drug-loaded colloidal systems are particularly of practical interest, such as micelles, polymersomes, nanogels, liposomes, nanocapsules, polymeric, inorganic nanoparticles, etc., with diameters in the range of 10 nm to 400 nm. The authors explained the correlations between the molecular characteristics of the starting polymers and colloidal characteristics, as well as the encapsulation efficiency and/or the drug release kinetics. In vitro biological analyses completed the physicochemical characterization of the polymeric systems. The primary audience is researchers (i.e., esearch scholars and scientists) pursuing research on drug delivery systems, chemistry, and biology. The secondary audience is represented by pharmaceutical manufacturers, healthcare personnel, and professors in Master's programs in chemistry/biomedical sciences.

The ten chapters of this reprint are proposed by authors from different countries: Romania, India, China, Russia, Saudi Arabia, Brazil, Spain, USA, and Jordan.

Leonard Atanase
Editor

Article

Drug Delivery Systems Based on Pluronic Micelles with Antimicrobial Activity

Corina Popovici [1], Marcel Popa [1,2,*], Valeriu Sunel [3], Leonard Ionut Atanase [2,4,*] and Daniela Luminita Ichim [4]

[1] "Cristofor Simionescu" Faculty of Chemical Engineering and Environmental Protection, "Gheorghe Asachi" Technical University of Iasi, 700050 Iasi, Romania; corinapopovici77@yahoo.com
[2] Academy of Romanian Scientists, 050045 Bucharest, Romania
[3] Faculty of Chemistry, "Alexandru Ioan Cuza" University of Iasi, 700506 Iasi, Romania; vsunel@uaic.ro
[4] Faculty of Medical Dentistry, "Apollonia" University of Iasi, 700511 Iasi, Romania; danielaluminitaichim@yahoo.com
* Correspondence: marpopa2001@yahoo.fr (M.P.); leonard.atanase@yahoo.com (L.I.A.)

Abstract: Bacterial oral diseases are chronic, and, therefore, require appropriate treatment, which involves various forms of administration and dosing of the drug. However, multimicrobial resistance is an increasing issue, which affects the global health system. In the present study, a commercial amphiphilic copolymer, Pluronic F127, was used for the encapsulation of 1-(5′-nitrobenzimidazole-2′-yl-sulphonyl-acetyl)-4-aryl-thiosemicarbazide, which is an original active pharmaceutical ingredient (API) previously synthesized and characterized by our group, at different copolymer/API weight ratios. The obtained micellar systems, with sizes around 20 nm, were stable during 30 days of storage at 4 °C, without a major increase of the Z-average sizes. As expected, the drug encapsulation and loading efficiencies varied with the copolymer/API ratio, the highest values of 84.8 and 11.1%, respectively being determined for the F127/API = 10/1 ratio. Moreover, in vitro biological tests have demonstrated that the obtained polymeric micelles (PMs) are both hemocompatible and cytocompatible. Furthermore, enhanced inhibition zones of 36 and 20 mm were observed for the sample F127/API = 2/1 against *S. aureus* and *E. coli*, respectively. Based on these encouraging results, it can be admitted that these micellar systems can be an efficient alternative for the treatment of bacterial oral diseases, being suitable either by injection or by a topical administration.

Keywords: Pluronic; F127; polymeric micelles; drug delivery system; active pharmaceutical ingredient; antimicrobial activity

1. Introduction

Polymeric micelles (PMs) represent a distinct class of carriers formed from amphiphilic block or graft copolymers [1,2]. The driving force of the self-assembly of these amphiphilic copolymers in aqueous media, at the nanometric scale, is the thermodynamic incompatibility between the different sequences (hydrophilic/hydrophobic) [3]. During the last two decades, PMs have attracted a great interest, in particular owing to their strong ability to solubilize water-insoluble active principles and thus to be effective drug nanocarriers [4–7]. Like surfactant micelles, polymer micelles have a core–corona structure capable of encapsulation various types of molecules. In addition, PMs show greater resistance to the effects of dilution than surfactant micelles due to their generally very low CMC (in the order of mg/L). Moreover, it is possible to optimize their colloidal characteristics by modulating the nature of the hydrophobic or hydrophilic copolymers sequences. Thus, the structure of the hydrophobic core can be adjusted in order to optimize its affinity for the incorporated active ingredient. Finally, due to their small size and their hydrophilic surface, PMs are poorly recognized by the immune system, which would make it possible to prolong the plasma half-life of their cargo. PMs therefore possess unique characteristics among all drug delivery systems [8–10].

Among the amphiphilic copolymers, one of the most studied systems are non-ionic poly(ethylene oxide)-poly(propylene oxide)-poly(ethylene oxide) (PEO-PPO-PEO) triblock copolymers, also called poloxamers or Pluronics®, which is a registered trademark of BASF [11]. In aqueous solution, these copolymers self-assemble to form spherical or elongated micelles, with a PPO core and PEO corona, but other structures may exist depending on the lengths of each block. F127, having a linear structure PEO_{97}-PPO_{69}-PEO_{97} (indexes are polymerization degree values), is one of the most suitable Pluronics for preparation of drug delivery systems as it has a hydrophilic/lipophilic balance (HLB) of 22, a molecular weight of 12,600 g/mol, low critical micellar concentration (CMC = 0.0031%, w/w), excellent biocompatibility and satisfactory safety [12,13]. In addition, owing to their small CMC, F127 micelles are stable to dilution in the presence of biological fluids.

Different types of active principles can be solubilized within the PPO hydrophobic core [14] or can be conjugated to the PEO corona [15]. The loading of hydrophobic molecules in this type of PMs, with sizes usually smaller than 100 nm, can increase their solubility, stability and improve their pharmacokinetics as well as biodistribution [14].

One of the most important advantage of these copolymers is that their hydrophilic PEO corona prevents aggregation, protein adsorption and especially recognition by the reticuloendothelial system (RES), leading thus to an increased blood circulation time. Due to these important advantages, the first micellar formulation approved for clinical trials was based on doxorubicin-loaded Pluronic F127 and L61 [16].

A growing problem of modern society is related to the uncontrolled use of antibiotics which have as a consequence increased multimicrobial resistance, affecting thus the global health system. An extremely sophisticated microbial environment is located in the oral cavity, in which can be found around 700 strains that can contribute to the rapid development of oral diseases [17]. Even if the synthesis of new drugs with antimicrobial activity has recorded remarkable progress in the last few years, their low solubility and bioavailability, high toxicity and inadequate release profile limits their clinical utilization [18]. In order to overcome these drawbacks, different types of PMs were studied for the preparation of formulation with antimicrobial activity [19–23]. Among those, we are interested in the use of Pluronics as nanocarriers for active principles with antimicrobial activity.

Hashemi et al. [24] encapsulated antimicrobial peptides in P127 micelles and observed, by ex vivo tests, that the obtained micellar formulation reduced the populations of fungal pathogens in tracheal and lung tissue.

In another study, Purro et at. [25] obtained a conjugate between Pluronic P127 and the siderophore desferrioxamine B (DFO) complexed to Ga, DFO:Ga^{III} (DG). These authors demonstrated that their conjugate was efficient against strains of *P. aeruginosa*.

Recently, it was proved that the inclusion of photosensitizers, based on cationic porphyrins, into the Pluronic polymer micelles (F-127) significantly increased the efficiency of the antimicrobial photodynamic inactivation [26].

Based on the evident advantages of the Pluronic micelles, we propose in this study the preparation and characterization of a micellar system with enhanced antimicrobial activity due to an encapsulated original active pharmaceutical ingredient (API), 1-(5′-nitrobenzimidazole-2′-yl-sulphonyl-acetyl)-4-aryl-thiosemicarbazide, which was previously synthesized [27]. The chemical structure of this active principle is provided in Figure 1.

The obtained micellar system was both physicochemical and biological characterized. Moreover, the antimicrobial activity was assessed on different Gram-positive and Gram-negative bacteria.

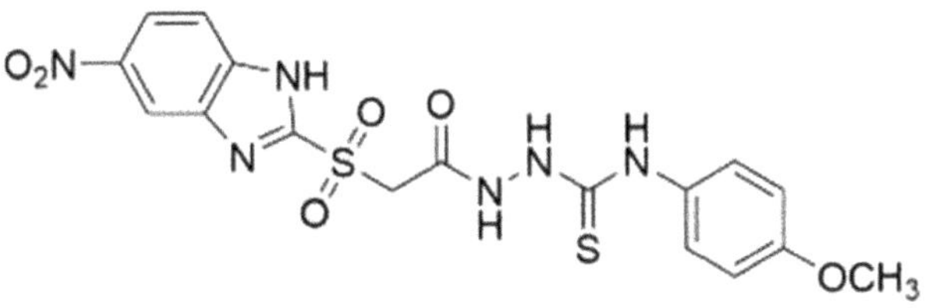

Figure 1. Chemical structure of the active pharmaceutical ingredient (API): 1-(5′-nitrobenzimidazole-2′-yl-sulphonyl-acetyl)-4-aryl-thiosemicarbazide.

2. Experimental Section

2.1. Materials

The active pharmaceutical ingredient 1-(5′-nitrobenzimidazole-2′-yl-sulphonyl-acetyl)-4-aryl-thiosemicarbazide was synthesized and fully characterized in a previous publication [27]. Pluronic F127 was purchased from Sigma Aldrich (Steinheim, Germany). All the solvents were used as received and without further purification. Human dermal fibroblasts cell line (HDFa) and the necessary supplies (antibiotic cocktail: penicillin and streptomycin, non-essential amino acids, trypsin solution and fetal bovine serum (FBS)) for in vitro cytotoxicity assay were purchased from Thermo Fisher Scientific (Waltham, MA, USA). Freeze-dried stains (*Escherichia coli*-ATCC^R 11775^{TM}, *Pseudomonas aeruginosa*-ATCC^R 10154^{TM}, *Klebsiella pneumoniae*-ATCC^R BAA-1705^{TM}, *Staphylococcus aureus*-ATCC^R 25 923^{TM} and *Porphyromonas gingivalis*-ATCC^R 33277^{TM}) were purchased from ATCC (Manassas, VA, USA). Chapman agar (mannitol salt agar) was purchased from Oxoid (Hampshire, UK) and MacConkey agar from G&M Procter Ltd. (Perth, UK).

2.2. Methods

2.2.1. Micelle's Preparation Procedure

Preparation of PMs was carried out by a dialysis method starting from a common solvent. In a typical procedure, 100 mg of Pluronic F127 was added in 10 mL of dimethylsulfoxide (DMSO) solution and stirred at room temperature until complete dissolution of copolymer. Afterwards, the solution was dialyzed against 1 L of ultrapure water using cellulose dialysis membranes (molecular weight cut off: 12 kDa, manufacturer, Sigma Aldrich, Steinheim, Germany). The water was changed eight times during 24 h of dialysis. The dry powder was collected after the freeze drying of the micellar solutions and then was stored at $-4°$ before further use.

A similar procedure was used for the preparation of API-loaded PMs with the difference that at the solution of the block copolymer in DMSO were added different amounts of API in order to have three weight ratios between the copolymer and the active principle, such as: 10:1; 5:1 and 2:1.

2.2.2. Physicochemical Characterization Methods

The micellar sizes were investigated by Dynamic Light Scattering (DLS) measurements in phosphate buffer solution (PBS; pH = 7.4), this medium being similar to the in vivo medium. DLS were carried out on a Malvern Zetasizer Pro (Malvern Panalytical, Worcestershire, UK), using the NIBS (Non-Invasive BackScattering) technology, equipped with a 4 mW He–Ne laser operating at a wavelength of 532 nm and at a scattering angle of 173°. The software package of the instrument calculates, by using the Stokes–Einstein equation, the hydrodynamic diameter (volume average) Dv, the Z-average diameter, which is an intensity-weighted size average and the polydispersity index (PDI) of the sample. In order to determine the mean diameter of the particles, the data were collected in automatic mode, typically requiring a measurement duration of 70 s. For each experiment, five consecutive measurements were carried out. The stability of the micellar solutions, stored at 4 °C, was assessed as a function of time for 30 days. The zeta potential values of PMs were determined by electrophoresis in phosphate buffer solution (PBS; pH = 7.4) using the same instrument.

In order to determine the encapsulation efficiency of API, calibration curves were constructed in DMSO, using different concentrations of API, and their absorbance values were recorded on a Nanodrop spectrophotometer (Thermo Scientific, Waltham, MA, USA) at the wavelength of 480 nm. A known amount of the API-loaded micelles, as powder, was dissolved in 1 mL of DMSO in order to completely destroy the micelles and to release the loaded active principle. The amount of API from the micelles was spectrophotometrically quantified, based on the calibration curve, using a UV spectrometer (Nanodrop One, Thermo Scientific, Waltham, MA, USA). Drug encapsulation efficiency (DEE) and drug loading efficiency (DLE) were calculated using Equations (1) and (2), respectively:

$$\text{DEE}(\%) = \frac{amount\ of\ drug\ in\ micelles}{amount\ of\ added\ drug} \times 100 \quad (1)$$

$$\text{DLE}(\%) = \frac{amount\ of\ drug\ in\ micelles}{amount\ of\ added\ polymer\ and\ drug} \times 100 \quad (2)$$

Three determinations were performed for each sample and the errors were ±0.3%.

2.2.3. Biological Characterization Methods

The hemolytic potential of the obtained PMs was evaluated using a spectrophotometric method adapted from Rata et al. [28]. These tests were started after obtaining the institutional ethical authorization (28/02.06.2022-The scientific research ethics committee) and the appropriate informed consent. The blood from the healthy non-smoking human volunteer was collected in vacutainer tubes and treated with PMs. In total, 5 mL anti-coagulated blood was centrifuged at 2000 rpm (RCF = 381× g) for 5 min and washed with normal saline solution several times to completely remove the plasma and obtain erythrocytes. After purification, erythrocytes were re-suspended in 25 mL normal saline solution. PMs saline solution with different concentrations (0.5 mL) was added to 0.5 mL of erythrocytes suspension (final concentrations were 10, 50, 100, and 200 mg PMs/mL erythrocytes suspension). Positive (100% lysis) and negative (0% lysis) control samples were prepared by adding equal volumes (0.5 mL) of Triton X-100 and a standard saline solution. The samples were incubated at 37 °C for 180 min. Once every 30 min the samples were gently shaken to re-suspend erythrocytes and PMs. After the incubation time, the samples were centrifuged at 2000 rpm (RCF = 381× g) for 5 min and 100 µL of supernatant was incubated for 30 min at room temperature to allow hemoglobin oxidation. Oxyhemoglobin absorbance in supernatants was measured at 540 nm using a Nanodrop One UV-Vis Spectrophotometer from Thermo Fischer Scientific, Waltham, MA, USA. All samples were analyzed in triplicate. The hemolytic percentage was calculated using Equation (3):

$$\text{Haemolysis}\ (\%) = \frac{(A_S - A_{NC})}{(A_{PC} - A_{NC})} \times 100 \quad (3)$$

where, A_s is the absorbance of the sample; A_{NC} and A_{PC} are the absorbance values of the negative and positive control, respectively.

The MTT method was applied to assess the in vitro cytotoxicity of free and API-loaded PMs by using adherent adult human fibroblast cells of dermal origin (HDFa). After thawing the fibroblast cells in the thermostatic bath at 37 °C (Digital thermostatic baths, DIGIBATH 2–BAD\2RAYPA, Spain) the cells (HDFa) were cultured in complete medium grow: DMEM (Dulbecco's Modified Eagle Medium) supplemented with 10% fetal bovine serum, 1% antibiotics and 1% non-essential amino acids, at 37 °C and a humidified atmosphere of 5% CO_2 (MCO-5AC CO_2 Incubator, Panasonic Healthcare Co., Ltd., Sakata Oizumi-Machi Ora-Gun Gunma, Japan). Cells were allowed to proliferate in culture flasks (NuncTM EasYFlask 25 cm^2 TM, ThermoFisher Scientific, Roskilde, Denmark) to reach 80% confluence, then were trypsinized with 0.05% trypsin solution at 37 °C, followed by the addition of complete medium to neutralize the trypsin, were centrifuged (Rotofix-32A, Hettich, Andreas GmbH Hettich & Co.KG, Tuttlingen, Germany) and re-suspended in complete medium DMEM. For performing the in vitro cytotoxicity assay, the reagents were purchased from Thermo Fisher Scientific. After centrifugation and re-suspension in fresh medium, the viable cells were plated in on flat-bottom 96-well plates (TPP Techno Plastic Products AG, Switzerland) and incubated for 24 h. After 24 h of incubation the culture grow medium was replaced with fresh medium, and the test material were put in direct contact with the fibroblast cells at three types of concentrations: 10, 50 and 100 µg/mL. Prior to performing the cytotoxicity test, the materials were sterilized with UV-VIS radiation for 3 min. Microscopic analysis was performed on the inverted optical microscope (CKX41, Olympus, Tokyo, Japan) with a built-in camera and QuickPHOTO camera 3.0 software. Cell viability determination was performed at 24 and 48 h after incubation of the cells treated by the quantitative colorimetric assay with tetrazolium salt (3-(4,5-dimethylthiazol-2-yl)-2,5-diphenyltetrazolium bromide (MTT)), (Merck Millipore, Darmstadt, Germany). After 24 h incubation, 100 µL of culture medium was replaced with 100 µL of fresh medium followed by adding 10 µL of MTT dye to each well and incubating for 4 h at 37 °C with 5% CO_2. After 4 h of incubation, 90 µL of medium was removed from each well and 100 µL of DMSO was added to dissolve the formazan crystals, followed by re-incubation for 10 min at 37 °C. The absorbance was measured at 570 nm using a Multiskan FC automatic plate reader (Thermo Fisher Scientific Oy, Finland) with Sknalt Software 4.1 software. Each sample was tested in triplicate, and cell viability was expressed as % of untreated cells (control) considered 100% viable.

Testing on the antimicrobial effect of the API, before and after incorporation into PMs based on F127, was performed by determining the diameter of the zone of inhibition by diffusion method, using three reference strains of Gram-negative bacteria (*Escherichia coli* ($ATCC^R$ 11775^{TM}), *Pseudomonas aeruginosa* ($ATCC^R$ 10154^{TM}), *Klebsiella pneumoniae* ($ATCC^R$ BAA—1705^{TM})), a Gram-pozitive bacteria (*Staphylococcus aureus* ($ATCC^R$ $25\,923^{TM}$)), and an anaerobic bacterium (*Porphyromonas gingivalis* ($ATCC^R$ 33277^{TM})). The inoculum used in the five test microorganisms used for seeding had a turbidity of 0.5 McFarland. On each plate of culture medium, in the wells made, 100 µL of the micellar suspension with a concentration of 200 µg/mL was placed. The diameters of the inhibition zones were measured after incubation at 37 °C, for 24 h in the case of *Staphylococcus aureus, Escherichia coli, Pseudomonas earuginosa, Klebsiella pneumoniae* and 48 h for *Porphyromonas gingivalis* (grown in Anaeroar, under anaerobic conditions). It was considered that the tested samples have an antimicrobial activity if around the wells there were areas of inhibition of the growth of the test microorganism, with a significantly large diameter, i.e., at least 15 mm.

The experiments were repeated three times, and the results (mm area of inhibition) were expressed as mean values.

3. Results

3.1. Micellar Sizes and Stability

Generally, in order to have the highest drug-loading efficiency during its circulation in blood, the micelles should be small enough to evade detection and destruction by the reticular endothelial system. The obtained micellar system loaded with API was designed in order to be administrated either by injection or by a topical application. Independent of the administration route, an important characteristic of this type of drug delivery systems is represented by their size and polydispersity index. In Table 1 are presented the colloidal characteristics of the free and API-loaded PMs.

Table 1. Diameter in volume (Dv), polydispersity index (PDI) and zeta potential (ZP) values of free and API-loaded PMs.

Sample	Dv (nm)	PDI	ZP (mV)
F127	20.3 ± 0.1	0.141	−3.0
F127/API = 10/1 (% w/w)	20.7 ± 0.2	0.198	−3.0
F127/API = 5/1 (% w/w)	20.3 ± 0.2	0.284	−2.8
F127/API = 2/1 (% w/w)	21.4 ± 0.3	0.308	−2.7

From this table it is possible to notice that the micellar Dv of free micelles is around 20 nm, which is in concordance with the literature data concerning the F127 micelles [29]. It can also be observed that only a small increase of the Dv is noticed when the API is loaded at different copolymer/API ratios. This behavior can be explained be a modification of the aggregation number by the fact that the encapsulation of a hydrophobic molecule, as it is the case for our active principle, leads to a more compact micellar core. In Figure 2 are shown the size distribution curves in volume for free and API-loaded PMs at 37 °C.

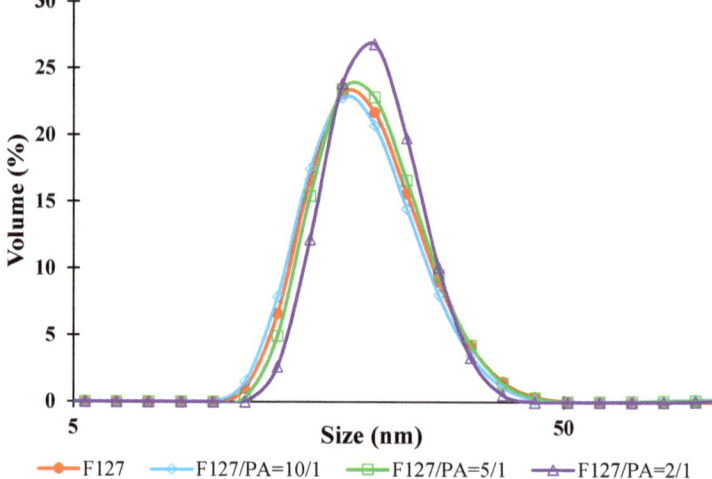

Figure 2. Size distribution curves in volume for free and API-loaded PMs at 37 °C.

From Figure 2 it appears that these curves have a monomodal distribution and that the PDI does not increase drastically with increasing the amount of API loaded which indicates the fact that there are no micellar aggregates. However, from data provided in Table 1, it can be observed that the PDI values increase slightly with the increase of the copolymer/API ratio but these values are always under 0.3. Moreover, from this table it appears that the ZP values are almost neutral, as expected for non-ionic copolymers such as Pluronics [30,31]. In this case, the micelles are sterically stabilized by the PEO sequences.

Among the colloidal properties which are important characteristic of a colloidal solution is the suspension stability as a function of time. Figure 3 shows the evolution of the Z-average as a function of time for all prepared micellar suspensions.

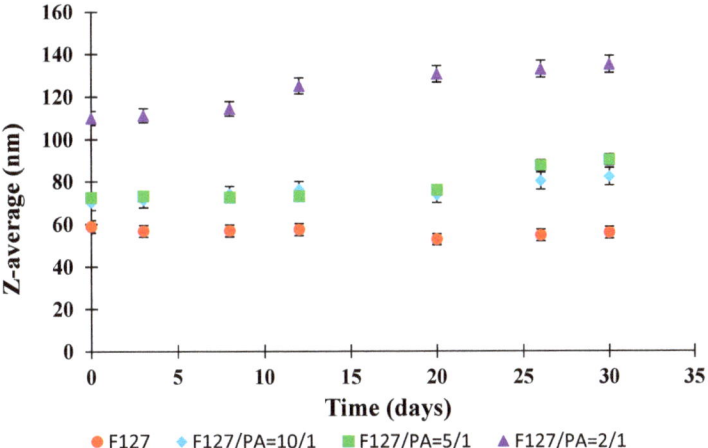

Figure 3. Evolution of the Z-average values as a function of time.

Figure 3 illustrates that only a slight increase of the Z-average values is noticed after 20 days for the sample F127/API = 2/1. For the other analyzed samples, the micellar suspension is stable, within the experimental error limits, during 30 days which, is an advantage for the long term usage and storage of these systems.

3.2. Drug Encapsulation Efficiencies

The drug encapsulation efficiency (DEE) and drug loading efficiency (DLE) of different samples are provided in Table 2.

Table 2. DEE and DLE values for API-loaded PMs.

F127/API Ratio (% w/w)	DEE (%)	DLE (%)
10/1	84.8	11.1
5/1	66.6	10.3
2/1	30.9	7.7

From Table 2, it appears that API was successfully incorporated into the PMs and that both the DEE and DLE values are increased by decreasing the initial copolymer/API ratio. At a F127/API ratio of 10/1, the DEE has the highest value of 84.8%, whereas the DLE is equal to 11.1%. These values are comparable with other values in the literature concerning the encapsulation of other active molecules in Pluronics micelles [31].

The main aim of this study was the preparation of a micellar system loaded with an active principle having an antimicrobial activity and which could be used for the treatment of oral diseases. As it was demonstrated that the API-loaded PMs are stable in time, it was of interest to study also the release kinetics of the API in PBS (pH = 7.4) at 37 °C. In Figure 4 are given the curves of the cumulative release kinetics for free API and also for the API-loaded PMs.

A first observation from Figure 4 is related to the fact that the release kinetics of free API is higher than that of the encapsulated active principle. This behavior is explained by the fact that the API loaded within the micelles must pass through a barrier formed by the copolymer core and corona. Comparing the micellar systems with loaded API, it appears that the release kinetics are higher for the sample F127/API = 10/1, where the amount of API is the lowest. Moreover, the release kinetic curves are typical for a release controlled by diffusion in which the equilibrium is reached after 24 h even if not all the amount of API was released.

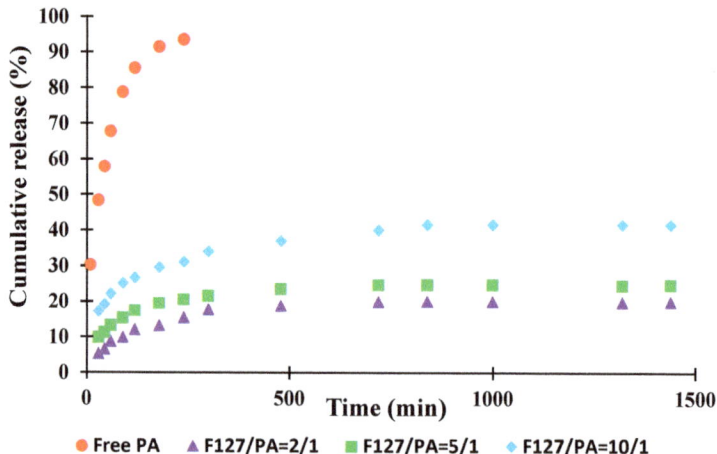

Figure 4. Evolution of the cumulative release kinetics in PBS (pH = 7.4) at 37 °C.

3.3. Assessment of the Haemolysis Degree

Hemolysis is the destruction of red blood cells along with the release of hemoglobin and other internal components into the surrounding fluid. If this destruction occurs in a significant number of red blood cells in the body, it can lead to dangerous pathological conditions [32,33]. Therefore, all biomedical products designed for intravenous administration should be evaluated for their hemolytic potential [34]. The small sizes of this micellar system make it suitable for administration by injection. In this case, it was worthy to study their interaction with the blood components. Figure 5 shows the evolution of the hemolysis degree as a function of time and concentration.

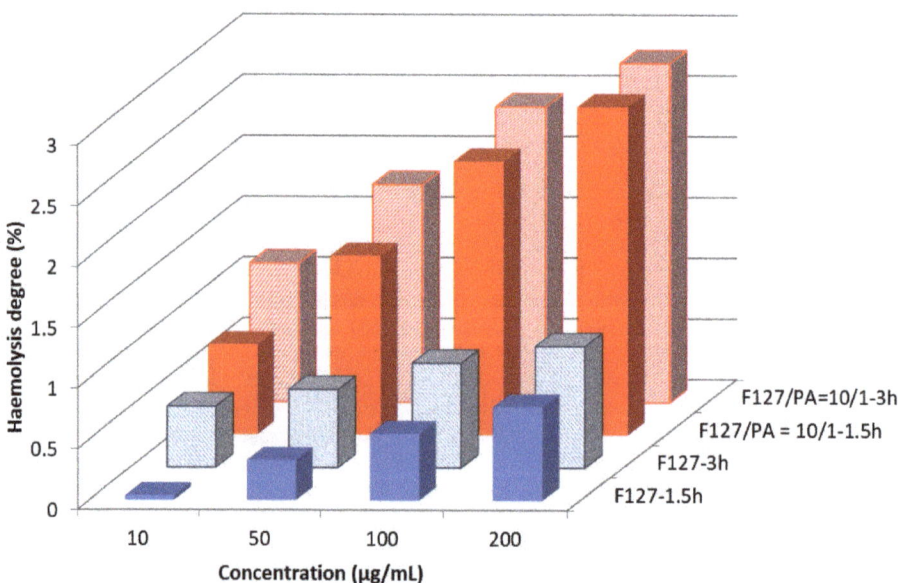

Figure 5. Evolution of the hemolysis degree for free F127 micelles and F127/API = 10/1 sample as a function of concentration and time.

It is well known that a material is considered as being hemotoxic if the hemolysis degree is higher than 5% [35]. From Figure 5 it can be noticed that even at the highest concentration of 200 μg/mL, the hemolysis degree is smaller than 3% which proves that our micellar systems are hemocompatible.

3.4. In Vitro Cytotoxicity Analysis

As these micellar systems are designed for biomedical applications, it was of interest to study their cytotoxicity in order to assess their biocompatibility. Cellular viability was evaluated on human fibroblast after 24 and 48 h, and the results are given in Figure 6.

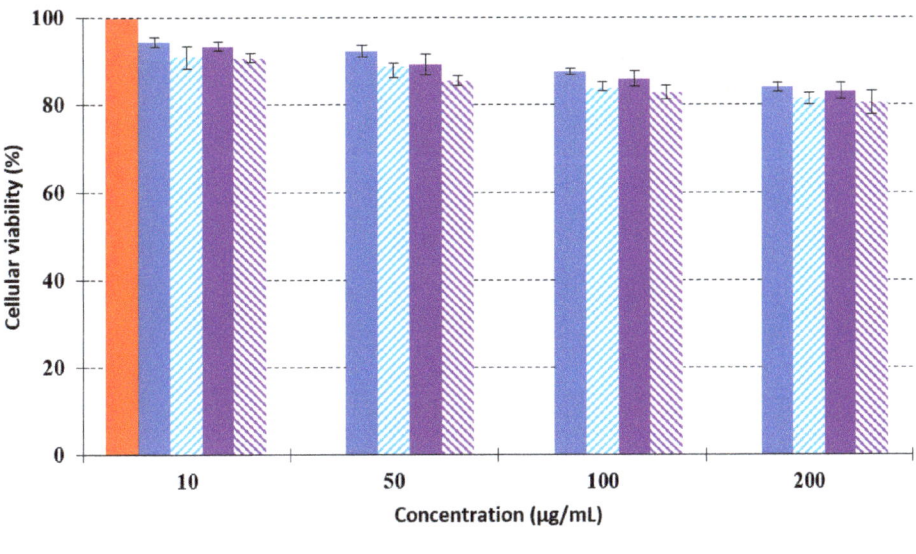

Figure 6. Evolution of the in vitro cellular viability after 24 and 48 h as a function of the concentration for sample F127/API = 10/1.

The data provided in Figure 6 show, first of all, that the tested PMs at all four established concentrations (10, 50, 100 and 200 μg/mL) did not show cytotoxic effects on fibroblast cells at both 24 and 48 h after incubation. Then, it can be observed that the cellular viability decreases with increasing the concentration, but, even at the highest concentration of 200 μg/mL and 48 h, the cellular viability remains higher than 80%, which is the threshold from which a material is considered to be cytotoxic. This biocompatible behavior is also illustrated by the photographs given in Table 3.

The photographs given in Table 3 are supplementary proof that the prepared micellar system has a high cellular viability, and, therefore, it can be considered as safe for biomedical applications.

Table 3. Micrographs of the fibroblast cells after 24 and 48 h.

Sample		Micrographs	
		24 h	48 h
	Control		
F127	10 µg/mL		
	50 µg/mL		
	100 µg/mL		
	200 µg/mL		

Table 3. *Cont.*

Sample	Micrographs	
	24 h	48 h
F127/PA = 10/1 (g/g) — 10 µg/mL		
50 µg/mL		
100 µg/mL		
200 µg/mL		

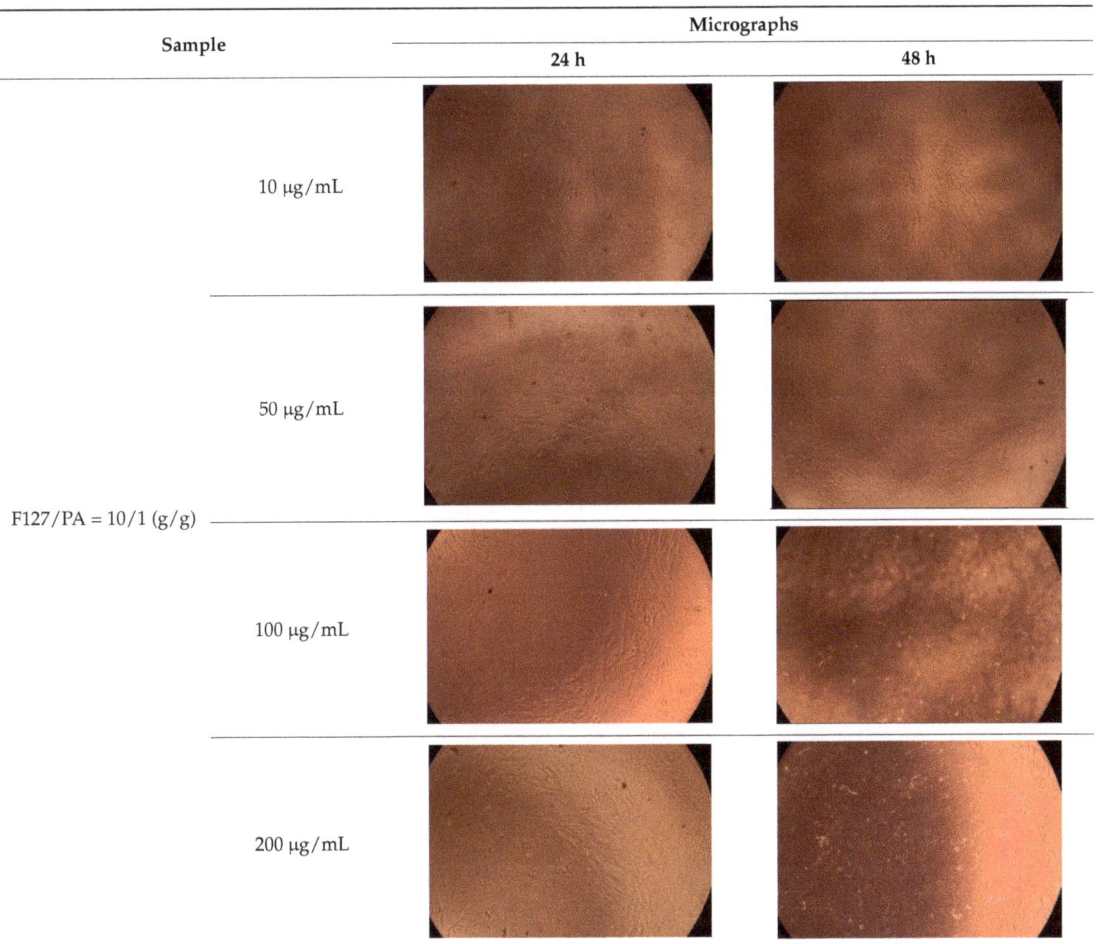

3.5. In Vitro Antimicrobial Activity

Finally, owing to the chemical structure of our original active principle, it was of interest to investigate the antimicrobial activity of the obtained micellar systems in order to determine if they can be potentially used for the treatment of oral diseases (especially the bacterial plaque). Different bacterial strains were selected based on their presence in the oral cavity and the obtained results are summarized in Table 4.

From this table, it is obvious that free micelles have the lowest antimicrobial activity. In the case of free API, it clearly appears that it has an enhanced antimicrobial activity on *E. coli* (25 ± 0.4 mm) and especially on *S. aureus* (39 ± 0.6 mm) with inhibition diameters much higher than 15 mm. In addition, a medium activity was observed for *K. Pneumonias* (14 ± 0.5 mm), whereas almost no activity was detected against *P. aeruginosa* and *P. gingivalis*. This tendency is also confirmed for API-loaded PMs. In fact, by increasing the F127/API ratio, the antimicrobial activity increases, approaching very much the value of free API. Our results suggest that the antibacterial activity is higher against Gram-positive bacteria than Gram-negative ones. A Gram-positive bacillus does not have an outer cell wall beyond the peptidoglycan membrane, which makes it more absorbent for different molecules [36]. On the contrary, Gram-negative bacteria have a thin peptidoglycan layer and have an outer lipid membrane. This can contribute to the antibiotic resistance of the bacterial cells and be involved in the interaction with host cells. The difference in the antibacterial activity against the studied bacterial strains can be correlated with the structural difference between the lipid layer of these Gram-negative bacteria, which affects the cell permeability. Moreover, the lack of the antibacterial activity against *P. aeruginosa* and *P. gingivalis* demonstrate that our system is highly selective in killing different strains of bacteria, and this can be considered an important advantage. Concerning the molecular mechanism of the antibacterial activity of these types of compounds, it is still under debate in the literature, but it seems that the inhibitory action towards bacterial topoisomerases could be the main explanation [37,38]. Undoubtedly, more in-depth research in this area is urgently needed in order to establish the exact mechanism of action of these compounds. The photographs from Figure 7 are a good indication of this antimicrobial activity.

Table 4. Inhibition zone diameter (mm) for different bacterial strains in the presence of free PMs, free API and API-loaded PMs. Concentration was equal to 200 µg/mL.

Strain	Negative Control (Free PMs)	Positive Control (Free API)	F127/PA = 2/1	F127/PA = 5/1	F127/PA = 10/1
Escherichia coli	8 ± 0.2	25 ± 0.4	20 ± 0.4	17 ± 0.3	9 ± 0.4
Pseudomonas aeruginosa	8 ± 0.3	9 ± 0.3	9 ± 0.3	8 ± 0.2	8 ± 0.2
Klebsiella pneumonia	8 ± 0.2	14 ± 0.5	10 ± 0.4	9 ± 0.3	8 ± 0.3
Staphylococcus aureus	8 ± 0.2	39 ± 0.6	36 ± 0.6	35 ± 0.5	32 ± 0.5
Porphyromonas gingivalis	8 ± 0.3	9 ± 0.3	9 ± 0.3	8 ± 0.3	8 ± 0.3

Gram-positive

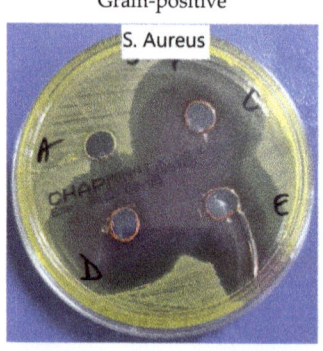

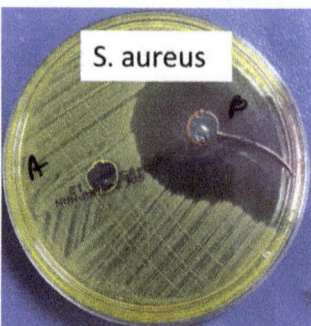

Gram-negative

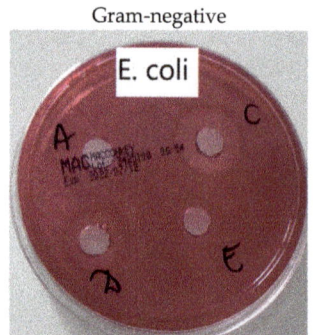

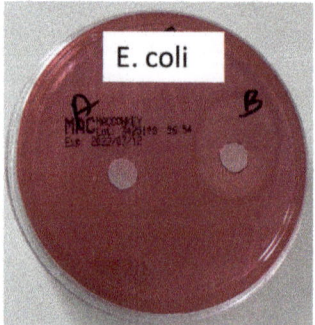

Figure 7. Antimicrobial activity of free and API-loaded PMs against Gram-positive and Gram-negative strains. A-negative control (free F127 micelles), B-positive control (API), C—F127/API = 2/1, D—F127/API = 5/1 and E—F127/API = 10/1.

4. Conclusions

In this paper it was demonstrated that the polymeric micelles (PMs) obtained using a commercial available copolymer, Pluronic F127, can be loaded with an active pharmaceutical ingredient (API) at different copolymer/API weight ratios. The free and API-loaded PMs were characterized, by DLS, in terms of their size, stability and zeta potential. It was found that the loading of API in the free F127 micelles, of around 20 nm in diameter, has no significant impact either on the micellar size or on the polydispersity index. The API-loaded PMs show high stability over time, proven by almost constant average sizes over 30 days. In vitro tests have proven both their non-cytotoxic behavior, with cellular viabilities higher than 80%, and low degree of hemolysis (smaller than 3% even at high concentrations such as 200 µg/mL). The in vitro testing was completed by antimicrobial tests on several types of bacterial strains (Gram-positive and Gram-negative bacteria). An enhanced antimicrobial activity was observed on *S. aureus* and *E. coli*, whereas for *K. Pneumonias* only a medium one was recorded. Multidrug-resistant bacterial infection is expected to be overcome by these effective nanoscale drug carriers. Nanoscale drug PMs have been shown to be able to release the API in a controlled manner, increase the penetration to biofilms, improve drug stability and enhance drug bioavailability; therefore, it can be admitted that these PMs might be recommended as safe systems for further in vivo tests.

Author Contributions: Conceptualization, C.P., M.P., V.S., D.L.I. and L.I.A.; methodology, C.P., M.P. and L.I.A.; investigation, C.P., D.L.I. and L.I.A. writing–original draft preparation, M.P. and L.I.A.; visualization, M.P. and L.I.A.; supervision, M.P. and L.I.A. All authors have read and agreed to the published version of the manuscript.

Funding: This work was supported by a grant of the Romanian Ministry of Education and Research, CNCS—UEFISCDI, project number PN-III-P1-1.1-TE-2019-0664, within PNCDI-III.

Institutional Review Board Statement: Not applicable.

Informed Consent Statement: Not applicable.

Data Availability Statement: The data presented in this study are available on request from the corresponding author.

Conflicts of Interest: The author declares no conflict of interest.

References

1. Riess, G. Micellization of block copolymers. *Prog. Polym. Sci.* **2003**, *28*, 1107–1170. [CrossRef]
2. Atanase, L.I.; Desbrieres, J.; Riess, G. Micellization of synthetic and polysaccharides-based graft copolymers in aqueous media. *Prog. Polym. Sci.* **2017**, *73*, 32–60. [CrossRef]
3. Atanase, L.I.; Riess, G. Block copolymers as polymeric stabilizers in non-aqueous emulsion polymerization. *Polym. Int.* **2011**, *60*, 1563–1573. [CrossRef]
4. Xu, W.; Ling, P.; Zhang, T. Polymeric micelles, a promising drug delivery system to enhance bioavailability of poorly water-soluble drugs. *J. Drug Deliv.* **2013**, *2013*, 340315. [CrossRef]
5. Ghezzi, M.; Pescina, S.; Padula, C.; Santi, P.; Del Favero, E.; Cantu, L.; Nicoli, S. Polymeric micelles in drug delivery: An insight of the techniques for their characterization and assessment in biorelevant conditions. *J. Contr. Rel.* **2021**, *332*, 312–336. [CrossRef]
6. Kalinova, R.; Dimitrov, I. Functional Polyion Complex Micelles for Potential Targeted Hydrophobic Drug Delivery. *Molecules* **2022**, *27*, 2178. [CrossRef]
7. Salkho, N.M.; Awad, N.S.; Pitt, W.G.; Husseini, G.A. Photo-Induced Drug Release from Polymeric Micelles and Liposomes: Phototriggering Mechanisms in Drug Delivery Systems. *Polymers* **2022**, *14*, 1286. [CrossRef]
8. Atanase, L.I.; Riess, G. Self-Assembly of Block and Graft Copolymers in Organic Solvents: An Overview of Recent Advances. *Polymers* **2018**, *10*, 62. [CrossRef]
9. Iurciuc-Tincu, C.-E.; Cretan, M.S.; Purcar, V.; Popa, M.; Daraba, O.M.; Atanase, L.I.; Ochiuz, L. Drug Delivery System Based on pH-Sensitive Biocompatible Poly(2-vinyl pyridine)-b-poly(ethylene oxide) Nanomicelles Loaded with Curcumin and 5-Fluorouracil. *Polymers* **2020**, *12*, 1450. [CrossRef] [PubMed]
10. Atanase, L.I. Micellar Drug Delivery Systems Based on Natural Biopolymers. *Polymers* **2021**, *13*, 477. [CrossRef]
11. Jarak, I.; Varela, C.L.; da Silva, E.T.; Roleira, F.F.M.; Veiga, F.; Figueiras, A. Pluronic-based nanovehicles: Recent advances in anticancer therapeutic applications. *Eur. J. Med. Chem.* **2020**, *206*, 112526. [CrossRef]

12. Sotoudegan, F.; Amini, M.; Faizi, M.; Aboofazeli, R. Nimodipine-loaded Pluronic® block copolymer micelles: Preparation, characterization, in-vitro and in-vivo studies. *Iran. J. Pharm. Sci.* **2016**, *15*, 641–661.
13. Meng, X.; Liu, J.; Yu, X.; Li, J.; Lu, X.; Shen, T. Pluronic F127 and D-α-Tocopheryl Polyethylene Glycol Succinate (TPGS) mixed micelles for targeting drug delivery across the blood brain barrier. *Sci. Rep.* **2017**, *7*, 2964. [CrossRef] [PubMed]
14. Yu, J.; Qiu, H.; Yin, S.; Wang, H.; Li, Y. Polymeric Drug Delivery System Based on Pluronics for Cancer Treatment. *Molecules* **2021**, *26*, 3610. [CrossRef]
15. Salay, L.C.; Prazeres, E.A.; Marín Huachaca, N.S.; Lemos, M.; Piccoli, J.P.; Sanches, P.R.S.; Cilli, E.M.; Santos, R.S.; Feitosa, E. Molecular interactions between Pluronic F127 and the peptide tritrpticin in aqueous solution. *Colloid Polym. Sci.* **2018**, *296*, 809–817. [CrossRef]
16. Ezhilrani, V.C.; Karunanithi, P.; Sarangi, B.; Joshi, R.G.; Dash, S. Hydrophilic-hydrophilic mixed micellar system: Effect on solubilization of drug. *SN Appl. Sci.* **2021**, *3*, 371. [CrossRef]
17. Zhang, M.; Yu, Z.; Lo, E.C.M. A new pH-responsive nano micelle for enhancing the effect of a hydrophobic bactericidal agent on mature *Streptococcus mutans* biofilm. *Front. Microbiol.* **2021**, *12*, 761583. [CrossRef] [PubMed]
18. Tănase, M.A.; Raducan, A.; Oancea, P.; Dițu, L.M.; Stan, M.; Petcu, C.; Scomoroșcenco, C.; Ninciuleanu, C.M.; Nistor, C.L.; Cinteza, L.O. Mixed Pluronic—Cremophor Polymeric Micelles as Nanocarriers for Poorly Soluble Antibiotics—The Influence on the Antibacterial Activity. *Pharmaceutics* **2021**, *13*, 435. [CrossRef]
19. Groo, A.C.; Matougui, N.; Umerska, A.; Saulnier, P. Reverse micelle-lipid nanocapsules: A novel strategy for drug delivery of the plectasin derivate AP138 antimicrobial peptide. *Int. J. Nanomed.* **2018**, *13*, 7565–7574. [CrossRef]
20. Song, H.; Yin, Y.; Peng, J.; Du, Z.; Bao, W. Preparation, Characteristics, and Controlled Release Efficiency of the Novel PCL-PEG/EM Rod Micelles. *J. Nanomater.* **2021**, *2021*, 8132868. [CrossRef]
21. Barros, C.H.N.; Hiebner, D.W.; Fulaz, S.; Vitale, S.; Quinn, L.; Casey, E. Synthesis and self-assembly of curcumin-modified amphiphilic polymeric micelles with antibacterial activity. *J. Nanobiotechnol.* **2021**, *19*, 104. [CrossRef] [PubMed]
22. Chen, X.; Guo, R.; Wang, C.; Li, K.; Jiang, X.; He, H.; Hong, W. On-demand pH-sensitive surface charge-switchable polymeric micelles for targeting *Pseudomonas aeruginosa* biofilms development. *J. Nanobiotechnol.* **2021**, *19*, 99. [CrossRef] [PubMed]
23. Chen, L.; Hong, Y.; He, S.; Fan, Z.; Du, J. Poly(ε-caprolactone)-Polypeptide Copolymer Micelles Enhance the Antibacterial Activities of Antibiotics. *Acta Phys. Chim. Sin.* **2021**, *37*, 1910059. [CrossRef]
24. Hashemi, M.M.; Holden, B.S.; Taylor, M.F.; Wilson, J.; Coburn, J.; Hilton, B.; Nance, T.; Gubler, S.; Genberg, C.; Deng, S.; et al. Antibacterial and Antifungal Activities of Poloxamer Micelles Containing Ceragenin CSA-131 on Ciliated Tissues. *Molecules* **2018**, *23*, 596. [CrossRef]
25. Purro, M.; Qiao, J.; Liu, Z.; Ashcraft, M.; Xiong, M.P. Desferrioxamine:gallium-pluronic micelles increase outer membrane permeability and potentiate antibiotic activity against Pseudomonas aeruginosa. *Chem. Commun.* **2018**, *54*, 13929–13932. [CrossRef] [PubMed]
26. Zhdanova, K.A.; Savelyeva, I.O.; Ignatova, A.A.; Gradova, M.A.; Gradov, O.V.; Lobanov, A.V.; Feofanov, A.V.; Mironov, A.F.; Bragina, N.A. Synthesis and photodynamic antimicrobial activity of amphiphilic *meso*-arylporphyrins with pyridyl moieties. *Dyes Pigm.* **2020**, *181*, 108561. [CrossRef]
27. Popovici, C.; Pavel, C.-M.; Sunel, V.; Cheptea, C.; Dimitriu, D.G.; Dorohoi, D.O.; David, D.; Closca, V.; Popa, M. Optimized Synthesis of New Thiosemicarbazide Derivatives with Tuberculostatic Activity. *Int. J. Mol. Sci.* **2021**, *22*, 12139. [CrossRef]
28. Rață, D.M.; Cadinoiu, A.N.; Atanase, L.I.; Bacaita, S.E.; Mihalache, C.; Daraba, O.M.; Popa, M. "In vitro" behaviour of aptamer-functionalized polymeric nanocapsules loaded with 5-fluorouracil for targeted therapy. *Mater. Sci. Eng. C* **2019**, *103*, 109828. [CrossRef]
29. Wang, H.; Williams, G.R.; Wu, J.; Wu, J.; Niu, S.; Xie, X.; Li, S.; Zhu, L.M. Pluronic F127-based micelles for tumor-targeted bufalin delivery. *Int. J. Pharm.* **2019**, *559*, 289–298. [CrossRef]
30. Butt, A.M.; Amin, M.C.I.M.; Katas, H.; Sarisuta, N.; Witoonsaridsilp, W.; Benjakul, R. In Vitro Characterization of Pluronic F127 and D-Tocopheryl Polyethylene Glycol 1000 Succinate Mixed Micelles as Nanocarriers for Targeted Anticancer-Drug Delivery. *J. Nanomater.* **2012**, *2012*, 916573. [CrossRef]
31. Niu, J.; Yuan, M.; Chen, C.; Wang, L.; Tang, Z.; Fan, Y.; Liu, X.; Ma, Y.J.; Gan, Y. Berberine-Loaded Thiolated Pluronic F127 Polymeric Micelles for Improving Skin Permeation and Retention. *Int. J. Nanomed.* **2020**, *15*, 9987–10005. [CrossRef] [PubMed]
32. Cadinoiu, A.N.; Peptu, C.A.; Fache, B.; Chailan, J.F.; Popa, M. Microparticulated systems based on chitosan and poly(vinyl alcohol) with potential ophthalmic applications. *J. Microencapsul.* **2015**, *32*, 381–389. [CrossRef] [PubMed]
33. Burlui, V.; Popa, M.; Cadinoiu, A.N.; Stadoleanu, C.; Mihalache, G.; Zamaru, V.; Dârtu, L.; Folescu, E.; Rata, D.M. Physico-chemical characterization and in vitro hemolysis evaluation of titanium dioxide nanoparticles. *Int. J. Med. Dent.* **2015**, *5*, 124.
34. Cadinoiu, A.N.; Rata, D.M.; Atanase, L.I. *Biocompatible Injectable Polysaccharide Materials for Drug Delivery in Polysaccharide Carriers for Drug Delivery*; Maiti, S., Jana, S., Eds.; Woodhead Publishing: Sawston, UK, 2019; pp. 127–154.
35. Alupei, L.; Lisa, G.; Butnariu, A.; Desbrieres, J.; Cadinoiu, A.N.; Peptu, C.A.; Calin, G.; Popa, M. New folic acid-chitosan derivative based nanoparticles–potential applications in cancer therapy. *Cellul. Chem. Technol.* **2017**, *51*, 631–648.
36. Zhang, Y.; Li, S.; Xu, Y.; Shi, X.; Zhang, M.; Huang, Y.; Liang, Y.; Chen, Y.; Ji, W.; Kim, J.R.; et al. Engineering of hollow polymeric nanosphere-supported imidazolium-based ionic liquids with enhanced antimicrobial activities. *Nano Res.* **2022**, *15*, 5556–5568. [CrossRef]

37. Kosikowska, U.; Wujec, M.; Trotsko, N.; Płonka, W.; Paneth, P.; Paneth, A. Antibacterial Activity of Fluorobenzoylthiosemicarbazides and Their Cyclic Analogues with 1,2,4-Triazole Scaffold. *Molecules* **2021**, *26*, 170. [CrossRef]
38. Janowska, S.; Khylyuk, D.; Andrzejczuk, S.; Wujec, M. Design, Synthesis, Antibacterial Evaluations and In Silico Studies of Novel Thiosemicarbazides and 1,3,4-Thiadiazoles. *Molecules* **2022**, *27*, 3161. [CrossRef]

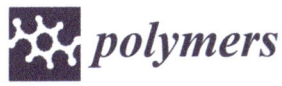

Article

Preparation of DNC Solid Dispersion by a Mechanochemical Method with Glycyrrhizic Acid and Polyvinylpyrrolidone to Enhance Bioavailability and Activity

Min Lu [1,†], Wei Wei [1,†], Wenhao Xu [1], Nikolay E. Polyakov [2,3], Alexandr V. Dushkin [1,3] and Weike Su [1,4,*]

1. National Engineering Research Center for Process Development of Active Pharmaceutical Ingredients, Collaborative Innovation Center of Yangtze River Delta Region Green Pharmaceuticals, Zhejiang University of Technology, Hangzhou 310000, China; yaolilumin@163.com (M.L.); 13619587845@163.com (W.W.); xuwenhao@zjut.edu.cn (W.X.); avd@ngs.ru (A.V.D.)
2. Institute of Chemical Kinetics and Combustion, 630090 Novosibirsk, Russia; polyakov@kinetics.nsc.ru
3. Institute of Solid State Chemistry and Mechanochemistry, 630090 Novosibirsk, Russia
4. Key Laboratory for Green Pharmaceutical Technologies and Related Equipment of Ministry of Education, College of Pharmaceutical Sciences, Zhejiang University of Technology, Hangzhou 310000, China
* Correspondence: pharmlab@zjut.edu.cn
† These authors contributed equally to this paper.

Abstract: To exploit aqueous-soluble formulation and improve the anticoccidial activity of 4,4′-dinitrocarbanilide (DNC, active component of nicarbazin), this paper prepared DNC/GA/PVP K30 solid dispersion (SD) with glycyrrhizic acid (GA) and polyvinylpyrrolidone (PVP) K30 by a mechanical ball milling method without using any organic solvent. Fourier transform infrared spectroscopy, X-ray diffraction, differential scanning calorimetry, and scanning electron microscopy were used for the solid state characterization. High performance liquid chromatography, critical micelle concentration, particle characterization, and transmission electron microscopy were used to evaluate the behavior in aqueous solution. In addition, the oral bioavailability, tissue distribution, and anticoccidial activity of DNC/GA/PVP K30 SD were investigated as well. Compared with free drug, the novel formulation not only improved the solubility and dissolution rate of DNC, but also inhibited the fecal output of oocysts and enhanced the therapeutic effect of coccidiosis. According to the experiment results, the DNC/GA/PVP K30 SD increased 4.64-fold in oral bioavailability and dramatically enhanced the concentration in liver which provided a basis for further research in schistosomiasis. In summary, our findings suggested that DNC/GA/PVP K30 SD may have promising applications in the treatment of coccidiosis.

Keywords: nicarbazin; DNC; glycyrrhizic acid; PVP; micelles; coccidiosis

1. Introduction

Chicken coccidiosis is one of the infectious parasite severe diseases and causes increasing morbidity and mortality every year in the poultry industry throughout the world, hampering the productivity and economic development [1]. As a leading parasite, *E. tenella* is the most pathogenic and fatal for seven species of coccidian parasite, which live inside the intestines' tract of chicken, cause malabsorption of nutrients, enteritis, and fluid loss destruction, resulting in extensive hemorrhage and death [2]. These coccidian parasites are highly widespread and can persist for long periods in the environment, such as feces and litter, causing many chickens to become infected with more than 50% mortality rate [3,4], and the cost of coccidiosis in chickens is estimated to $3 billion USD per annum [1]. Therefore, control of the coccidiosis in poultry is essential for food security and economic development.

Nicarbazin (NIC) is a broad-spectrum anticoccidial agent with high security and low resistance, the equimolar complex of 4,4′-dinitrocarbanilide (DNC) with 2-hydroxy-4,6-dimethylpyrimidine (HDP), which has been synthesized and utilized since the 1950s [5]. Previous contribution has reported that NIC [6] inhibited the second generation schizont of *E. tenella*, which was essential for killing and restraining coccidia growth; therefore, NIC has been proposed as a promising preventing strategy for the therapy of coccidiosis. The active ingredient of NIC is DNC (a highly insoluble drug), and the other part HDP had no coccidioidal activity [7], hydrogen bonding between them greatly improving the anticoccidial effect compared to that of the administration of DNC alone [8]. However, the further application of DNC is restricted due to its poor solubility in aqueous, and improving the water solubility may enhance the drug efficacy. Formation of Solid dispersion (SD) has been widely used to increase aqueous solubility of hydrophobic drugs, which provided an important route to improve dissolution rate and bioavailability and reduce toxicity. Various manufacture approaches have been employed to produce solid dispersion, such as solvent evaporation, melting, etc. [9]. However, solvent evaporation will leave the product with residual organic solvent and high temperature of the melting process may lead to decomposition or degradation of the drug [10]. Recently, a novel strategy, mechanochemistry preparing SD widely applied as drug solubilization, has the remarkable characteristics of changing particle structure and enhancing the activity of drug by simple operation with high efficiency and less solvent residual [11]. Researchers prepared curcumin SD using the mechanochemical method; the bioavailability of curcumin increased 10 times, the corresponding water solubility has also been greatly improved; meanwhile, the SD had strong lipid-lowering ability and anticancer capacity compared with curcumin [12,13]. Xu et al. conducted a mechanochemical ball milling technique to achieve astaxanthin solid dispersion [14] and 5-amino salicylic acid pH-sensitive hydrogel [15] with higher aqueous solubility than pure drugs. Therefore, mechanical ball milling will be a great choice to enhance the water solubility of drugs under the conditions of being environmentally friendly and having high-performance productivity.

Glycyrrhizic acid (GA) is a kind of naturally occurring sweetener, as a triterpenoid glycoside had been extracted from glycurrhiza root with antiviral, anti-inflammatory, anticancer, and hepatoprotective activities et al. [16]. In a drug delivery system, GA has strong complexation ability for drugs and self-association in aqueous solution as an amphiphilic molecule could enhance the solubility and permeability of hydrophobic compounds, rendering the drug dispersed in aqueous solutions to form a stable and homogenous solution [17]. Yang et al. [18] had prepared Paclitaxel-loaded glycyrrhizic acid micelles to increase the bioavailability of paclitaxel. Zheng et al. reported a method to prepare inositol hexanicotinate SD with GA and arabic gum by mechanical ball milling to enhance the solubility and bioavailability. Polyvinylpyrrolidone (PVP) was obtained by radical polymerization of the monomer, N-vinyl-pyrrolidone, and proved to be a safe water-soluble polymer with a variety of molecular weight and viscosity [19,20]. The degree of polymerization and molecular weight of PVP determines its viscosity, which is represented by the K value. A great number of reports claimed that the application of PVP could enhance dissolution rate and bioavailability of drug, such as phenytoin, sulfathiazole, hydrocortisone, disulfiram, etc. [21]. Frizon et al. [22] prepared loratadine SD with PVP K30 to enhance its dissolution rate. Previous reports pointed out that PVP and tannic acid (TA) combined stabilization through hydrogen-bonding interactions between carbonyl groups (PVP) and hydroxyl groups (TA) [23]. However, whether glycyrrhizic acid, an excipient containing hydroxyl groups just like TA, is combined with PVP to change the physicochemical properties of drugs by mechanical ball milling has not been reported, which is a significant study for verifying this point.

In this study, DNC (the active compound of nicarbazin) was used as the model drug, using mechanical ball milling to prepare DNC SD with GA and PVP K30, then characterizing the solid state and evaluating the behavior in aqueous solution. Furthermore, the permeability and tissue distribution studies of DNC SD were conducted. The anticoccidial activity of DNC SD was investigated in vivo as well. Our finding indicated that DNC SD could self-assemble to form micelles when dissolved in water, meanwhile showing the improvement of aqueous solubility, dissolution rate, and oral bioavailability, which may have potential application in coccidiosis therapy.

2. Materials and Methods

2.1. Materials

Nicarbazin (NIC) was purchased from Wuhan Changcheng Chemical Technology Development Co., Ltd. (Wuhan, China, purity: 98%). Glycyrrhizic acid (GA) was purchased from Shaanxi Pioneer Biotechnology Co., Ltd. (Xi'an, China, purity ≥ 98%). PVP K30 was acquired from Meryer Co., Ltd. (Shanghai, China, purity: 100%). All other chemicals used were analytical grade. Figure 1 shows the chemical structures of NIC, GA, and PVP.

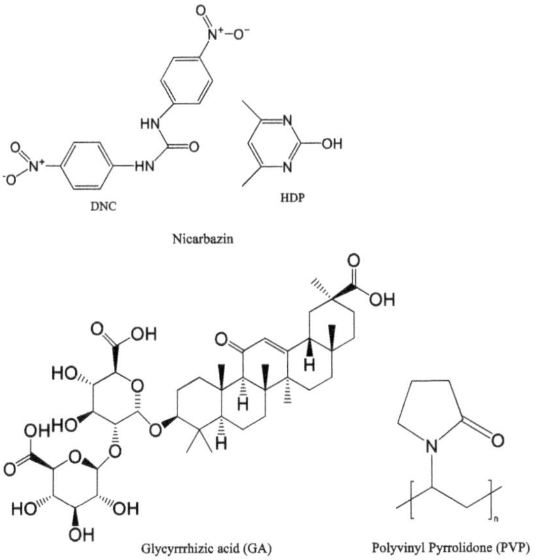

Figure 1. Structure of NIC, GA, and PVP.

2.2. Preparation of the DNC Solid Dispersions (DNC SDs)

DNC solid dispersions were prepared by a planetary ball mill (PM 400, Retsch, Haan, Germany). Briefly, 0.44 g DNC and 1.76 g GA (mass ratio 1/4), or 0.37 g DNC, 1.46 g GA, and 0.37 g PVP K30 (mass ratio 1/4/1) were added to a 50 mL ball mill tank with 82 g steel balls (diameter 9 mm) with rotation speed 300 rpm, 30 min, were called DNC/GA SD and DNC/GA/PVP K30 SD, respectively. Physical mixing (PM): the ingredients were mixed (with same SD mass ratio) and called DNC/GA PM and DNC/GA/PVP K30 PM.

2.3. Analysis of DNC by HPLC

The amount of DNC was determined with a high-performance liquid chromatography (HPLC) system (Agilent 1200, Santa Clara, CA, USA) with a diode-array detector. Chromatography was performed on an Agilent C18 column (3.0 × 250 mm, 5 µm) at 25 °C. The mobile phase consisted of acetonitrile-water (60:40, v/v); the flow rate was 0.8 mL/min. The injection volume was 10 µL and the detection wavelength was 350 nm.

2.4. Content Test for DNC in Solid Dispersion

To determine the content of DNC in SDs after ball milling, a certain weight of DNC SDs samples completely dissolved in 10 mL of dimethyl sulfoxide (DMSO) and was then analyzed by HPLC.

2.5. Powder X-ray Diffraction (PXRD)

An X-ray diffractometer (DNC/GA, Bruker D2 PHASER, Karlsruhe, Germany; DNC/GA/PVP K30, D/max-Ultima IV, Rigaku, Japan) was used to collect the wide-angle XRD profiles. The samples were exposed to CuKa radiation under 30 kV and 10 mA. The 2-theta of a wide angle XRD was in the range of 3°–40° with a speed of 4°/min, with a 0.02° step size.

2.6. Differential Scanning Calorimetry (DSC)

Thermal analysis (TA) of the accurately weighted amounts samples (5 mg) were carried out by DSC with the TA Instruments (SERIES 2000, Mettler-Toledo, Columbus, OH, USA) in Atmosphere. Samples were heated from 40 °C to 350 °C (heating rate: 10 °C/min).

2.7. Fourier Transform Infrared (FT-IR)

To measure the functional groups spectra of the samples used, a Nicolet iS50 Fourier spectrophotometer (Thermo Fisher Technology Co., Ltd., Waltham, MA, USA), in the range of 400 to 4000 cm^{-1} and the resolution was 2 cm^{-1}. All samples were dried and prepared as thin tablets with KBr.

2.8. Nuclear Magnetic Resonance (NMR) Relaxation Study

The NMR spectrum of DNC SDs was recorded on a Bruker NMR spectrometer, and samples were completely dissolved in D_2O solution, and the spin–spin relaxation time (T_2) was determined by the standard sequence Carr–Purcell–Meiboom–Gill (CPMG) of the Avance version of the Bruker pulse sequence library.

2.9. Scanning Electron Microscopy (SEM)

A Zeiss Gemini 500 field SEM (Carl Zeiss AG, Germany) was used to acquire the electronic images of samples. Before observation, the coating of samples with gold was performed.

2.10. Particle Size Distribution and Zeta Potential

Dissolving 1 mg sample in 1 mL pure water, the particle size, polydispersity index (PDI), and zeta potential of samples were measured by dynamic light scattering (DLS, Nano ZS90, Malvern Instruments, Malvern, UK).

2.11. Drug Encapsulation Efficiency (EE) and Loading Capacity (LC) Determination

The accurately weighed DNC SDs were dissolved in water, centrifuged, and then the supernatant was analyzed by HPLC. The EE and LC of DNC SDs were calculated according to Equations (1) and (2):

$$EE\ (\%) = (\text{weight of drug in nanoparticle})/(\text{total weight of drug}) \times 100\% \quad (1)$$

$$LC\ (\%) = (\text{weight of drug in micelles})/(\text{total weight of micelles}) \times 100\% \quad (2)$$

2.12. Solubility Determination

Excess DNC SD samples were added into a 20 mL Erlenmeyer flask with 10 mL distilled water, shaken (200 r) at 37 °C for 12 h. Then, the samples solution was filtered and analyzed by HPLC.

2.13. Transmission Electron Microscope (TEM)

DNC micelles were placed on a copper grid covered with nitrocellulose. TEM (HT7700, Hitachi Co., Ltd., Tokyo, Japan) was used to evaluate the morphology of DNC micelles in water.

2.14. Determination of Critical Micelle Concentration (CMC)

Nile red (NR) was used as the fluorescent probe to determine the CMC of micelles. First, 0.4 mg NR was added to a 10 mL acetone solution; then, 30 µL NR solution were added to 10 brown vials and evaporated, with each vial added in a different concentration gradient of DNC micelles solution, avoiding light shaking for 24 h. The samples solution detected the fluorescence intensity at the wavelength of 620 nm and excited at 579 nm.

2.15. Dissolution Determination

Dissolution tests of DNC SDs samples were performed using a dissolution tester (RC-6, Tianjin Jingtuo Instrument Technology Co., Ltd., Tianjin, China) at the paddle rotation speed of 100 rpm in 900 mL of distilled water at 37 ± 0.5 °C. Samples equivalent to 10 mg DNC were added to 900 mL of dissolution medium. At 5, 10, 15, 30, 45, 60, 90, and 120 min, 1 mL dissolution medium was withdrawn and an equal volume of distilled water was added. The collected samples were filtered and then analyzed by HPLC.

2.16. Parallel Artificial Membrane Permeability Assay (PAMPA)

The PAMPA experiment was used to predict the passive intestinal absorption of DNC SDs with transwell inserts (polycarbonate membrane, 6.5 mm, 0.4 µm pore size, Corning Incorporated). For the preparation of artificial membrane, each donor plate hole had 60 µL 5% added (2% DOPC in hexadecane) hexadecane/hexane solution (v/v), which evaporated completely. Then, each acceptor plate had 1 mL distilled water added, and the samples were added to the donor plate hole, shaken (25 °C, 200 rpm), taken out at 0.5, 1, 1.5, 2, 2.5, 3, 3.5, 4, 4.5, and 5 h, and evaluated by HPLC [13].

2.17. Animal Experiments

2.17.1. Bioavailability Study

Healthy male Sprague–Dawley (SD) rats with the weight of 200 ± 20 g were provided by the Zhejiang Academy of Medical Science. All the experiments involving SD rats were performed in accordance with protocols approved by the Ethics Committee of Zhejiang University of Technology (Certificate number: 201907).

SD rats were fasted 12 h before operation and randomly divided into 5 groups (n = 6 per group) as follows: DNC, NIC, NIC commercial product (NIC CP), DNC/GA, and DNC/GA/PVP K30 group with oral administration 0.5% CMC-Na suspension of DNC (90 mg/kg) equivalent weight. Blood was collected from the eyelids at different time points: 0.5, 1, 2, 4, 8, 12, 24, 48, and 72 h, then centrifuged (4 °C, 10,000 rpm for 10 min) with supernatant collected, and the plasma samples were stored at −80 °C and applied for determination of the DNC content. In addition, 1 mL methanol was added to a 100 µL plasma sample, and the solution was vortexed and centrifuged at 10,000 rpm for 10 min. The supernatant was collected and the content of DNC in the plasma was determined by ultra-high performance liquid chromatography-tandem mass spectrometry (UPLC-MS, ACQUITY H-Class/Xevo TQS, Waters Corporation, Milford, MA, USA). The UPLC system was as follows: chromatography was performed on a C18 column (2.1 × 50 mm, 1.7 µm, Acquity UPLC), column temperature was 30 °C, with methanol–water (90:10, v/v) as the mobile phase, and the flow rate was 0.2 mL/min. MS analysis: electrospray ionization (ESI) with positive mode; capillary voltage, 3 kV; source temperature, 150 °C desolvation temperature, 350 °C; gas flow rate, 650 L/H.

2.17.2. Tissue Distribution Experiment

Thirty SD rats were fasted 12 h and randomly divided into 5 groups (n = 6 per group) as follows: DNC, NIC, and NIC commercial product, DNC/GA and DNC/GA/PVP K30 group with oral administration 0.5% CMC-Na suspension of DNC (90 mg/kg) equivalent weight for 7 days in a row. After the experiment was over, all animals were euthanized using CO_2 following IACUC guidelines. The heart, liver, spleen, lungs, and kidneys were dissected and washed clean by normal saline (NS), wiped with filter paper. In addition,

0.2 g tissues were weighed and mixed with 1 mL NS and homogenized using a tissue grinder to obtain tissue homogenate. Furthermore, 0.5 mL tissue homogenate was added to 1 mL methanol and then homogenized and centrifuged at 4 °C at 10,000 rpm for 10 min; the supernatant was used to determine the content of DNC in the organs by UPLC-MS.

2.17.3. Anti-Occidial Activity

Twelve-day-old male broilers were weighed and assigned to 5 groups (n = 15 per group): negative control group (NC) with normal diet, positive control group (PC) with normal diet, DNC group (90 mg DNC /kg diet), NIC commercial product (NIC CP) group (125 mg NIC/kg diet) and DNC/GA/PVP K30 groups (540 mg DNC/GA/PVP K30 SD/L water). On day 1, each chick was orally challenged with 50,000 Eimeia tenella live oocysts except the NC group. The NC group was gavaged with distilled water at an equal volume. The clinical symptoms were recorded such as bloody stools, anorexia, huddling together, and disheveled feathers. The feces from days 7 to 9 were collected, and the amount of oocyst per gram (OPG) was measured according to the McMaster technique [24]. On the 9th day of the test, the chicks were weighed and sacrificed, the cecum dissected and scored, and the anticoccidial index (ACI) was calculated according to Equation (3) as well.

Survival rate (*100%) is the ratio of the number of surviving chickens to the initial number of chickens. The relative weight gain rate is the ratio of the average weight gain of the infected group to the average weight gain of the NC group. The lesions score included hemorrhages, thickening of the cecum wall and mucoid discharge, ranging from 0 to 4. Oocysts percent gram converts to oocysts value according to Table 1. The ACI value [24,25] was ordered in different grades: ACI < 120 (inactive), 120 < ACI < 140 (mild), 140 < ACI < 160 (moderate), and ACI > 180 (excellent).

Table 1. Conversion between oocysts percent gram and oocysts value.

OPG/PC OPG	<1%	1–25%	25–50%	50–75%	75–100%
Oocysts value	0	5	10	20	40

OPG/PC OPG = 100*Oocysts/g output of per group/Oocysts/g output of the PC group.

ACI = survival rate + relative weight gain rate (RWGR) − (lesions score + oocysts value) (3)

2.18. Stability Test of DNC SDs

The same net weight of DNC/GA/PVP K30 SD was packed in aluminum foil bags, stored in a drug stability test chamber at 40 °C, 75% humidity, and the drug content, particle size, zeta potential, and X-ray diffraction were measured at 1, 2, and 3 months.

2.19. Statistical Analysis

PKSolver (China Pharmaceutical University, Nanjing, China) was used to calculate the parameters of bioavailability experiments. The data statistical analyses were performed with IBM SPSS Statistics 26. After one-way analysis of variance (ANOVA), a two-tailed Student's t-test was used. Results were reported as the means ± S.E.M., and it is significantly different when the value of $p < 0.05$.

3. Results and Discussion

3.1. DSC

DSC thermograms of all solid samples were shown in Figure 2. The endothermic peak of pure DNC exhibited approximately 329.63 °C with a ΔH value of 708.13 J·g^{-1}. After adding GA and PVP K30, the endothermic peak of crystalline DNC still existed in the physical mixture because of the dilution effect. After ball milling, the intensity of DNC peak completely disappeared and ΔH values dropped to nearly 0 J·g^{-1}, certifying that DNC had converted to a partly amorphous state by the mechanical ball milling.

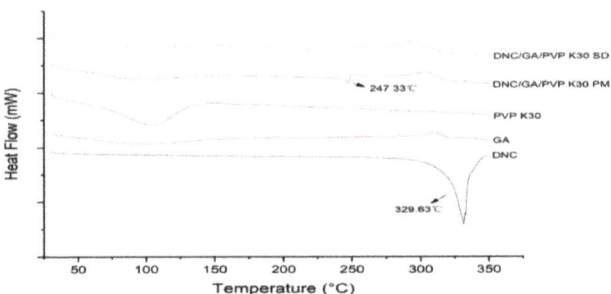

Figure 2. DSC curves of DNC, GA, PVP K30, DNC/GA/PVP K30 PM, and DNC/GA/PVP K30 SD.

3.2. PXRD

Powder X-ray diffractograms of all solid samples were shown in Figure 3. DNC has characteristic peaks at 2θ angles of 7.06, 9.42, 12.4, 16.5, 19.1, 24.6, 25.9, etc. DNC/GA PM and DNC/GA/PVP K30 PM showed several characteristic peaks as well, but the crystalline peaks of DNC SDs were drastically decreased or even disappeared. Both diffractograms and thermograms of DNC SDs' results illustrated the conversion of the DNC to an amorphous dispersion by mechanical ball milling. Previous studies had showed that the amorphous state of drug could enhance the water solubility and dissolution [21]. Zhang et al. prepared amorphous camptothecin solid dispersion to increase the solubility of camptothecin by 178-fold and the cumulative release amount reached 46% [13].

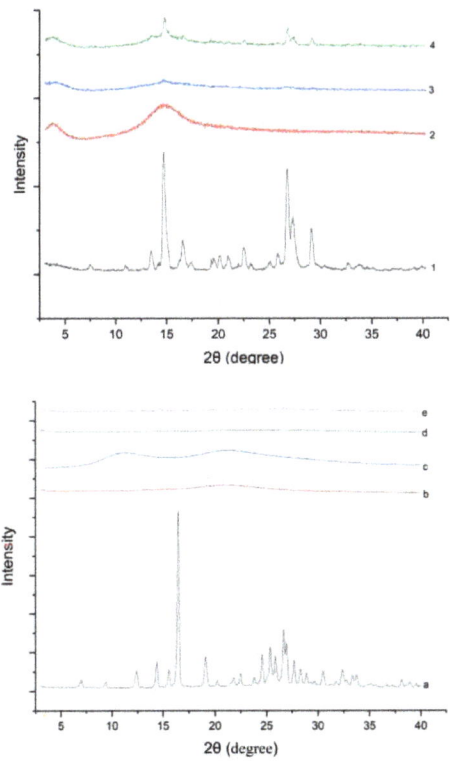

Figure 3. PXRD curves of (1, a) DNC, (2, b) GA, (3) DNC/GA SD, (4) DNC/GA PM, (c) PVP K30, (d) DNC/GA/PVP K30 SD, (e) DNC/GA/PVP K30 PM.

3.3. FT-IR

FT-IR spectroscopy was used to estimate the possible by-products in the ball milled DNC SDs, with the infrared spectrum of pure DNC, GA, PVP K30, and DNC PMs above, Figure 4. Pure DNC has strong tensile vibration peaks at 1731 cm^{-1}, 1654 cm^{-1}, 1497 cm^{-1}, 1114 cm^{-1}, 848 cm^{-1}, and 750 cm^{-1} observed the same characteristic peaks at DNC PM as well. Although the peak of DNC/GA/PVP K30 SD has weakened, there was no shift of the characteristic peaks, indicating that the coordination interaction may not have taken place and DNC were not destructed during the process of ball milling [26,27]. The most likely reason for the weakening of characteristic peaks is that the DNC SD has less components of DNC.

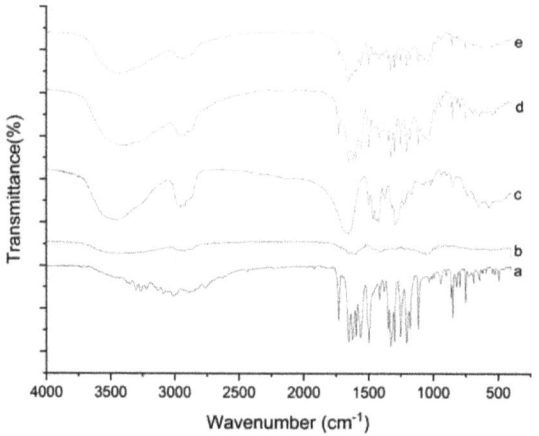

Figure 4. FT-IR spectra of (**a**) DNC, (**b**) GA, (**c**) PVP K30, (**d**) DNC/GA/PVP K30 SD, and (**e**) DNC/GA/PVP K30 SD.

3.4. SEM

Microscopic surface morphology of all solid samples was shown in Figure 5. The pure GA (Figure 5b) and PVP K30 (Figure 5c) without processing have a spherical and porous surface structure. When GA, PVP K30, and DNC were milled in a planetary ball mill for 30 min, the original shape of spherical GA and PVP K30 was destroyed, and the formulation of DNC SDs had irregular smaller particles. The results indicated that mechanical ball milling could reduce the particles and have more irregular shapes.

3.5. Solubility

The water solubility results of DNC, DNC SDs, and corresponding physical mixture were shown in Table 2. The solubility of physical mixtures was slightly higher than that of DNC because of the solubilizing capacity of GA and PVP K30. The operation and equipment may cause the slight fluctuation of drug content, which was measured. In the binary system, the solubility of DNC was 451.33 μg/mL, which was 22,566 times as much as that of pure DNC; however, the ternary system (DNC/GA/PVP K30 SD, 921.44 μg/mL) showed a 46,072-fold increase. These results indicated that GA might self-assemble to form micelles in water to increase solubility, and become significantly outstanding when PVP K30 was added, which preliminarily confirmed that the possible combination of GA and PVP K30 plays a role in solubilization.

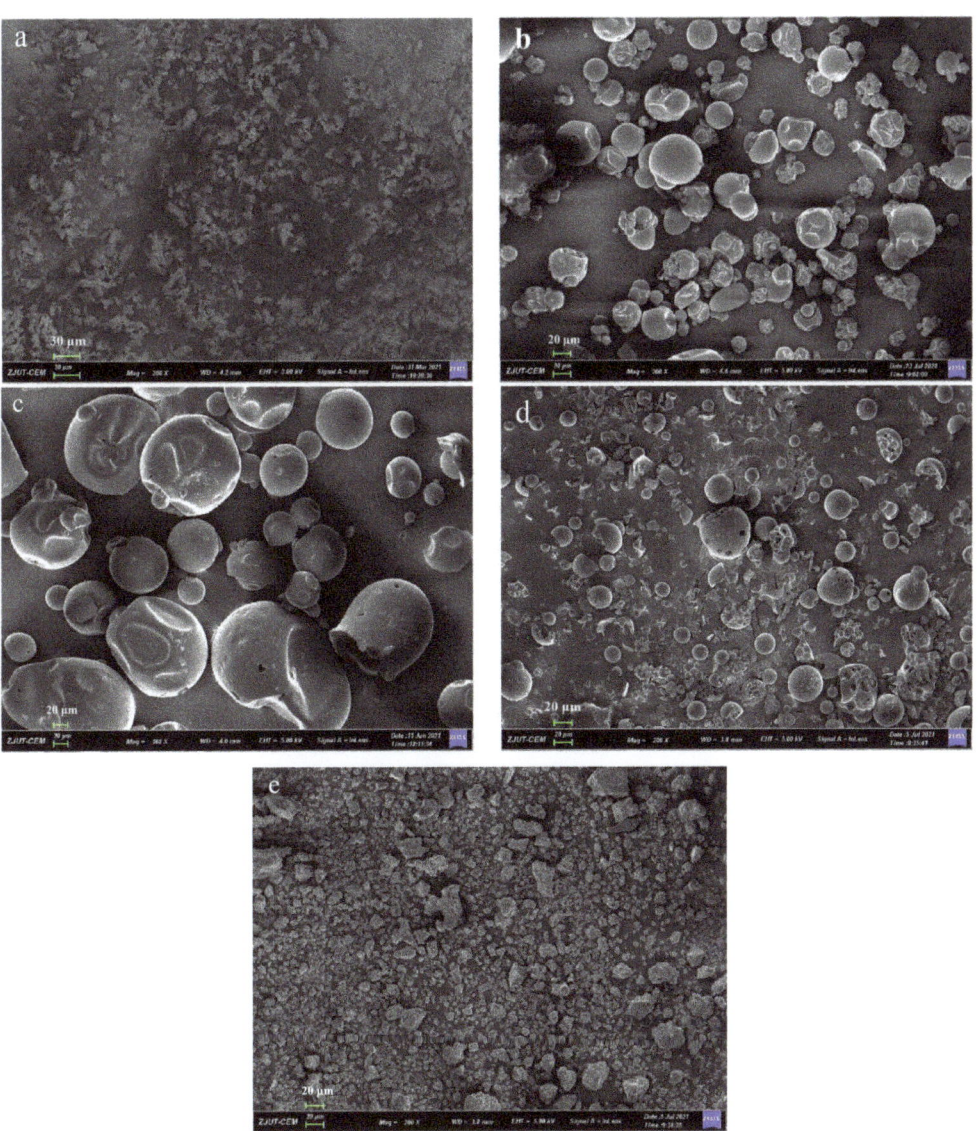

Figure 5. SEM images of (**a**) DNC, (**b**) GA, (**c**) PVP K30, (**d**) DNC/GA/PVP K30 PM, (**e**) DNC/GA/PVP K30 SD.

Table 2. The water solubility of DNC SDs and DNC PMs.

Sample	Solubility (μg/mL)	Increasing Times	Content (%)
DNC	0.02	-	100
DNC/GA SD	451.33	22,566.50	100.91
DNC/GA PM	0.10	5.00	105.18
DNC/GA/PVP K30 SD	921.44	46,072.00	96.89
DNC/GA/PVP K30 PM	1.55	77.50	105.46

3.6. Particle Size, PDI, Zeta Potential, EE and LC

Table 3 presents the results of particle size, PDI, zeta potential, encapsulation efficiency, and loading capacity of DNC SDs. The particle size of DNC/GA and DNC/GA/PVP K30 micelles were 101.01 nm and 101.27 nm (Figure 6), while performing with the same dispersion index. Compared with the binary system, DNC/GA/PVP K30 micelles showed less zeta potential (−38.27), high EE (99.76%) and LC (18.13%). Therefore, it was manifested that encapsulation of DNC into GA polymeric micelles resulted in a homogenous and stable dosage form in aqueous solution with high drug loading. The advantage was more obvious by adding PVP K30 and being consistent with water solubility, which makes DNC water-soluble administration possible.

Table 3. The particle size, PDI, zeta potential, EE, and LC of DNC SDs.

Sample	DNC/GA SD	DNC/GA/PVP K30 SD
Size (nm)	101.01 ± 0.07	101.27 ± 0.64
PDI	0.24 ± 0.01	0.236 ± 0.01
Zeta potential (mW)	−31.15 ± 0.68	−38.27 ± 0.21
EE (%)	69.39%	99.76%
LC (%)	13.87%	18.13%

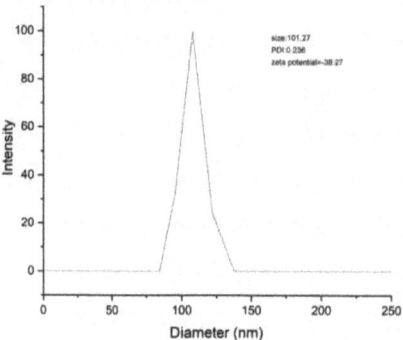

Figure 6. The particle size of DNC/GA/PVP K30 SD.

3.7. TEM

The morphological characters of DNC micelles in water are as shown in Figure 7. DNC SD dissolves in water to form smooth spheroid micelles by self-assembly (Appendix A).

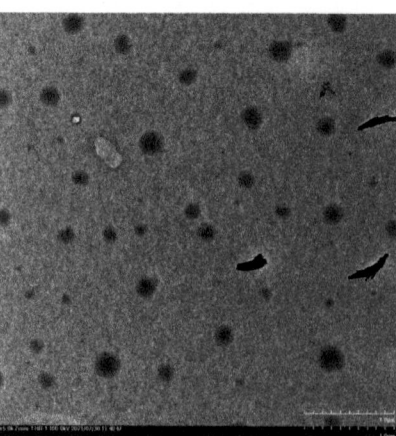

Figure 7. TEM image of DNC/GA/PVP K30 micelles.

3.8. CMC

The CMC value of DNC/GA micelles in an aqueous solution was 0.429 mg/mL, and the value of DNC/GA/PVP K30 micelles was 0.221 mg/mL, suggesting that the self-assembled micelles were extremely stable in the water solution (Figure 8). The increase of hydrophilicity could induce the enhancement of water solubility and decrease the CMC value further, which was consistent with the above results of water solubility.

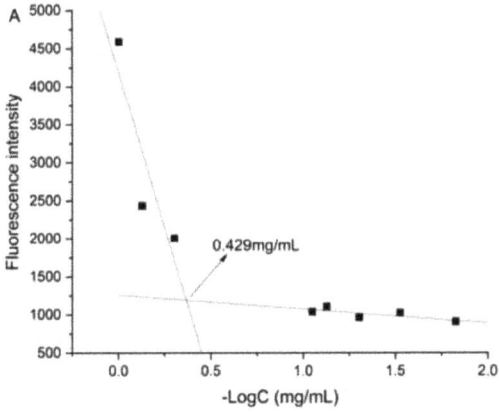

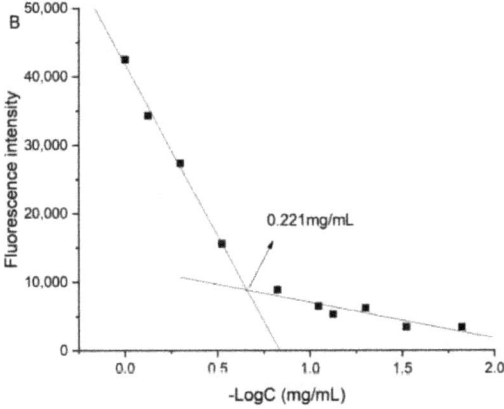

Figure 8. CMC value of (**A**) DNC/GA micelles; (**B**) DNC/GA/PVP K30 micelles.

3.9. The Relaxation Time

The method for the relaxation time of DNC and DNC solid dispersions by using an NMR relaxation study was discussed. The results of T_2 relaxation times of the DNC binary and ternary system were 637.50 ms and 79.63 ms, Figure 9, shorter than DNC. Spin–spin relaxation times are related to molecular motions and binding state, the slow molecular mobility, and the closer the intermolecular interaction, the smaller the T_2. It showed that the stabilization of DNC/GA/PVP K30 SD in D_2O was clearly higher than that of the binary system, indicating that it accumulated in water to form more stable micelles [27].

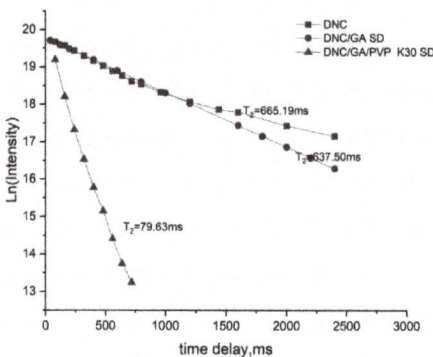

Figure 9. The relaxation time of DNC, DNC/GA SD, and DNC/GA/PVP K30 SD.

3.10. In Vitro Drug Release

The dissolution situation of DNS SDs was shown in Figure 10. In aqueous solution, the final release rates of the DNC/GA SD and DNC/GA/PVP K30 SD within 120 min were 55.6% and 79.38%. In general, the increase of solubility can improve the dissolution of drugs, while the amorphous drugs have a higher apparent solubility, which makes it easy to release drugs. For DNC/GA/PVP K30 SD, DNC was released very fast during the first five minutes and continued to increase its concentration up to 120 min. The pH of chicken intestine is almost 6.8 and digestion is fast, so it is desirable that the drug can be released rapidly in vivo and then exert drug efficacy. Experiment results revealed that DNC SDs could self-assemble into polymeric micelles in water and release in the intestine quickly.

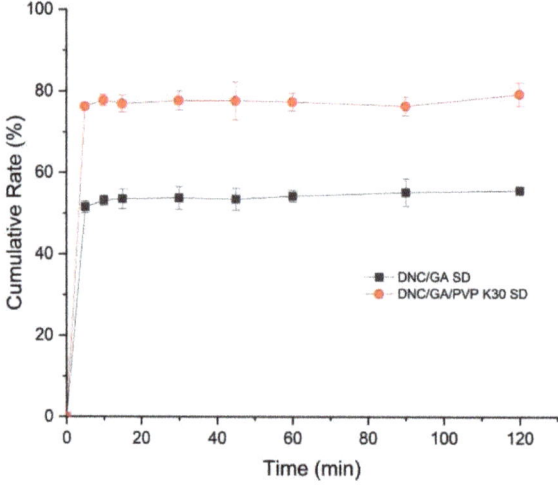

Figure 10. The cumulative release of DNC/GA SD and DNC/GA/PVP K30 SD.

3.11. Permeability

The cumulative penetration of DNC/GA/PVP K30 SD was 0.152 µg and increased 2.5 times that of DNC, and the results were statistically significant, presented in Figure 11. Researchers conduct the PAMPA to predict the passive transmembrane ability of compounds. In this study, our finding indicated that the cooperation of GA and PVP K30 stimulated the passive transport of DNC in the intestine.

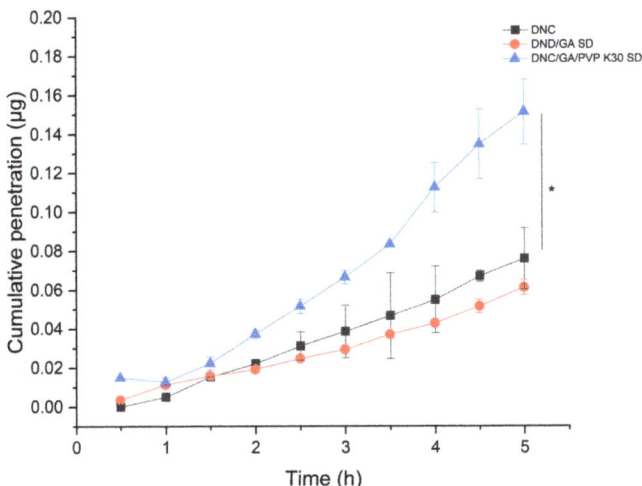

Figure 11. The cumulative penetration of DNC, DNC/GA SD, and DNC/GA/PVP K30 SD versus DNC, * $p < 0.05$.

In the binary system, GA dramatically improved the solubility and release of DNC in aqueous solution by the mechanical process. When DNC/GA SD dissolved in water solution, the iminos of DNC and the carboxyl group of GA could form intermolecular hydrogen bonds to change the phenomenon of insoluble water. In order to select a better SD system, the DNC/GA/PVP K30 SD was prepared. The above results indicated that the ternary system surprised us more compared with the binary system, although it was also a dispersed amorphous state, and the water solubility and dissolution rate of it were significantly prominent. Our hypothesis on the DNC/GA/PVP K30 SD was as follows: firstly, there was not a chemical reaction during the mechanochemical ball milling process but only decreased the size to an amorphous state. Secondly, PVP K30 could interact with GA. The hydroxyl group of GA combined with the carboxyl group of PVP K30 to form hydrogen bonds when DNC/GA/PVP K30 SD was dissolved in water; the PVP K30 contained a hydrophilic end and the nitro group of DNC as a lipophilic group aggregated and self-assembled to form micelles (Scheme 1). Certainly, DNC may also have hydrogen bonding with PVP K30 or GA to produce DNC-GA and DNC-PVP micelles, and these micelles might present at the same time.

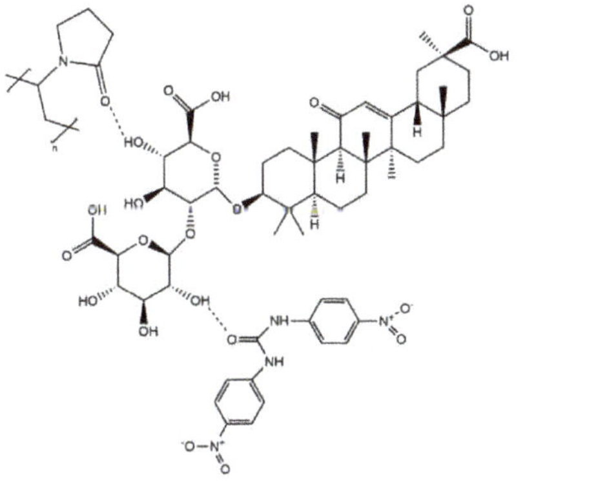

Scheme 1. The hydrogen bonding of DNC, GA, and PVP K30 in water.

3.12. Pharmacokinetic

As shown in Figure 12, profiles of plasma concentration time were determined after the oral administration of DNC and two species of DNC micelles in rats, respectively. After oral administration of DNC/GA/PVP K30 micelles, the T_{max} (time of maximum concentration), C_{max} (peak of plasma concentration), and AUC_{0-72} (area under the time curve) were 2.86 h, 14.46 μg/mL, and 238.35 μg/mL.h; however, the results of DNC were 4 h, 7.09 μg/mL, and 51.37 μg/mL.h, respectively (Table 4). The C_{max} and AUC_{0-72} in DNC/GA/PVP K30 micelles group were 2.04 times and 4.64 times higher than that for pure DNC ($p < 0.01$). It significantly proved that DNC/GA/PVP K30 SD self-assembled into micelles in water, enhanced the release rate, and improved the absorption of DNC in vivo by mechanical ball milling procession. Meanwhile, previous studies showed that the HDP possessed the property of facilitating the absorption [7]; in this research, the nicarbazin possessed higher oral bioavailability than DNC, which was consistent with it. Furthermore, the C_{max} of DNC/GA micelles was dramatically lower than pure DNC ($p < 0.01$), $T_{1/2}$ was extended, and AUC_{0-72} was 309 μg/mL.h; it was speculated that DNC/GA micelles have a sustained-release effect and could maintain the blood concentration for a longer time to a certain extent, but the low drug concentration may have difficulty exerting the activity of DNC. The oral bioavailability of DNC/PVP K30 SD was increased, which may be down to certain factors. Firstly, the pharmaceutical dosage form changed, the DNC had strong hydrophobic compounds, and the solubility of DNC could significantly increase in water after being prepared DNC SDs by mechanochemical grinding to enhance the oral bioavailability. With the development of pharmacy, the emergence of new drug dosage forms, on the one hand, to ensure the long-term efficacy of drugs, and, on the other hand, greatly facilitates patients. The same dose lipophilic compounds revealed different oral bioavailability, and the solution is higher than suspension and powder is the worst. Secondly, DNC and the hydrophilic carrier material (GA and PVP K30) form a better micellar system in aqueous solution. The ternary system showed higher water solution, drug release, and permeation compared with a binary system, promoting the increase of blood concentration when DNC/GA/PVP K30 micelles enter into blood.

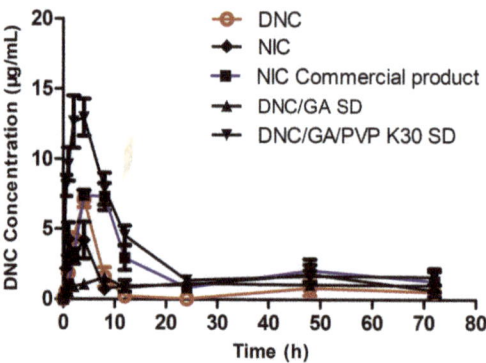

Figure 12. Concentration of DNC in plasma after oral administration.

Table 4. Pharmacokinetic parameters of DNC by oral administration (n = 6).

Parameter	DNC	NIC	NIC CP	DNC/GA SD	DNC/GA/PVP K30 SD
C_{max} (μg/mL)	7.09 ± 0.97	5.99 ± 2.13	8.29 ± 0.82 *	1.89 ± 0.58 **	14.46 ± 1.41 **
T_{max} (h)	4.00 ± 0.00	4.00 ± 0.00	5.2 ± 2.68	7.20 ± 1.79 **	2.86 ± 0.99 *
$T_{1/2}$ (h)	25.28 ± 4.01	22.91 ± 4.54	17.68 ± 6.51	95.60 ± 31.06	24.55 ± 3.85
$AUC_{0\to72}$ (μg/mL.h)	46.49 ± 11.83	79.82 ± 23.35 *	107.32 ± 28.45 **	93.78 ± 13.67 *	238.35 ± 27.73 **
$AUC_{0\to\infty}$ (μg/mL.h)	67.11 ± 20.83	87.14 ± 27.06 *	116.018 ± 34.61 **	239.30 ± 114.05 *	287.03 ± 72.75 **

Compared with DNC, * $p < 0.05$, ** $p < 0.01$.

3.13. Tissue Distribution Study

Figure 13 presented the tissue drug distribution profiles of DNC after oral administration of pure DNC and two species of DNC micelles in rats, respectively. After continuous oral administration for one week, DNC was distributed into heart, liver, spleen, lungs, and kidneys, compared with that in pure DNC. The DNC concentration in DNC/GA/PVP K30 micelles group was significantly higher in these tissues. This study was consistent with bioavailability results; when the drug entered into the bloodstream, it was absorbed by the liver and distributed throughout the body, finally excreted in the kidneys.

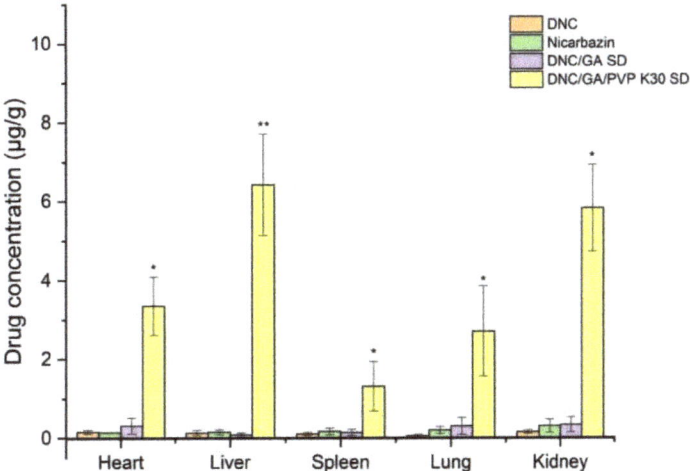

Figure 13. DNC concentration in tissue after oral administration (n = 6) versus DNC, * $p < 0.05$, ** $p < 0.01$.

The figure showed that the accumulation of DNC in the liver tissue was 6.42 μg/mL, which was the largest. Warren [28] found that nicarbazin had inhibited the activity of hepatosplenic shistosomiasis mansoni in mice and there was an improvement in sustainability after terminal treatment. Antunes et al. [29] reported the suppressive activity on oviposition in schistosoma mansoni infection. Due to the liver-targeting aggregation effect of DNC/GA/PVP K30 micelles, it could hypothesize that the DNC ternary system may have a strong inhibitory effect for hepatosplenic shistosomiasis mansoni and markedly ameliorates early disease. At the beginning of the experiment, the NIC commercial product was administrated as well; however, the study animals had unexplained deaths, and all of them died before the experiment was over. The results showed that the NIC commercial product was toxic to rats, which indicated a safety hazard for human beings and the importance of new product development.

3.14. Assessment of Anticoccidial Activity

3.14.1. Clinical Symptoms

After being infected with *E. tenella*, the PC showed disheveled feathers, dullness, and anorexia. These symptoms in DNC, NIC commercial product, and DNC/GA/PVP K30 groups were much milder. From the 5th day on, the appearance of hemorrhaging in feces were noted except NS; the 6th day was most serious, indicating that the model was successfully established. The DNC/GA/PVP K30 group had the least bloody stools, suggested that DNC SD could alleviate the appearance of hemorrhages, as showed in Figure 14-1. There was no dead chick throughout the experiment, thus the mortality rate was 0% in each group.

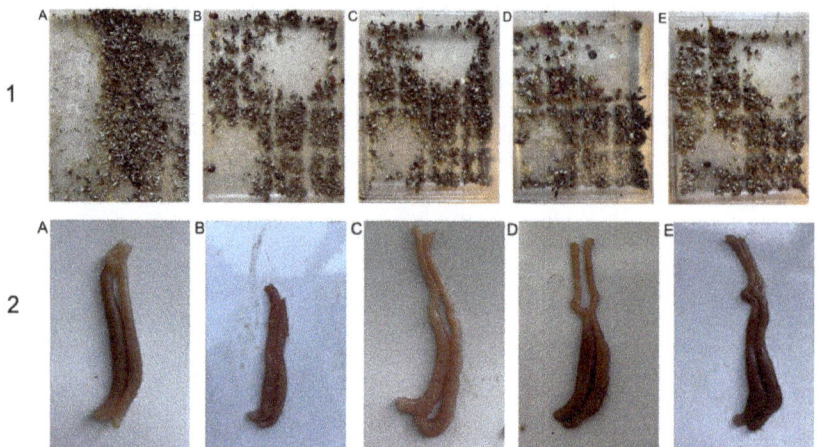

Figure 14. The images of hematochezia (1) and cecum (2) of chicken. (**A**) NC, (**B**) PC, (**C**) DNC, (**D**) NIC commercial product, (**E**) DNC/GA/PVP K30 SD.

3.14.2. Lesion Scores

Figure 14-2 presented the pathological lesion scores of study groups. PC observed severe pathologic lesions, which included atrophy, wall thickening, erosion, and dark blood clotting. The ceca in DNC and NIC commercial products displayed various degrees of morphology. According to pathologic performance, the cecum was scored, and the more severe, the higher the grades. It was clear that the lesion scores of NC and PC were 0 and 28, and DNC/GA/PVP K30 obtained a lower score than PC. Results suggested that both cecal lesions symptoms were relieved with DNC/GA/PVP K30.

3.14.3. Relative Weight Gain Rate

The average gain weight of chicks in PC and DNC were 44.8 g and 47.2 g, respectively, and relative weight gain rates were 82.05% and 86.45%, as shown in Table 5. After being infected with coccidiosis, the PC showed a decrease in food intake induced decreased weight gain, with a lower RWG than the other groups; although DNC could alleviate this to some extent, the effect was minimal without HDP. After prophylactic administration with DNC/GA/PVP K30, the RWG up to 96.66% was better than the NIC commercial product.

Table 5. The relative weight gain rate of each group.

Sample	NC	PC	DNC	NIC CP	DNC/GA/PVP K30 SD
The average weight gain (g)	54.60 ± 8.04	44.80 ± 16.18	47.20 ± 6.28	50.00 ± 6.14	52.78 ± 6.14
Relative weight gain rate (%)	100	82.05	86.45	91.57	96.66

3.14.4. Oocysts per Gram Output and ACI

Except for NC, the oocysts in feces could be detected, and the number of oocysts significantly decreased after the administration of DNC/GA/PVP K30. The relative inhibition rate of oocysts of DNC, NIC commercial product, and DNC/GA/PVP K30 SD were 49.01%, 34.95%, and 77.36%, as shown in Table 6. The PC showed a non-anticoccidial effect, and the ACI index was 114.05. ACI indices of DNC and NIC commercial product groups were 127.45 and 135.67 (Table 7), which were regarded as inefficient anticoccidial agents. The DNC SD treated group showed moderate anticoccidial effects, and the ACI index was 167.67, Table 7.

Table 6. The oocysts percent gram of each group.

Sample	NC	PC	DNC	NIC CP	DNC/GA/PVP K30 SD
Oocysts percent gram/($\times 10^6$)	0 ± 0	30.33 ± 1.16	15.47 ± 0.66	19.76 ± 1.11	6.88 ± 0.74
Relative inhibition rate (%)	-	0	49.01	34.95	77.36
Oocysts value	0	40	20	20	5

Table 7. ACI.

Sample	NC	PC	DNC	NIC CP	DNC/GA/PVP K30 SD
Survival rate (%)	100	100	100	100	100
RWG rate (%)	100	82.05	86.45	91.57	96.67
Cecum score	0	28	19	16	9
Oocysts value	0	40	20	20	5
ACI	200	114.05	147.45	155.57	182.67

When chickens eat infective sporozoites, it reaches the cecum in the digestive tract and infects cecal epithelial cells into capillaries, with blood in the intestinal gland to complete the fission and sexual reproduction stage, and then penetrating the intestinal villous epithelial cells into the intestinal lumen to destruct the cecum with some symptoms occurring such as hemorrhaging in feces, disheveled feathers, dullness, anorexia, and inducing mass fatality in severe cases [30]. In order to reduce the risk of infection, farms usually for sustainability overused anticoccidial drugs. Due to adding them long-term to the diet, the oocysts used in the experiment may have developed some resistance to nicarbazin active ingredients and may be the reason for the low ACI value of the experimental NIC commercial product. Although HDP was an inactive substance and enhanced the activity of DNC 10-fold, researchers presumed, by increasing absorption, that the water solubility has not been improved, which was still inconvenient for drug administration. In this research, we used DNC (NIC active compound), GA, and PVP K30 to prepare solid dispersion without any organic solution, not only enhancing water solubility and oral bioavailability, but also the anticoccidioidal capability being better than NIC commercial products. Moreover, there was a relationship wherein the ACI increased with an enhancement in bioavailability, and the increased order as follows: DNC, NIC commercial product, and DNC/GA/PVP K30. Based on previous results, HDP enhanced 10-fold the anticoccidioidal capability of DNC by facilitating absorption, and we speculated that the DNC ternary solid dispersion system prepared by mechanochemistry could enhance the efficacy in the same certain, exceeding that of the co-crystal compound nicarbazine composed of HDP and DNC.

3.15. Stability Study

Figure 15 showed the results of DNC/GA/PVP K30 SD of PXRD, drug content, particle size, and zeta potential after three months of storage. As shown below, DNC/GA/PVP K30 SD has no significant crystal transformation and is still in a relatively amorphous state. Meanwhile, the DNC content of SDs, particle size, and zeta potential remained relatively steady compared with the original state. For DNC/GA/PVP K30 SD, which was packed with an aluminum foil bag, the size of particles increased after 30 days conjecturing that the package of that batch was not sealed completely, causing moisture absorption and agglomeration, and then decreased, indicating that DNC solid dispersion could keep the drug stable for a certain time. This evidence manifested that DNC SDs that were prepared by a mechanochemical force could reduce the generation of outgrowth, maintaining stability during the storage process.

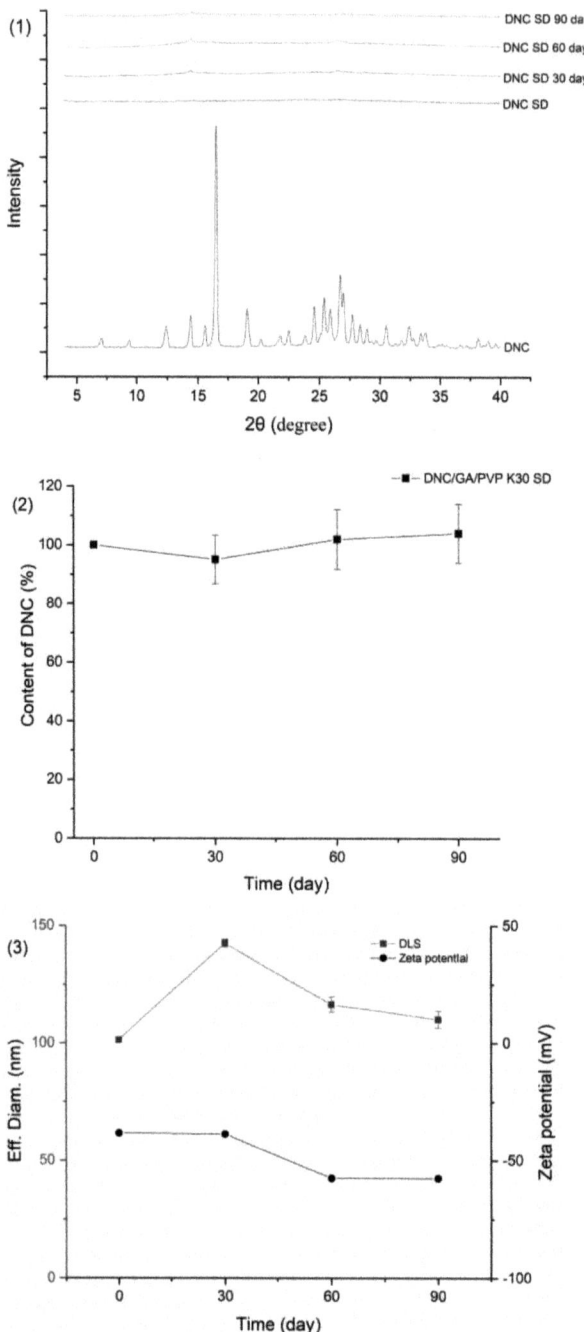

Figure 15. Stability test of (**1**) PXRD, (**2**) Drug content, (**3**) Particle size, and zeta potential of DNC/GA/PVP K30 SD.

4. Conclusions

In this investigation, DNC/GA/PVP K30 solid dispersion was successfully prepared via mechanical ball milling, which self-assembled to form DNC micelles in water and assigned the therapy of coccidiosis in vivo. The characterization of solid state results demonstrated that DNC was dispersed in GA/PVP K30, the crystal disappeared into an amorphous state, there was no chemical reaction during the ball milling process, and stability can be maintained in a certain period. When DNC/GA/PVP K30 SD dissolved in aqueous solution, the solubility of DNC enhanced 46,072 times that of pure DNC, and the drug release behavior in vitro was significantly improved. After the DNC was entrapped to form micelles, the concentration of DNC in blood and tissue enhanced, and the anticoccidial effect was correspondingly increased compared with free drugs, which makes it possible for animals to be administrated DNC drug through drinking water. In addition, the accumulation of DNC micelles in the liver tissue may have potential therapeutic and preventive effects on schistosomiasis. Thus, the research elucidated an innovative preparation of an amorphous DNC SD formulation by mechanical ball milling and might be a promising formulation to increase the therapeutic.

Author Contributions: Conceptualization, W.S.; methodology, M.L.; software, W.W.; formal analysis, N.E.P.; resources, W.S.; data curation, W.X.; writing, M.L.; project administration, A.V.D.; funding acquisition, W.S. All authors have read and agreed to the published version of the manuscript.

Funding: This research was funded by the Science and Technology Agency of Zhejiang Province of China, Grant No. 2019C04023.

Institutional Review Board Statement: The study was conducted in accordance with the Declaration of Helsinki, and approved by the Ethics Committee of Zhejiang University of Technology (Certificate Number: 201907, Date of Approval: 28/5/2021).

Informed Consent Statement: Not applicable.

Data Availability Statement: Not applicable.

Conflicts of Interest: The authors declare no conflict of interest.

Appendix A

When DNC/GA/PVP K30 SD was dissolved in water, the Tyndall effect appeared, indicating that solid dispersion was evenly dispersed in water to form micelles.

Figure A1. DNC/GA/PVP K30 SD was dissolved in water solution.

References

1. Blake, D.P.; Tomley, F.M. Securing poultry production from the ever-present Eimeria challenge. *Trends Parasitol.* **2013**, *30*, 12–19. [CrossRef]
2. Chapman, H.D. Milestones in avian coccidiosis research: A review. *Poult. Sci.* **2014**, *93*, 501–511. [CrossRef] [PubMed]
3. Oljira, D.; Melaku, A.; Bogale, B. Prevalence and Risk Factors of Coccidiosis in Poultry Farms in and Around Ambo Town, Western Ethiopia. *Am.-Eurasian J. Sci. Res.* **2012**, *7*, 146–149.
4. Karaer, Z.; Guven, E.; Akcay, A.; Kar, S.; Nalbantoglu, S.; Cakmak, A. Prevalence of subclinical coccidiosis in broiler farms in Turkey. *Trop. Anim. Health Prod.* **2012**, *44*, 589–594. [CrossRef] [PubMed]
5. Da Costa, M.J.; Bafundo, K.W.; Pesti, G.M.; Kimminau, E.A.; Cervantes, H.M. Performance and anticoccidial effects of nicarbazin-fed broilers reared at standard or reduced environmental temperatures. *Poult. Sci.* **2017**, *96*, 1615–1622. [CrossRef]
6. Mcloughlin, D.K.; Wehr, E.E. Stages in the Life Cycle of *Eimeria tenella* Affected by Nicarbazin. *Poult. Sci.* **1960**, *39*, 534–538. [CrossRef]
7. Porter, C.C.; Gilfillan, J.L. The Absorption and Excretion of Orally Administered Nicarbazin by Chickens. *Poult. Sci.* **1955**, *34*, 995–1001. [CrossRef]
8. Rogers, E.; Brown, R.; Brown, J.; Kazazis, D.; Leanza, W.; Nichols, J.; Ostlind, D.; Rodino, T. Nicarbazin complex yields dinitrocarbanilide as ultrafine crystals with improved anticoccidial activity. *Science* **1983**, *222*, 630–632. [CrossRef]
9. Bhujbal, S.V.; Mitra, B.; Jain, U.; Gong, Y.; Zhou, Q.T. Pharmaceutical amorphous solid dispersion: A review of manufacturing strategies. *Acta Pharm. Sin. B* **2021**, *11*, 2505–2536. [CrossRef]
10. Zheng, L.; Sun, C.; Zhu, X.; Xu, W.; Su, W. Inositol Hexanicotinate Self-micelle Solid Dispersion is an Efficient Drug Delivery System in the Mouse Model of Non-alcoholic Fatty Liver Disease. *Int. J. Pharm.* **2021**, *602*, 120576. [CrossRef]
11. Dushkin, A.V.; Tolstikova, T.G. Complexes of Polysaccharides and Glycyrrhizic Acid with Drug Molecules—Mechanochemical Synthesis and Pharmacological Activity. In *The Complex World of Polysaccharides*; IntechOpen: London, UK, 2012.
12. Zhang, Q.; Feng, Z.; Wang, H.; Su, C.; Lu, Z.; Yu, J.; Dushkin, A.V.; Su, W. Preparation of camptothecin micelles self-assembled from disodium glycyrrhizin and tannic acid with enhanced antitumor activity. *Eur. J. Pharm. Biopharm.* **2021**, *164*, 75–85. [CrossRef] [PubMed]
13. Zhang, Q.; Wang, H.; Feng, Z.; Lu, Z.; Su, W. Preparation of pectin-tannic acid coated core-shell nanoparticle for enhanced bioavailability and antihyperlipidemic activity of curcumin. *Food Hydrocoll.* **2021**, *119*, 106858. [CrossRef]
14. Su, W.; Polyakov, N.E.; Xu, W.; Su, W. Preparation of Astaxanthin Micelles Self-Assembled by a Mechanochemical Method from Hydroxypropyl β-Cyclodextrin and Glyceryl Monostearate with Enhanced Antioxidant Activity. *Int. J. Pharm.* **2021**, *605*, 120799. [CrossRef] [PubMed]
15. Xu, W.; Su, W.; Xue, Z.; Pu, F.; Xie, Z.; Jin, K.; Polyakov, N.E.; Dushkin, A.V.; Su, W. Research on Preparation of 5-ASA Colon-Specific Hydrogel Delivery System without Crosslinking Agent by Mechanochemical Method. *Pharm. Res.* **2021**, *38*, 693–706. [CrossRef]
16. Selyutina, O.Y.; Polyakov, N.E. Glycyrrhizic acid as a multifunctional drug carrier—From physicochemical properties to biomedical applications: A modern insight on the ancient drug. *Int. J. Pharm.* **2019**, *559*, 271–279. [CrossRef]
17. Gong, C.; Deng, S.; Wu, Q.; Xiang, M.; Wei, X.; Li, L.; Xiang, G.; Wang, B.; Lu, S.; Chen, Y. Improving antiangiogenesis and anti-tumor activity of curcumin by biodegradable polymeric micelles. *Biomaterials* **2013**, *34*, 1413–1432. [CrossRef]
18. Yang, F.H.; Zhang, Q.; Liang, Q.Y.; Wang, S.Q.; Zhao, B.X.; Wang, Y.T.; Cai, Y.; Li, G.F. Bioavailability enhancement of paclitaxel via a novel oral drug delivery system: Paclitaxel-loaded glycyrrhizic acid micelles. *Molecules* **2015**, *20*, 4337–4356. [CrossRef]
19. Koczkur, K.K.; Mourdikoudis, S.; Polavarapu, L.; Skrabalak, S.E. Polyvinylpyrrolidone (PVP) in nanoparticle synthesis. *Dalton Trans.* **2015**, *44*, 17883–17905. [CrossRef]
20. Bühler, V. *Polyvinylpyrrolidone Excipients for Pharmaceuticals*; Springer: Berlin/Heidelberg, Germany, 2005.
21. Kurakula, M.; Rao, G. Pharmaceutical Assessment of Polyvinylpyrrolidone (PVP): As Excipient from Conventional to Controlled Delivery Systems with a Spotlight on COVID-19 Inhibition. *J. Drug Deliv. Sci. Technol.* **2020**, *60*, 102046. [CrossRef]
22. Frizon, F.; Eloy, J.; Donaduzzi, C.M.; Mitsui, M.L.; Marchetti, J.M. Dissolution rate enhancement of loratadine in polyvinylpyrrolidone K-30 solid dispersions by solvent methods. *Powder Technol.* **2013**, *235*, 532–539. [CrossRef]
23. Erel, I.; Sukhishvili, S.A. Hydrogen-Bonded Multilayers of a Neutral Polymer and a Polyphenol. *Macromolecules* **2008**, *41*, 3962–3970. [CrossRef]
24. Long, P.L.; Millard, B.J.; Joyner, L.P.; Norton, C.C. A guide to laboratory techniques used in the study and diagnosis of avian coccidia. *Folia Vet. Lat.* **1976**, *6*, 201–217. [PubMed]
25. Morisawa, Y.; Kataoka, M.; Kitano, N.; Matsuzawa, T. Studies on anticoccidial agents. 10. Synthesis and anticoccidial activity of 5-nitronicotinamide and its analogs. *J. Med. Chem.* **1977**, *20*, 129–133. [CrossRef] [PubMed]
26. Zhang, Q.H.; Suntsova, L.; Chistyachenko, Y.S.; Evseenko, V.; Khvostov, M.V.; Polyakov, N.E.; Dushkin, A.V.; Su, W.K. Preparation, physicochemical and pharmacological study of curcumin solid dispersion with an arabinogalactan complexation agent. *Int. J. Biol. Macromol.* **2019**, *128*, 158–166. [CrossRef]
27. Zhang, Q.; Polyakov, N.E.; Chistyachenko, Y.S.; Khvostov, M.V.; Frolova, T.S.; Tolstikova, T.G.; Dushkin, A.V.; Su, W. Preparation of curcumin self-micelle solid dispersion with enhanced bioavailability and cytotoxic activity by mechanochemistry. *Drug Deliv.* **2018**, *25*, 198–209. [CrossRef]

28. Warren, K.S. Suppression of Hepatosplenic Shistosomiasis Mansoni in Mice by Nicarbazin, a Drug That Inhibits Egg Production by Schistosomes. *J. Infect. Dis.* **1970**, *121*, 514–521. [CrossRef]
29. Antunes, C.M.F.; Katz, N.; Dias, E.P.; Pellegrino, J. An attempt to obtain infertile live worms in mice infected with *Schistosoma mansoni*. *Ann. Trop. Med. Parasitol.* **1974**, *68*, 237–238. [CrossRef]
30. Kawazoe, U.; Di Fabio, J. Resistance to diclazuril in field isolates of *Eimeria* species obtained from commercial broiler flocks in Brazil. *Avian Pathol.* **1994**, *23*, 305–311. [CrossRef]

Article

Development and Characterization of PEGylated Fatty Acid-*Block*-Poly(ε-caprolactone) Novel Block Copolymers and Their Self-Assembled Nanostructures for Ocular Delivery of Cyclosporine A

Ziyad Binkhathlan [1,2,*,†], Abdullah H. Alomrani [1,2,†], Olsi Hoxha [1], Raisuddin Ali [1,2], Mohd Abul Kalam [1,2] and Aws Alshamsan [1,2]

1. Department of Pharmaceutics, College of Pharmacy, King Saud University, P.O. Box 2457, Riyadh 11451, Saudi Arabia; aomrani@ksu.edu.sa (A.H.A.); 436108203@student.ksu.edu.sa (O.H.); ramohammad@ksu.edu.sa (R.A.); makalam@ksu.edu.sa (M.A.K.); aalshamsan@ksu.edu.sa (A.A.)
2. Nanobiotechnology Research Unit, College of Pharmacy, King Saud University, P.O. Box 2457, Riyadh 11451, Saudi Arabia
* Correspondence: zbinkhathlan@ksu.edu.sa
† These authors contributed equally to this work.

Abstract: Low aqueous solubility and membrane permeability of some drugs are considered major limitations for their use in clinical practice. Polymeric micelles are one of the potential nano-drug delivery systems that were found to ameliorate the low aqueous solubility of hydrophobic drugs. The main objective of this study was to develop and characterize a novel copolymer based on poly (ethylene glycol) stearate (Myrj™)-*block*-poly(ε-caprolactone) (Myrj-*b*-PCL) and evaluate its potential as a nanosystem for ocular delivery of cyclosporine A (CyA). Myrj-*b*-PCL copolymer with various PCL/Myrj ratios were synthesized *via* ring-opening bulk polymerization of ε-caprolactone using Myrj (Myrj S40 or Myrj S100), as initiators and stannous octoate as a catalyst. The synthesized copolymers were characterized using ^{1}H NMR, GPC, FTIR, XRD, and DSC. The co-solvent evaporation method was used to prepare CyA-loaded Myrj-*b*-PCL micelles. The prepared micelles were characterized for their size, polydispersity, and CMC using the dynamic light scattering (DLS) technique. The results from the spectroscopic and thermal analyses confirmed the successful synthesis of the copolymers. Transmission electron microscopy (TEM) images of the prepared micelles showed spherical shapes with diameters in the nano range (<200 nm). Ex vivo corneal permeation study showed sustained release of CyA from the developed Myrj S100-*b*-PCL micelles. In vivo ocular irritation study (Draize test) showed that CyA-loaded Myrj S100-b-PCL$_{88}$ was well tolerated in the rabbit eye. Our results point to a great potential of Myrj S100-*b*-PCL as an ocular drug delivery system.

Keywords: cyclosporine A; ethoxylated fatty acid; block copolymer; polymeric micelles; ocular

---

1. Introduction

Ethoxylated fatty acids e.g., PEG stearates (sold under the trademark Myrj™) are non-ionic surfactants widely used in various drug delivery systems (Figure 1). When intravenously injected, the PEG component of Myrj™ extends the circulation of the drug in plasma, while the fatty acid enhances the solubility of the fat-soluble drug. Depending on the length of PEG used, the ethoxylated fatty acid products have hydrophilic-lipophilic balance (HLB) values in the range of 11–18.8 [1].

The United States Food and Drug Administration (US FDA) has approved the use of Myrj™ products as safe excipients used in cosmetics, pharmaceutical formulations, and food additives [1]. Recently, there has been a growing interest in the use of Myrj™ products in the pharmaceutical industry. Several of these Myrj™ products have been used as absorption enhancers, emulsifiers, solubilizers, permeation enhancers, and stabilizers.

Some derivatives of Myrj™ have also been utilized as inhibitors of P-gp to increase intestinal permeability and enhance the oral bioavailability of P-gp substrates [2–4].

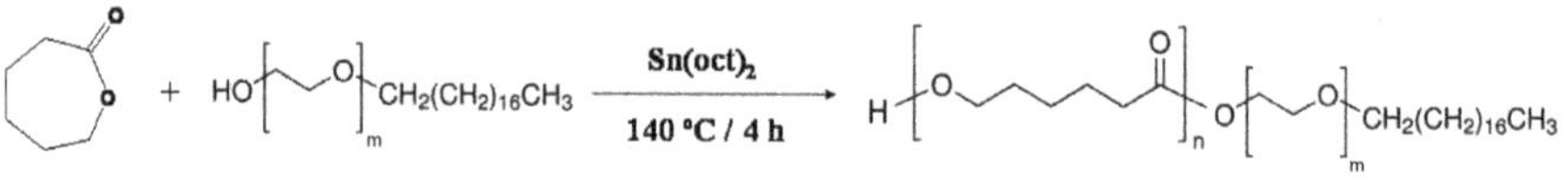

Figure 1. Synthetic scheme for Myrj-*b*-PCL copolymers.

Poly (ε-caprolactone) (PCL) is an eco-friendly polyester and has been widely utilized for tissue engineering and controlled drug delivery applications [5]. It allows homogenous distribution of drug molecules within the polymer matrix due to better compatibility with a variety of drugs [6]. Moreover, PCL exhibit, very slow degradation leading to prolonged drug release lasting for a few months [7]. The mechanical and physicochemical properties of PCL can easily be modified by blending or co-polymerization with different polymers. It has been found that co-polymerization of PCL with different polymers alters several intrinsic properties of PCL such as solubility, ionization, degradation pattern, and crystallinity resulting in a custom-made polymer with desired features for efficient drug delivery [8].

Polymeric micelles fabricated from PCL-based polymer were successful candidates for the delivery of lipophilic drugs such as valspodar, CyA, and paclitaxel [9–11]. Moreover, we recently showed that copolymerization of PCL with D–α–tocopheryl polyethylene glycol succinate (TPGS; with PEG MW > 2 kDa) was capable to self-assemble into nanocarriers with diameters smaller than 200 nm and critical association concentrations (CAC) in the nanomolar range [6]. More importantly, several derivatives of TPGS-*b*-PCL significantly improved the water solubility of paclitaxel (from ca. 0.3 µg/mL up to 88.4 µg/mL) and sustained the release of the loaded drug in vitro [6]. We demonstrated that paclitaxel-loaded TPGS$_{5000}$-*b*-PCL$_{15000}$ micellar formulation exhibited less than 10% drug release during the first 12 h, and approximately 36% cumulative drug release during 72 h in contrast to 61 and 100% paclitaxel release, respectively, from the marketed formulation (Ebetaxel®) [6].

Cyclosporine A (CyA) is a cyclic peptide and potent immunosuppressant agent prescribed for organ transplant patients (including heart, lung, and kidney) to prevent organ rejection. It is also very effective in the treatment of bone marrow transplants, systemic immune disorders, and various dermatological diseases (alopecia areata, psoriasis, pyoderma gangrenosum) [12]. Moreover, CyA is used to treat dry eye disease and uveitis [13]. However, CyA has poor corneal permeation owing to its rigid cyclic structure, high molecular weight, and barrier properties of the cornea [14]. For instance, after administration of a common eye-drop solution containing 0.05% CyA to treat dry eye syndrome, more than 95% of the drug reaches systemic circulation through trans-nasal or conjunctival absorption [15].

The main objective of the current study was to develop and characterize a novel copolymer based on poly (ethylene glycol) stearate (Myrj™)-*block*-poly(ε-caprolactone) (Myrj-*b*-PCL) and evaluate its potential as a nanosystem for ocular delivery of CyA.

2. Materials and Methods

2.1. Materials

Ethoxylated fatty acids [PEG-40 Stearate (commercially known as Myrj™ 52 or Myrj™ S40) and PEG-100 Stearate (commercially known as Myrj™ 59 or Myrj™ S100)], stannous octoate (95%), ε-Caprolactone, and HPLC-grade tetrahydrofuran (THF) were purchased from Sigma-Aldrich (St. Louis, MO, USA). Deuterated chloroform (CDCl$_3$, 99.8%) was purchased from Cambridge Isotope Laboratories Inc. (Tewksbury, MA, USA). Acetonitrile (HPLC grade) was supplied by Fisher Scientific Co. (Leicestershire LE/15 RG, UK). Cyclosporine A (CyA) was purchased from Molekula Group LLC (Irvine, CA, USA). Potassium dihydrogen orthophosphate, dipotassium hydrogen orthophosphate, and potassium chloride were obtained from BDH Chemical Ltd. (Poole, England). All other chemicals were of analytical grade. Deionized water was prepared in-house using a Millipore system.

2.2. Methods

2.2.1. Synthesis of Myrj-b-PCL Copolymers

Ring-opening polymerization of ε-caprolactone was the approach used to synthesize the copolymers [6,16]. For Myrj-b-PCL copolymer, either Myrj S40 (MW ~ 2000 Da) or Myrj S100 (MW ~ 4700 Da) were used as a macroinitiator and stannous octoate was used as the catalyst. Monomer (ε-caprolactone) to catalyst molar ratio was always kept at 1:500. Different ε-caprolactone to Myrj feed ratios were used to synthesize Myrj-b-PCL block copolymers with varying degrees of ε-caprolactone polymerization. Briefly, either Myrj S40 or Myrj S100, ε-caprolactone, and stannous octoate were added to an ampoule that was previously flamed and purged with nitrogen gas, which was then sealed under vacuum. The reaction was conducted at 140 °C for 4 h. Then, it was terminated by removing the reaction vessel (ampoule) from the oven and storing it at room temperature overnight.

2.2.2. Characterization of the Myrj-b-PCL Copolymers

^{1}H Nuclear Magnetic Resonance (^{1}H NMR) Spectroscopy

The products of the chemical reaction were evaluated by ^{1}H NMR (Bruker Ultra shield 500.133 MHz spectrometer) in CDCL$_3$. Tetramethylsilane (TMS) was used as an internal standard, and Topsin software was used to process the data and obtain the spectra. The number average molecular weight of all synthesized copolymers was determined from ^{1}H NMR spectra by comparing the peak intensity of (–CH$_2$CH$_2$O–, δ = 3.65 ppm) present in the PEO segment of each copolymer to that of PCL (H–O–CH$_2$–, δ = 4.07 ppm). The calculation used the integration area of the peaks of methylene protons of PCL at 4.07 ppm and of PEG at 3.65 ppm, respectively [6,16].

GPC Chromatography

Gel permeation chromatography system GPCmax (Malvern Panalytical Ltd., Malvern, OH, USA) and Viscotek Triple Detector Array 305 (Malvern, OH, USA) were used to determine the weight-average and number-average molecular weights as well as the molecular weight distribution of Myrj-b-PCL copolymers. A sample of 100 µL of the polymer from the prepared solution was injected into PLgel 5 µm MIXED-D, 7.5 × 300 mm (Agilent Technologies Inc., Santa Clara, CA, USA). The sample was pumped at a rate of 1 mL/min using THF as a mobile phase. The analysis was performed at 35 °C temperature. The data acquisition and processing were carried out by OmniSEC^{TM} (Version 4.7.0.406, Malvern Panalytical Ltd., Malvern, OH, USA) software. Prior to running the samples, GPC calibration was performed using EasiVial PS-M standards (Varian, Palo Alto, CA, USA).

Fourier-Transform Infrared (FTIR) Spectroscopy

The FTIR spectra of Myrj-b-PCL copolymers were recorded using an Alpha FTIR spectrophotometer (Bruker, Karlsruhe, Germany) equipped with Platinum Attenuated Total Reflectance (ATR) Module and diamond hemisphere. The instrument control as well as the data recording and processing are performed using OPUS version 7.8 (Bruker Optik GmbH, Ettlingen, Germany) software. Briefly, a small portion of the sample was kept on the sample holder covering the diamond hemisphere and the infrared (IR) beam was allowed to pass through the sample. The data were recorded in mid-infrared range starting from 4000 to 375 cm^{-1} with a spectral resolution of 2 cm^{-1}. The software recorded the percentage transmittance with changing wave numbers to provide the infrared spectra.

Differential Scanning Calorimetry (DSC)

Thermograms of the Myrj-b-PCL copolymers were obtained utilizing DSC-60 (Shimadzu, Tokyo, Japan). Samples (3–5 mg) of the copolymers were loaded into aluminum pans. The samples were heated from 25–200 °C under nitrogen gas at a heating rate of

10 °C/min and purging at 40 mL/min along with scanning. Data analysis was conducted using the TA60 Version 2.10 (Shimadzu, Tokyo, Japan) thermal analysis software.

X-ray Diffraction (XRD)

The crystallinity state of the synthesized Myrj-*b*-PCL copolymers was studied using X-ray Diffractometer. Samples from the copolymers were loaded in the Ultima IV XRD instrument (Rigaku, Tokyo, Japan). The X-ray diffraction data of the samples were collected over the 3.0–50.0 degrees 2θ range at a scan speed of 0.50 degrees/min. The scanning process was done at room temperature.

Preparation of Drug-Free and CyA-Loaded Myrj-b-PCL Micelles

Micelles of Myrj-*b*-PCL block copolymers were prepared by the co-solvent evaporation method. Certain quantity of block copolymers (30 mg) was dissolved in 0.5 mL acetone. This preparation was added to a 3 mL purified water dropwise under stirring. The preparation was left on stirring overnight to evaporate the organic solvent. The particle size and polydispersity index of the self-assembled micelles were evaluated by dynamic light scattering (DLS) using Malvern Zetasizer™ 3000 (Malvern Instruments Ltd., Malvern, UK).

2.2.3. Characterization of Myrj-b-PCL Micelles

Size, Polydispersity, and ζ-Potential

The particle size, polydispersity index, and ζ-potential of the self-assembled micelles were evaluated by dynamic light scattering (DLS) using Malvern Zetasizer™ 3000 (Malvern Instruments Ltd., Malvern, UK).

Critical Micelle Concentration (CMC)

The critical micelle concentration (CMC) of each block copolymer was evaluated by DLS as previously described [17,18]. The newly invented block copolymers were used to prepare micelles starting with the concentration of 500 µg/mL. For each block copolymer, subsequent two-fold serial dilutions by distilled water were applied to get various successive concentrations of a block copolymer. The intensity of the scattered light for the diluted samples of the prepared micelles was evaluated until reaching the levels where the intensity of the light scattering of the copolymer solution is similar to that of water. Zetasizer Nano ZS analyzer (Malvern Instruments Ltd., Malvern, UK) was used to measure the light scattering intensity as kilo counts per second (kcps) of each concentration of the block copolymers. The principle of this test is based on the fact that the readings of kcps remain constant at concentrations below CMC and start to abruptly increase at concentrations equal to CMC. All measurements were performed at 25 °C.

Drug Encapsulation Efficiency % and Drug Loading %

Samples (100 µL) of each prepared drug-loaded polymeric micelles after centrifugation at 13,000 rpm were diluted 100 times in acetonitrile and vortexed to disassemble the micelles and free the drug. The concentration of CyA in the supernatant was quantified using a previously published HPLC-UV assay [19]. Briefly, the HPLC system (Waters™ 1500 series controller, Boston, MA, USA) is equipped with a wavelength detector (Waters™ 2489a Dual™ Absorbance detector, Boston, MA, USA), pump (Waters™ 1525a Binary pump, Boston, MA, USA), and an automated sampling system (Waters™ 2707 Plus Autosampler, Boston, MA, USA) monitored by "Breeze (Waters™)" software. Sample (90 µL) containing CyA was analyzed using a mobile phase made of acetonitrile and purified water at a ratio of 75:25 running over C_{18} stationary phase (Macherey-Nagel, 4.6–150 mm, 10 µm particle size) maintained at 60 °C and 1.0 mL/min flow rate. The UV detector was set at 230 nm.

The encapsulation efficiency (%*EE*) and drug loading (%*DL*) of CyA in the prepared micelles were calculated using the following equations: Equations (1) and (2).

$$EE(\%) = \frac{\textit{Amount of drug loaded } (\mathbf{mg})}{\textit{Amount of drug added } (\mathbf{mg})} \times 100, \tag{1}$$

$$DL(\%) = \frac{Amount\ of\ drug\ loaded\ (\mathbf{mg})}{Amount\ of\ polymer\ (\mathbf{mg}) + drug\ added\ (\mathbf{mg})} \times 100, \qquad (2)$$

Morphology

Morphology of the self-assembled structures was characterized by transmission electron microscopy (TEM). An aqueous droplet of micellar solution (20 uL) was placed on a copper-coated grid (Ted Pella, Inc., Redding, CA, USA). The grid was kept horizontally for 20 s to allow the colloidal aggregates to settle. A drop of 2% solution of phosphotungstic acid (PTA) in PBS (pH = 7.0) was then added to give the negative stain. After 1 min, the excess fluid was removed by using a strip of filter paper. The samples were left to get dry at room temperature and loaded into a JEM-1010 Transmission electron microscope (JEOL, Tokyo, Japan) operating at an acceleration voltage of 80 kV. Images were recorded with a high-speed read-out side-mounted MegaView^{G2} (Olympus, Hamburg, Germany) camera and processed with iTEM (Olympus Soft Imaging Solutions GmbH, Münster, Germany) software.

Ex Vivo Corneal Permeation

New Zealand rabbits weighing 2–3 kg were used in this study, and the protocol was approved by the Research Ethics Committee (REC) at King Saud University (No. SE-19-133). The cornea of rabbit eyes was excised and fixed between the donor and receptor components of the fabricated double jacketed transdermal diffusion cells (sampling system-SFDC 6, Logan, NJ, USA). The fixing of the cornea was done in such a way that the epithelial surface faced the donor compartment. Simulated-tear fluid (pH 7.4) with 0.5% (w/v) Tween-80 was filled in the receptor component of the diffusion cells. The water (37 °C) was allowed to flow in the outer jacket of the diffusion cells. The diffusion cells were placed on different stations of the LOGAN instrument. Continuous magnetic stirring could remove any air bubbles in the receptor component during sampling. Of each group (in triplicate), 500 µL CyA-containing formulations (CyA-micelles and Restasis®, 0.05%, w/v) were placed in the donor compartments. The samples from the receptor component were taken at predetermined time intervals for 4 h.

The CyA content was analyzed by the previously reported LC-MS/MS method with minor modifications [20–22]. Briefly, chromatographic separation of CyA was achieved with the help of Waters Acquity UPLC H-Class (Waters, Milford, MA, USA) equipped with the quaternary solvent manager, degasser, and column heater, sample manager having standard flow-through-needle and maximum injection volume of 10 µL. A Waters Acquity UPLC BEH™ C18 column (1.7 µm, 2.1 mm × 50 mm, Waters, Milford, MA, USA), maintained at 45 °C was utilized for drug elution. The mobile phase was eluted in isocratic mode at a flow rate of 0.3 mL/min, consisting of a mixture of 20 mm ammonium acetate and acetonitrile (20:80). Total run time was fixed to 3.5 min.

Mass spectrometric detection was performed using Waters Acquity TQD detector (Waters, Milford, MA, USA), equipped with an electrospray ionization (ESI) source, and operated in positive ionization mode. Selected ion recording (SIR) was used for the identification and quantification of CyA. For CyA, the SIR was performed at m/z 1202.85. The source and desolvation temperatures were set at 150 °C and 350 °C, respectively. Nitrogen was used as desolvation and nebulizing gas at a 600 L/h flow rate. The capillary and cone voltages were set at 3 kV and 95 V, respectively. The MassLynx software (Version 4.1) was used to control the UPLC–MS/MS system as well for data acquisition and processing.

Samples were centrifuged at 15,000 rpm for 10 min and filtered (0.45 µm) to remove any particulate matter. Precisely 100 µL of the filtered sample was mixed with 900 µL of mobile phase composition and vortexed for 30 s. The diluted samples were transferred to UPLC vials and 5 µL of the samples were injected into the UPLC-MS/MS system.

The permeation parameters including the flux (J) and the apparent permeability (Papp) were calculated from the slopes obtained by plotting the permeated amount of CyA ($\mu g \cdot cm^{-2}$) against time (h), using the following equations: Equations (3) and (4).

$$dQ/dt = J\left(\mu gcm^{-2}\cdot h^{-1}\right), \quad (3)$$

$$J/c_0 = Papp\left(cmh^{-1}\right), \quad (4)$$

where (Q) signifies the amount of drug crossing the cornea, (dQ/dt) is the linear portion of the slope, (t) is the contact time of the formulation with the corneal surface, and (C_0) is the initial concentration of the drug.

2.2.4. In Vivo Ocular Irritation Test

All the procedures in this experiment were performed according to the guidelines of the Association for Research in Vision and Ophthalmology (ARVO) for animal use in ophthalmic and vision research. In accordance with the ARVO guidelines, only one eye (left eye) of all rabbits were chosen for testing purpose (the right eye was instilled with 0.9% NaCl as a negative control), and Draize's eye test was used for assessing the ocular safety of CyA-loaded Myrj-b-PCL micelles [23]. Generally, for one test formulation, a maximum of six animals (rabbits) are required; however, this number can be decreased to three if there might be a chance of any severe ocular damage [24]. Thus, in this study, nine rabbits were divided into three groups each containing three (n = 3) for the irritation testing of the three formulations. Forty microliters (40 µL) of each formulation including blank were instilled into the lower conjunctival sac of each rabbit of the respective groups. All the rabbits in the conscious state received 3 consecutive instillations in the conjunctival sac of the left eye at an interval of 10 min for a short-term eye irritation test. After 1 h of the last dosing, eyes were periodically observed for any injuries or signs and symptoms in the cornea, iris and conjunctiva, photographs were captured, and scoring was done.

2.2.5. Data Analysis

The collected data were reported as mean ± standard deviation (SD). Based on the number of groups being compared, statistical significance was analyzed either by Student's t-test or one-way analysis of variance (ANOVA) followed by a post hoc test (LSD). The level of significance was set at α = 0.05.

3. Results and Discussion

3.1. Synthesis and Characterization of Myrj-b-PCL Copolymers

Table 1 shows the different Myrj-b-PCL copolymers synthesized. The copolymers were characterized by several analytical techniques including ^{1}H NMR, GPC, FTIR, DSC, and XRD.

Table 1. Characteristics of the synthesized Myrj-b-PCL copolymers.

Block Copolymer [a]	Theoretical MW (g/mol)	Mn [b] (g/mol)	Mn [c] (g/mol)	Mw [d] (g/mol)	Dispersity [e] (Ð)
Myrj S40-b-PCL$_{18}$	4100	3500	3007	4037	1.34
Myrj S40-b-PCL$_{35}$	6040	5700	4116	5745	1.40
Myrj S100-b-PCL$_{44}$	9700	9400	6192	8628	1.39
Myrj S100-b-PCL$_{88}$	14,650	14,600	10,150	15,724	1.55
Myrj S100-b-PCL$_{131}$	19,700	19,600	12,019	28,880	1.67

[a] The number shown as a subscript indicates the polymerization degree of each block determined by ^{1}H NMR. [b] Number-average molecular weight measured by ^{1}H NMR. [c] Number-average molecular weight measured by GPC. [d] Weight-average molecular weight measured by GPC. [e] Dispersity (Mw/Mn) determined by GPC.

3.1.1. ^{1}H NMR

Representative ^{1}H NMR spectra of Myrj S40 and Myrj S100 and the synthesized copolymers are shown in Figures 2 and S1. Myrj-b-PCL copolymers show the following signals: δ = 4.07, 2.32, 1.67, and 1.38 ppm, which were assigned to (H–O–C$\underline{H}_2$–), (–CO–C$\underline{H}_2$–), (–CO–CH$_2$–C$\underline{H}_2$–CH$_2$–CH$_2$–CH$_2$–OH), and (–CO–CH$_2$–CH$_2$–C$\underline{H}_2$–CH$_2$–CH$_2$–OH) of the PCL segment, respectively. The peak at δ = 3.65 ppm was assigned to the methylene protons of the PEO in Myrj segment (CH$_3$–O–C$\underline{H}_2$–C$\underline{H}_2$–O–). The lower peaks in the aliphatic region (δ = 0.87–1.28 ppm) belong to various moieties of stearate tails. The molecular weight and the composition of the synthesized copolymers were determined by ^{1}H NMR based on the intensity ratio between the peaks at δ = 4.07 ppm and 3.65 ppm. ^{1}H NMR data were found to be consistent with the theoretical values, which confirms the successful synthesis of Myrj-b-PCL copolymers.

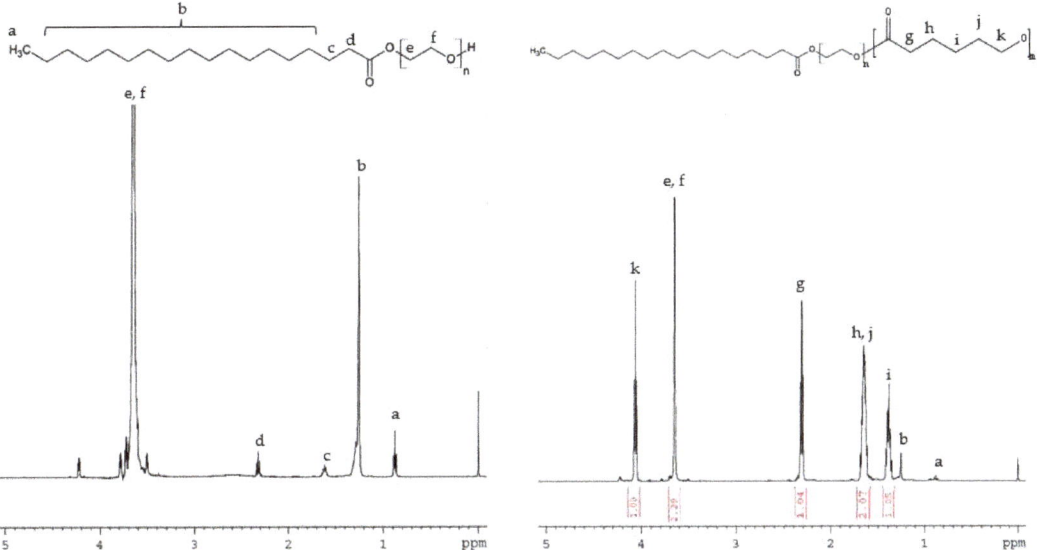

Figure 2. Representative ^{1}H NMR spectra of Myrj S100 (**left**) and Myrj S100-b-PCL$_{88}$ (**right**).

3.1.2. GPC

GPC analysis further confirmed the successful synthesis of Myrj-b-PCL with dispersity (Đ) ranging from 1.34–1.67. Figures 3 and S2 show the GPC chromatograms of Myrj S40-b-PCL and Myrj S100-b-PCL copolymers, respectively. The retention volumes for Myrj S100 appeared at 8.0 mL. However, as expected, the peaks for all synthesized Myrj S100-b-PCL copolymers shifted to lower retention volumes (higher molecular weights). Each copolymer appeared as a single peak, which confirms that the polymerization reaction was successful. The molecular weights of the synthesized copolymers were calculated based on the prepared calibration curve using polystyrene standards. The GPC data for all the synthesized copolymers are presented in Table 1.

3.1.3. FTIR

FTIR spectra of Myrj S40 and Myrj S100 as well as their corresponding copolymers were shown in Figures 4 and S3, respectively. For Myrj moieties, the characteristic bands for carbonyl groups (of stearate) appeared at 1737–1738 cm^{-1}. For the synthesized Myrj-b-PCL copolymers, the carbonyl band was shifted to 1724–1734 cm^{-1} with stronger intensity due to the formation of PCL. Moreover, it was noticed that with increasing the molecular weight of PCL, the aliphatic CH stretching band of ε-CL at 2944–2945 cm^{-1} increased.

On the other hand, the absorption band of CH stretching vibration in PEO and stearate moieties of Myrj at 2884–2885 cm^{-1} decreased. These observations verify the presence of intermolecular interactions and point to the formation of Myrj-*b*-PCL diblock copolymers. Indeed, a similar trend was previously reported for PEO-*b*-PCL copolymers [25].

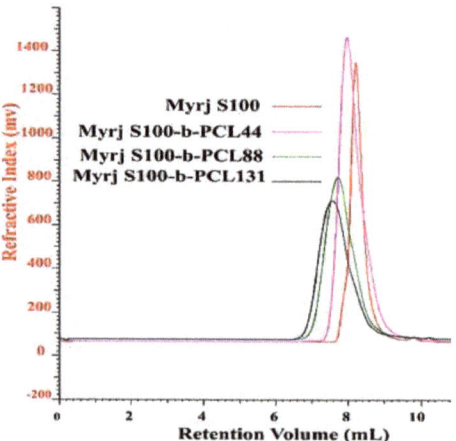

Figure 3. GPC chromatograms of unmodified Myrj S100 and the synthesized Myrj S100-*b*-PCL copolymers.

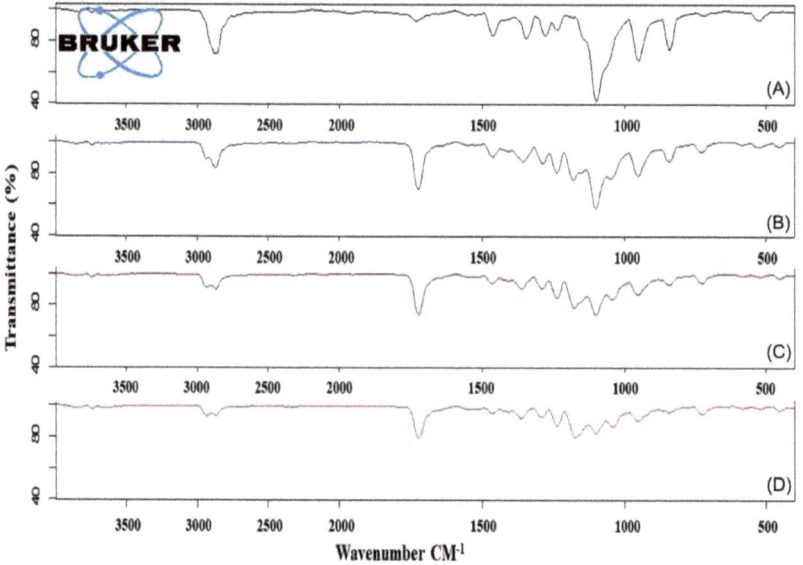

Figure 4. FTIR spectra of Myrj S100 (**A**), Myrj S100-*b*-PCL$_{44}$ (**B**), Myrj S100-*b*-PCL$_{88}$ (**C**), and Myrj S100-*b*-PCL$_{131}$ (**D**).

3.1.4. XRD

The X-ray diffractograms of Myrj and the synthesized Myrj-*b*-PCL copolymers are shown in Figures 5 and S4. The diffraction pattern of unmodified Myrj S40 and Myrj S100 exhibited good crystallinity with two sharp peaks appearing at 2θ = 19.3° and 23.4°, which are the characteristic crystalline peaks of PEO [26]. The synthesized copolymers showed different crystallinity behavior according to their PCL content. In addition to

PEO peaks (of Myrj), the Myrj-*b*-PCL copolymers showed strong peaks at 2θ = 21.5° and 23.8° corresponding to the PCL crystalline units [25]. Furthermore, in all the synthesized Myrj-*b*-PCL copolymers, the sharp crystalline peak of PEO at 2θ = 19.3° was reduced while the intensity of the characteristic peaks of PCL increased with the increase in PCL molecular weight. These may be attributed to a conformational change, which affected the crystallinity of the PEG segment of Myrj upon increasing the PCL molecular weight.

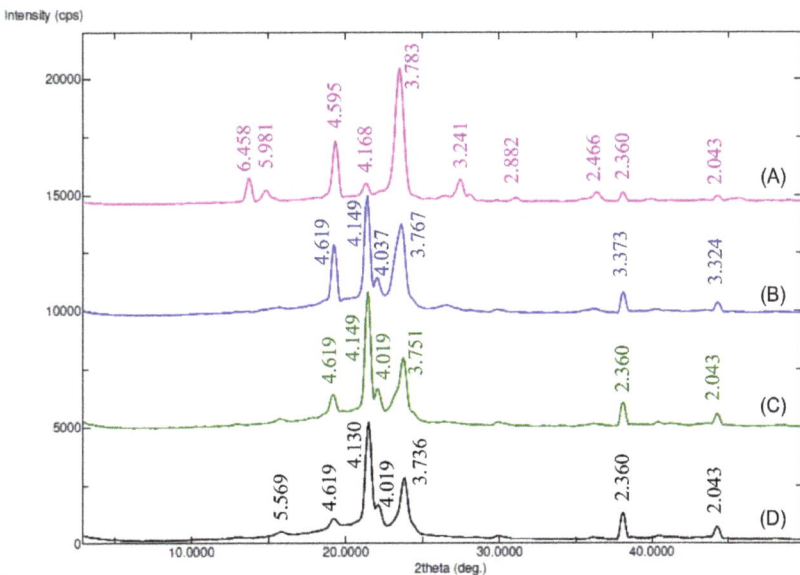

Figure 5. XRD diffractograms (peaks with d-values) of Myrj S100 (**A**), Myrj S100-*b*-PCL$_{44}$ (**B**), Myrj S100-*b*-PCL$_{88}$ (**C**), and Myrj S100-*b*-PCL$_{131}$ (**D**).

3.1.5. DSC

Thermal properties of Myrj-*b*-PCL copolymers and their Myrj precursors were investigated by DSC. As shown in Figures 6 and S5, DSC thermograms of Myrj S40, Myrj S40-*b*-PCL$_{18}$, and Myrj S40-*b*-PCL$_{35}$ exhibited strong peaks at 51.5 °C, 39.0 °C, and 50.5 °C, respectively, corresponding to their melting points. Moreover, DSC thermograms of Myrj S100, Myrj S100-*b*-PCL$_{44}$, Myrj S100-*b*-PCL$_{88}$, and Myrj S100-*b*-PCL$_{131}$ exhibited strong peaks at 59.0 °C, 48.5 °C, 58.0 °C, 59.0 °C, respectively, corresponding to their melting points.

It was noticed that the melting temperature of Myrj increases with the increase in molecular weight of PEO (i.e., Myrj S40 *vs.* Myrj S100). Moreover, the melting temperatures of the Myrj-*b*-PCL copolymers correlate well with the PCL content. Specifically, the melting temperatures of the copolymers increased with the increase in the PCL molecular weight. Moreover, it was noticed that unmodified Myrj S40 and Myrj S100 each had a single sharp peak demonstrating a crystal pattern typical for the PEO crystal phase reported with PEO homopolymer [26]. Myrj-*b*-PCL copolymers, on the other hand, showed either broader and/or bimodal peaks (Figure 6). The broad or bimodal peaks are likely due to overlapped or separate endothermic peaks of Myrj (PEO) and PCL blocks. This is in agreement with the XRD results, where Myrj-*b*-PCL copolymers demonstrated a summation of both the PEO and PCL diffraction peaks. This indicates that both blocks could crystallize and form separate crystals, which were also previously observed with PEO-*b*-PCL copolymers [26].

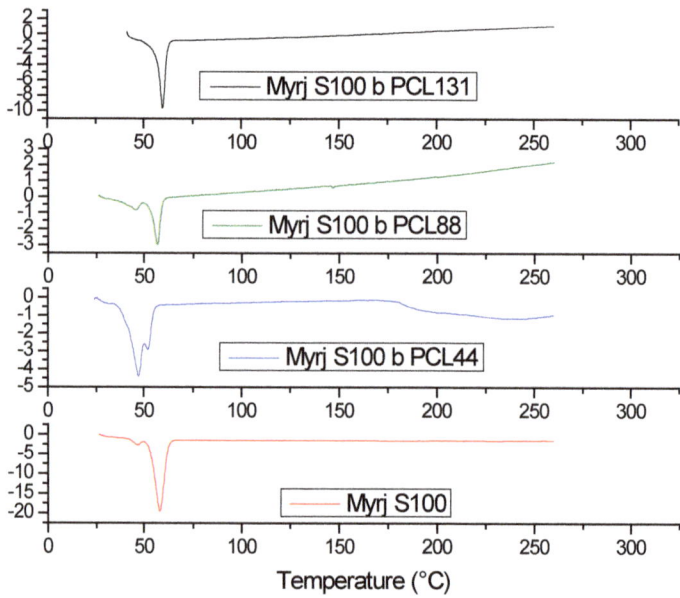

Figure 6. DSC thermograms of Myrj S100, Myrj S100-*b*-PCL$_{44}$, Myrj S100-*b*-PCL$_{88}$, and Myrj S100-*b*-PCL$_{131}$.

3.2. Preparation and Characterization of Drug-Free and CyA-Loaded Myrj-b-PCL Micelles

Several water-miscible organic solvents were used (including acetonitrile, tetrahydrofuran, and acetone) to prepare micelles using the cosolvent evaporation method. Despite several attempts to prepare micelles using different organic solvent-to-water ratios as well as other methods of preparation (e.g., film-hydration method), none of the drug-free or CyA-loaded Myrj S40-*b*-PCL copolymers formed micelles with an acceptable yield or size. On the other hand, both Myrj S100-*b*-PCL$_{88}$ and Myrj S100-*b*-PCL$_{131}$ copolymers formed micelles with mean diameters <200 nm and with good yield (minimal or no precipitation). The only difference between Myrj S40 and Myrj S100 is the molecular weight PEG. Myrj S40 has 40 repeating units of PEG (i.e., PEG molecular weight = 1760 Da), whereas Myrj S100 comprised 100 repeating units of PEG (i.e., PEG molecular weight = 4400 Da). It seems that a PEG chain longer than 40 (MW > 1760 Da) was needed to provide the necessary folding to bring the stearate moiety (hydrophobic) in proximity to PCL, the core-forming block, during micelle formation in water. It is believed that this would be the thermodynamically favorable orientation for Myrj-*b*-PCL copolymer during the micellization process in the aqueous phase (Figure 7), which would be similar to the one reported for flower-like micelles formed from A-B-A triblock copolymers, where B is the hydrophilic block (e.g., Pluronic-R) [27]. In fact, this is similar to what we have observed and recently reported with α-tocopheryl polyethylene glycol succinate-*b*-PCL copolymers, where the micelles only formed when the molecular weight of PEG was higher than 2000 Da [6]. Nonetheless, this is only a hypothetical model that needs to be verified. Further studies are needed to investigate the conformation of these micelles in water (e.g., by using ^{1}H NMR in D$_2$O) [28,29] and to examine the morphology of the self-assembled structures in water (e.g., by using cryogenic TEM or atomic force microscopy) [30,31].

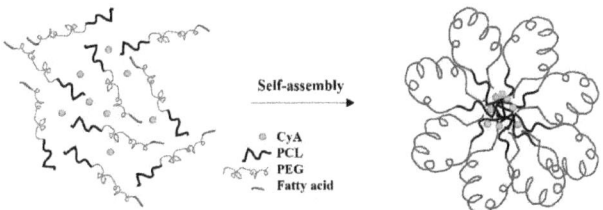

Figure 7. Proposed model for Myrj-b-PCL micelle formation in water.

3.2.1. Size, Polydispersity Index, and ζ-Potential

The particle size, polydispersity index (PDI), and ζ-potential of unloaded Myrj S100-b-PCL micelles are presented in Table 1. The mean particle size of the unloaded Myrj S100-b-PCL$_{44}$, Myrj S100-b-PCL$_{88}$, and Myrj S100-b-PCL$_{131}$ micelles was 49.8, 70.3, and 72.6 nm, respectively. The PDI values of Myrj S100-b-PCL$_{88}$ and Myrj S100-b-PCL$_{131}$ were both found to be $\leq$0.21, which reflects good homogeneity of particle size distribution. Myrj S100-b-PCL$_{44}$ micelles, however, showed a significantly higher PDI value (0.38 $\pm$ 0.14) compared to the other two. This is likely due to the presence of a bimodal population (Figure S6). It is also worth mentioning that there was significant polymer precipitation during the preparation of the micelles and after centrifugation. In the field of nano-drug delivery, a PDI $\leq$ 0.3 is considered optimal and indicates a homogenous population of the carrier system [32]. The ζ-potential values of all the Myrj S100-b-PCL micelles were nearly neutral to slightly negative, which is typical of polymeric micelles.

The particle size, PDI, and ζ-potential CyA-loaded Myrj S100-b-PCL are presented in Table 2. The size of the prepared micelles did not change significantly after drug loading except for Myrj S100-b-PCL$_{88}$, where the CyA-loaded micelle size significantly increased (p < 0.05, paired Student's t-test) compared to the unloaded counterpart. While CyA-loaded Myrj S100-b-PCL$_{88}$ showed a uniform size with a relatively low PDI value (Table 2, Figure S7), Myrj S100-b-PCL$_{44}$ and Myrj S100-b-PCL$_{131}$ showed high variability in size and a relatively higher PDI value (Table 2).

Table 2. Physical characterization of micelles before and after drug loading. Results are expressed as mean with $\pm$ SD, n = 3.

Polymeric Micelles	Particle Size (nm)	Polydispersity Index	ζ-Potential (mV)	EE (%)	DL (%)
Micelles before CyA loading					
Myrj S100-b-PCL$_{44}$	49.8 $\pm$ 14.9	0.38 $\pm$ 0.14	$-$7.0 $\pm$ 1.8	–	–
Myrj S100-b-PCL$_{88}$	70.3 $\pm$ 2.4	0.20 $\pm$ 0.04	$-$4.7 $\pm$ 0.3	–	–
Myrj S100-b-PCL$_{131}$	72.6 $\pm$ 1.3	0.21 $\pm$ 0.02	$-$9.4 $\pm$ 0.7	–	–
Micelles after CyA loading					
Myrj S100-b-PCL$_{44}$	41.1 $\pm$ 5.6	0.44 $\pm$ 0.06	$-$4.9 $\pm$ 3.2	15.23 $\pm$ 0.38	1.38 $\pm$ 0.03
Myrj S100-b-PCL$_{88}$	81.9 $\pm$ 2.9	0.18 $\pm$ 0.01	$-$6.8 $\pm$ 2.1	54.15 $\pm$ 3.03	5.62 $\pm$ 0.27
Myrj S100-b-PCL$_{131}$	63.5 $\pm$ 13.1	0.28 $\pm$ 0.05	$-$1.6 $\pm$ 1.7	23.04 $\pm$ 1.35	2.25 $\pm$ 0.13

3.2.2. CMC

The calculated CMC values for Myrj S100-b-PCL$_{88}$ and Myrj S100-b-PCL$_{131}$ micelles were 2.08 $\pm$ 0.08 and 2.70 $\pm$ 0.59 μM, respectively. The difference between the two CMC values was not statistically significant (Student's t-test, p > 0.05). These values are much lower than the values reported for polymeric surfactants, which are usually in the millimolar range.

3.2.3. Drug Encapsulation Efficiency (EE%) and Drug Loading (DL%)

The drug loading% (DL%) and encapsulation efficiency% (EE%) of CyA-loaded micelles are presented in Table 2. Myrj S100-b-PCL$_{88}$ showed a significantly higher EE% and DL% of CyA compared to Myrj S100-b-PCL$_{131}$ (p < 0.05, Student's t-test). Specifically, Myrj S100-b-PCL$_{88}$ had an EE% of over 54% and DL% of 5.62%, which translates to an aqueous solubility of 540 µg/mL (i.e., a 2348-fold increase in CyA solubility). Although this loading level is lower than that previously reported for CyA in methoxy poly(ethylene oxide)-*block*-poly(ε-caprolactone) (PEO-b-PCL) micelles [19,21], it is sufficient to prepare the required concentration for ocular administration (0.05% w/v) [33]. Based on these results, only Myrj S100-b-PCL$_{88}$ was selected for the next studies.

3.2.4. Morphology

The morphology of the unloaded, as well as CyA-loaded Myrj S100-b-PCL micelles, was studied by using transmission electron microscopy (TEM). The images revealed that these micelles were of spherical shape as shown in Figure 8.

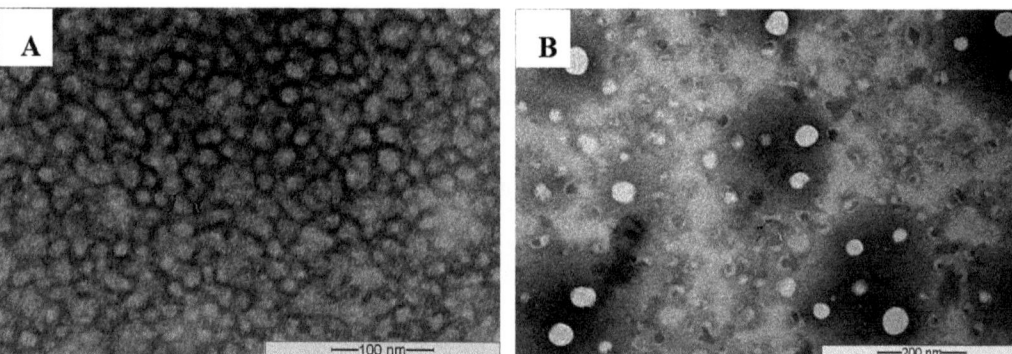

Figure 8. Representative TEM images obtained from unloaded Myrj S100-b-PCL$_{88}$ (**A**) and CyA-loaded Myrj S100-b-PCL$_{88}$ micelles (**B**) using JEOL JEM-1100 Transmission electron microscope (JAPAN) operating at an acceleration voltage of 80 kV.

3.3. Transcorneal Permeation of CyA-Loaded Myrj-b-PCL Micelles

The steady-state flux (*J*) and apparent permeability (P_{app}) of CyA was calculated by considering the involved corneal area (0.5024 cm^2), the volume of permeation medium (5.2 mL), and initial drug concentrations (500 µg/mL). In STF (pH 7.4), Tween-80 (0.5%, w/v) was added to improve the solubility of the highly lipophilic drug (CyA). From the plot shown in Figure 9, and the calculated values as summarized in Table 3, Myrj S100-b-PCL$_{88}$ micelles exhibited a smooth and linear permeation of CyA as compared to Restasis®. The cumulative amounts of CyA permeated were 51.14 ± 5.23 and 59.49 ± 8.67 µg.cm^{-2} (at 4 h) from CyA-Micelles and Restasis®, respectively. The difference was not statistically significant (p > 0.05, Student's t-test).

The permeation of CyA from Restasis® seemed higher initially (up to 1 h) as compared to CyA-Micelles, but overall, the total permeated amount of CyA through the excised rabbit cornea was comparable for the two formulations (p > 0.05, Student's t-test). From the permeation profile, it was concluded that CyA-Micelles may provide sustained delivery of the loaded CyA from the micelles compared to that of Restasis®.

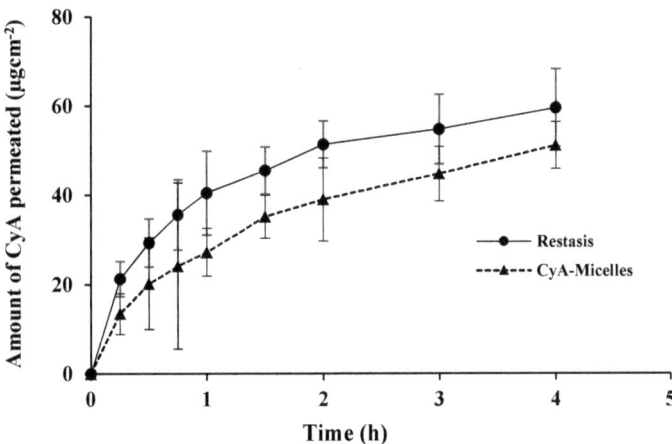

Figure 9. Transcorneal permeation of CyA from Myrj S100-*b*-PCL$_{88}$ micelles and Restasis®. (Mean ± SD, n = 3).

Table 3. Transcorneal permeation parameters for CyA from Myrj S100-*b*-PCL$_{88}$ micelles and Restasis® (Mean ± SD, n = 3).

Parameters	CyA-Micelles	Restasis®
Cumulative amount of CyA permeated (μg cm^{-2}) at 4th h	51.14 ± 5.23	59.49 ± 8.67
Steady-state flux, J (μg cm^{-2} h^{-1})	18.89 ± 2.99	17.14 ± 2.79
Permeability coefficient, P_{app} (cm h^{-1})	(3.78 ± 0.59) × 10^{-2}	(3.43 ± 0.56) × 10^{-2}

3.4. In Vivo Ocular Irritation Test

The irritation potential (if any) of the blank and CyA-loaded Myrj S100-*b*-PCL$_{88}$ micelles towards the anterior ocular segment was investigated for 24 h by following Draize's test [23]. The observations were noted after the instillation of 40 µL (three times at an interval of 10 min) into the left eyes of all the New Zealand rabbits (n = 3) in comparison to the marketed CyA ocular formulation (Restasis®). Any changes were observed by visual examination of the cornea, iris, and conjunctiva of the treated eyes of all rabbits [34]. On the basis of signs and symptoms of ocular irritation (including redness, swelling, chemosis, edema, cloudiness, edema, hemorrhage, and or any discharge other than normal) which may arise in the treated eyes, the scoring was done as per the theoretical scores for grading system (Table S1) [35]. The irritation potential was categorized according to a classification system (Table S2) [36]. The obtained scores and signs of discomfort (if any) during the experiment for all tested formulations were recorded in Table 4.

No significant signs or symptoms of discomfort were found during the irritation testing in the treated animals with blank micelles (unloaded), CyA-loaded micelles, and Restasis®. Figure 10a,a′,a″ represent images of NaCl treated (right) eyes of rabbits of the respective group (green arrows). Figure 10b,b′ represent the mild redness (red arrows) of the conjunctiva and abnormal discharge at 1 h post instillation of blank micelles and CyA-Micelles, respectively. This abnormal discharge might be due to the surfactant property of the di-block copolymer and physiological secretion, which is in agreement with our previous report [21]. No redness or inflammation was observed in the Restasis® treated eyes even at the initial hour (1 h) of the experiment as shown in Figure 10b″ (green arrow). In fact, in the present investigation, the Restasis® treated animals did not show any signs and symptoms of ocular irritation at any time points as represented in Figure 10c″,d″,e″ (green arrow). The redness of conjunctiva was decreased at 3 h post instillation of CyA-Micelles (Figure 10c′) (black arrow) and at 6 h it disappeared (Figure 10d′) (green arrow) and the eyes were recovered their normal conditions at 24 h (Figure 10e″) (green arrow). Redness of conjunctiva was also decreased at 3 h in the blank micelles treated eye (Figure 10c) (black arrow), it disappeared at 6 h (Figure 10d) (green arrow) and the eyes regained their normal condition at 24 h (Figure 10e) (green arrow), that might be due to the natural defensive mechanism of the animal and lower irritation potential of the unloaded Myrj S100-b-PCL$_{88}$ polymeric micelles.

Table 4. Weighted scores for eye irritation test by Restasis®, blank micelles, and CyA-Micelles.

Lesions in the Treated Eyes	Individual Scores for Eye Irritation								
	Restasis®			Blank Micelles			CyA-Micelles		
	Animal #			Animal #			Animal #		
	1st	2nd	3rd	1st	2nd	3rd	1st	2nd	3rd
Cornea									
I. Opacity (Degree of density)	0	1	0	0	1	1	0	1	0
II. Area of cornea	4	4	4	4	4	4	4	4	4
Total scores = (I × II × 5) =	0	20	0	0	20	20	0	20	0
Iris									
I. Lesion values	0	1	0	0	1	1	0	1	0
Total scores = (I × 5) =	0	5	0	0	5	5	0	5	0
Conjunctiva									
I. Redness	0	0	1	1	1	1	0	1	0
II. Chemosis	0	0	0	0	0	0	0	0	0
III. Mucoidal discharge	0	0	0	0	1	0	0	0	0
Total scores = (I + II + III) × 2 =	0	0	2	2	4	2	0	2	0

As a result of blank micelles instillation, a slight irritation was found in the treated eye of one rabbit with some watery discharge (slightly different from normal but not mucoidal) and it was given score 1. No corneal opacity or lesions in the eye structures was noted in the treated animals. For cornea, iris, and conjunctiva grade-0 was given for all the tested products. Based on a reported classification system for eye irritation scoring [36], the calculated maximum mean total score (MMTS) during the visual examination after 24 h of first dosing for the blank micelles was 19.33 (which is greater than 15.1 but less than 25). The MMTS for CyA-loaded micelles and Restasis® was 9.00 (which is greater than 2.6 but less than 15) (Table 5). Thus, the blank polymeric micelles were considered "mildly irritating" while the CyA-loaded polymeric micelles and the marketed Restasis® were "minimally irritating" to the rabbit eyes in the present investigation.

All the involved animals were active and healthy without any abnormal signs and symptoms of overall toxicity throughout the experiment, except for the observed signs as scored and mentioned in Table 4. No severe eye irritation was found that may arise due to any corneal abrasion related to the treatment or any obstruction in the lacrimal drainage by the CyA-micelles in comparison to Restasis®. This indicated that the developed Myrj S100-b-PCL$_{88}$ micellar formulation was non-irritant to rabbit eyes. Additionally, during the visual examination after 24 h of the treated eyes, no remaining formulations were observed in the eyes, suggesting a complete degradation or disposition of the applied formulations at 24 h. The overall obtained "minimally irritating" CyA-Micelles in this investigation were in agreement and substantiate with the previous reports, where the block copolymers such (PEO-b-PCL) [21] and methoxy poly(ethylene glycol)-hexylsubstituted poly(lactide) (MPEG-hexPLA) [37] micelle carriers were used for topical ocular delivery of CyA. Similarly, polyhydroxyethylaspartamide-polyethylene glycol with hexadecylamine [PHEA-PEG-C$_{16}$] copolymer-based micelles were used for topical ocular delivery of dexamethasone and no irritation was reported [38]. All together, we can conclude that CyA-Micelles were well tolerated by rabbit eyes.

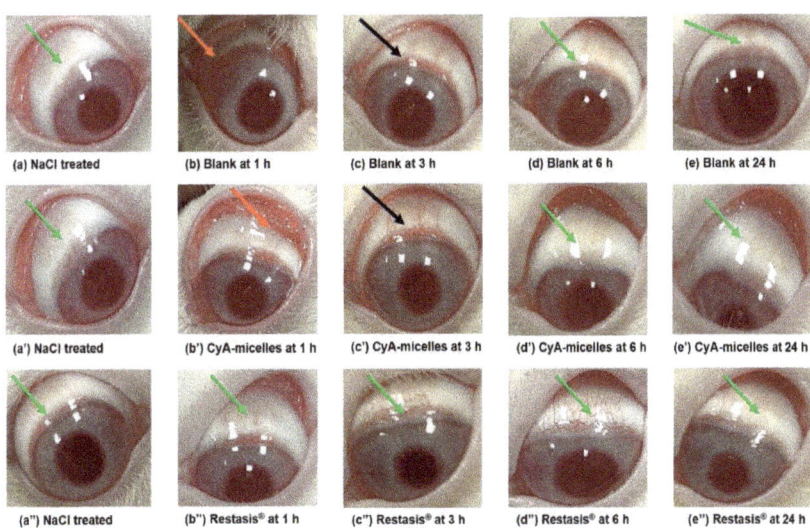

Figure 10. Photographs captured during irritation study in rabbit eyes. NaCl treated eyes of respective groups (a,a′,a″) (green arrows). Post instillation of blank Myrj S100-b-PCL$_{88}$ micelles at 1 h (b) (red arrow); at 3 h (c) (black arrow); at 6 h (d) and at 24 h (e). Post instillation of CyA-loaded Myrj S100-b-PCL$_{88}$ micelles (CyA-Micelles) at 1 h (b′) (red arrow); at 3 h (c′) (black arrow); at 6 h (d′) and at 24 h (e′). Post instillation of Restasis® at 1 h (b″); at 3 h (c″); at 6 h (d″) and at 24 h (e″). All other images did not show any redness or abnormal discharge (i.e., indicating the normal features) were represented by green arrows.

Table 5. Maximum Mean Total Score (MMTS) calculations for the tested formulations as per the obtained scores in Table S2.

	Restasis®				
Animal #→	1st	2nd	3rd	SUM	Average (SUM/3)
Cornea	0	20	0	20	6.67
Iris	0	5	0	5	1.67
Conjunctiva	0	0	2	2	0.66
SUM total =	0	25	2	27	9.00
	Blank micelles				
Animal #→	1st	2nd	3rd	SUM	Average (SUM/3)
Cornea	0	20	20	40	13.33
Iris	0	5	5	10	3.33
Conjunctiva	2	4	2	8	2.67
SUM total =	2	29	27	58	19.33
	CyA-Micelles				
Animal #→	1st	2nd	3rd	SUM	Average (SUM/3)
Cornea	0	20	0	20	6.67
Iris	0	5	0	5	1.67
Conjunctiva	0	2	0	2	0.66
Total =	0	27	0	27	9.00

4. Conclusions

Myrj S40-*b*-PCL and Myrj S100-*b*-PCL copolymers were successfully synthesized, but only Myrj S100-*b*-PCL formed micelles with optimum size, PDI, and no precipitation. The mean diameters of the prepared self-assembled structures were in the nano-range ($\leq$200 nm). Moreover, Myrj S100-*b*-PCL micelles significantly increased the aqueous solubility of CyA from ca. 23 µg/mL to over 540 µg/mL. The developed micelles showed a transcorneal permeation comparable to Restasis®, the leading commercial CyA ocular formulation. The in vivo ocular irritation study demonstrated that CyA-loaded Myrj S100-*b*-PCL$_{88}$ micelles were well tolerated in the rabbit eye. Our results point to a great potential for Myrj-*b*-PCL micelles to serve as an efficient solubilizing and delivery system for CyA and potentially other hydrophobic drugs.

Supplementary Materials: The following supporting information can be downloaded at: https://www.mdpi.com/article/10.3390/polym14091635/s1, Figure S1. Representative ^{1}H NMR spectra of Myrj S40 (A) and Myrj S40-*b*-PCL$_{35}$ (B); Figure S2. GPC chromatograms of unmodified Myrj S40 and the synthesized Myrj S40-*b*-PCL copolymers; Figure S3. FTIR spectra of Myrj S40 (A), Myrj S40-*b*-PCL$_{18}$ (B), and Myrj S40-*b*-PCL$_{35}$ (C); Figure S4. XRD diffractograms of Myrj S40 (A), Myrj S40-*b*-PCL$_{18}$ (B), and Myrj S40-*b*-PCL$_{35}$ (C); Figure S5. DSC thermograms of Myrj S40, Myrj S40-*b*-PCL$_{18}$, and Myrj S40-*b*-PCL$_{35}$; Figure S6. Representative size distribution profiles of (A) unloaded Myrj S100-*b*-PCL$_{44}$ micelles and (B) CyA-loaded Myrj S100-*b*-PCL$_{44}$ nanocarriers obtained by dynamic light scattering (Zetasizer Nano ZS, Malvern Instrument Ltd., Malvern, UK) showing bimodal size population; Figure S7. Representative size distribution profiles of (A) unloaded Myrj S100-*b*-PCL$_{88}$ micelles and (B) CyA-loaded Myrj S100-*b*-PCL$_{88}$ nanocarriers obtained by dynamic light scattering (Zetasizer Nano ZS, Malvern Instrument Ltd., Malvern, UK) showing bimodal size population; Table S1. Grading system for ocular lesions in irritation tests; Table S2. Classification of eye irritation scoring system.

Author Contributions: Conceptualization, Z.B., A.H.A. and A.A.; Data curation, O.H., R.A. and M.A.K.; Formal analysis, Z.B., A.H.A., O.H., R.A. and M.A.K.; Funding acquisition, Z.B.; Investigation, A.H.A., O.H. and M.A.K.; Methodology, Z.B., A.H.A., R.A. and M.A.K.; Project administration, Z.B. and A.H.A.; Supervision, Z.B. and A.H.A.; Writing—original draft, Z.B., O.H. and M.A.K.; Writing—review and editing, A.H.A., R.A. and A.A. All authors have read and agreed to the published version of the manuscript.

Funding: The authors extend their appreciation to the Deanship of Scientific Research at King Saud University for funding this work through research group No. RG-1441-416.

Institutional Review Board Statement: The animal study protocol was approved by Research Ethics Committee (REC) at King Saud University (protocol code SE-19-133; 13 February 2020) for studies involving animals.

Data Availability Statement: Data are available upon request from the corresponding author.

Conflicts of Interest: We declare that we have no known competing financial interest or personal relationships that could have appeared to influence the work reported in this paper.

References

1. Rowe, R.C.; Sheskey, P.J.; Owen, S.C.; Association, A.P. *Handbook of Pharmaceutical Excipients*; Pharmaceutical Press: London, UK, 2006.
2. Liu, C.; Wu, J.; Shi, B.; Zhang, Y.; Gao, T.; Pei, Y. Enhancing the Bioavailability of Cyclosporine A Using Solid Dispersion Containing Polyoxyethylene (40) Stearate. *Drug Dev. Ind. Pharm.* **2006**, *32*, 115–123. [CrossRef] [PubMed]
3. Lo, Y.L. Relationships between the hydrophilic-lipophilic balance values of pharmaceutical excipients and their multidrug resistance modulating effect in Caco-2 cells and rat intestines. *J. Control. Release* **2003**, *90*, 37–48. [CrossRef]
4. Wang, S.-W.; Monagle, J.; McNulty, C.; Putnam, D.; Chen, H. Determination of P-glycoprotein inhibition by excipients and their combinations using an integrated high-throughput process. *J. Pharm. Sci.* **2004**, *93*, 2755–2767. [CrossRef]
5. Dash, T.K.; Konkimalla, V.B. Poly-epsilon-caprolactone based formulations for drug delivery and tissue engineering: A review. *J Control. Release* **2012**, *158*, 15–33. [CrossRef] [PubMed]

6. Yusuf, O.; Ali, R.; Alomrani, A.H.; Alshamsan, A.; Alshememry, A.K.; Almalik, A.M.; Lavasanifar, A.; Binkhathlan, Z. Design and Development of D–α–Tocopheryl Polyethylene Glycol Succinate–block–Poly(ε-Caprolactone) (TPGS-b-PCL) Nanocarriers for Solubilization and Controlled Release of Paclitaxel. *Molecules* **2021**, *26*, 2690. [CrossRef]
7. Sinha, V.R.; Bansal, K.; Kaushik, R.; Kumria, R.; Trehan, A. Poly-epsilon-caprolactone microspheres and nanospheres: An overview. *Int. J. Pharm.* **2004**, *278*, 1–23. [CrossRef]
8. Woodruff, M.A.; Hutmacher, D.W. The return of a forgotten polymer—Polycaprolactone in the 21st century. *Prog. Polym. Sci.* **2010**, *35*, 1217–1256. [CrossRef]
9. Aliabadi, H.M.; Brocks, D.R.; Lavasanifar, A. Polymeric micelles for the solubilization and delivery of cyclosporine A: Pharmacokinetics and biodistribution. *Biomaterials* **2005**, *26*, 7251–7259. [CrossRef]
10. Binkhathlan, Z.; Hamdy, D.A.; Brocks, D.R.; Lavasanifar, A. Development of a polymeric micellar formulation for valspodar and assessment of its pharmacokinetics in rat. *Eur. J. Pharm. Biopharm.* **2010**, *75*, 90–95. [CrossRef]
11. Shuai, X.; Merdan, T.; Schaper, A.K.; Xi, F.; Kissel, T. Core-cross-linked polymeric micelles as paclitaxel carriers. *Bioconjugate Chem.* **2004**, *15*, 441–448. [CrossRef]
12. Amber, T.; Tabassum, S. Cyclosporin in dermatology: A practical compendium. *Dermatol. Ther.* **2020**, *33*, e13934. [CrossRef] [PubMed]
13. Airody, A.; Heath, G.; Lightman, S.; Gale, R. Non-Infectious Uveitis: Optimising the Therapeutic Response. *Drugs* **2016**, *76*, 27–39. [CrossRef] [PubMed]
14. Schultz, C. Safety and Efficacy of Cyclosporine in the Treatment of Chronic Dry Eye. *Ophthalmol. Eye Dis.* **2014**, *6*, 37–42. [CrossRef] [PubMed]
15. Peng, C.C.; Bengani, L.C.; Jung, H.J.; Leclerc, J.; Gupta, C.; Chauhan, A. Emulsions and microemulsions for ocular drug delivery. *J. Drug Deliv. Sci. Technol.* **2011**, *21*, 111–121. [CrossRef]
16. Binkhathlan, Z.; Alomrani, A.H.; Alshamsan, A.; Aljuffali, I.I.; Ali, R. Poly e-Caprolactone-Ethoxylated Fatty Acid Copolymers. U.S. Patent 9,622,973, 18 April 2017.
17. Topel, Ö.; Çakır, B.A.; Budama, L.; Hoda, N. Determination of critical micelle concentration of polybutadiene-block-poly(ethyleneoxide) diblock copolymer by fluorescence spectroscopy and dynamic light scattering. *J. Mol. Liq.* **2013**, *177*, 40–43. [CrossRef]
18. Soleymani Abyaneh, H.; Vakili, M.R.; Zhang, F.; Choi, P.; Lavasanifar, A. Rational design of block copolymer micelles to control burst drug release at a nanoscale dimension. *Acta Biomater.* **2015**, *24*, 127–139. [CrossRef] [PubMed]
19. Binkhathlan, Z.; Ali, R.; Qamar, W.; Lavasanifar, A. Pharmacokinetics of Orally Administered Poly(Ethylene Oxide)-block-Poly(ε-Caprolactone) Micelles of Cyclosporine A in Rats: Comparison with Neoral®. *J. Pharm. Pharm. Sci.* **2018**, *21*, 192s–199s. [CrossRef]
20. Al-Jenoobi, F.I.; Alam, M.A.; Alkharfy, K.M.; Al-Suwayeh, S.A.; Korashy, H.M.; Al-Mohizea, A.M.; Iqbal, M.; Ahad, A.; Raish, M. Pharmacokinetic interaction studies of fenugreek with CYP3A substrates cyclosporine and carbamazepine. *Eur. J. Drug Metab. Pharmacokinet.* **2014**, *39*, 147–153. [CrossRef]
21. Alshamsan, A.; Abul Kalam, M.; Vakili, M.R.; Binkhathlan, Z.; Raish, M.; Ali, R.; Alturki, T.A.; Safaei Nikouei, N.; Lavasanifar, A. Treatment of endotoxin-induced uveitis by topical application of cyclosporine a-loaded PolyGel in rabbit eyes. *Int. J. Pharm.* **2019**, *569*, 118573. [CrossRef]
22. Kalam, M.A.; Alshamsan, A. Poly (D, L-lactide-co-glycolide) nanoparticles for sustained release of tacrolimus in rabbit eyes. *Biomed. Pharmacother.* **2017**, *94*, 402–411. [CrossRef]
23. Draize, J.H.; Woodard, G.; Calvery, H.O. Methods for the study of irritation and toxicity of substances applied topically to the skin and mucous membranes. *J. Pharmacol. Exp. Ther.* **1944**, *82*, 377–390.
24. Lee, M.; Hwang, J.H.; Lim, K.M. Alternatives to In Vivo Draize Rabbit Eye and Skin Irritation Tests with a Focus on 3D Reconstructed Human Cornea-Like Epithelium and Epidermis Models. *Toxicol. Res.* **2017**, *33*, 191–203. [CrossRef] [PubMed]
25. Gyun Shin, I.L.; Yeon Kim, S.; Moo Lee, Y.; Soo Cho, C.; Yong Kiel, S. Methoxy poly(ethylene glycol)/ε-caprolactone amphiphilic block copolymeric micelle containing indomethacin.: I. Preparation and characterization. *J. Control. Release* **1998**, *51*, 1–11. [CrossRef]
26. Sun, J.; He, C.; Zhuang, X.; Jing, X.; Chen, X. The crystallization behavior of poly(ethylene glycol)-poly(ε-caprolactone) diblock copolymers with asymmetric block compositions. *J. Polym. Res.* **2011**, *18*, 2161–2168. [CrossRef]
27. Sharma, R.; Murali, R.; Murthy, C.N. Clouding and Aggregation Behavior of PPO-PEO-PPO Triblock Copolymer (Pluronic®25R4) in Surfactant Additives Environment. *Tenside Surfactants Deterg.* **2012**, *49*, 136–144. [CrossRef]
28. Mahmud, A.; Xiong, X.-B.; Lavasanifar, A. Novel Self-Associating Poly(ethylene oxide)-*block*-poly(ε-caprolactone) Block Copolymers with Functional Side Groups on the Polyester Block for Drug Delivery. *Macromolecules* **2006**, *39*, 9419–9428. [CrossRef]
29. Atanase, L.I.; Winninger, J.; Delaite, C.; Riess, G. Micellization and demicellization of amphiphilic poly(vinyl acetate)-graft-poly(N-vinyl-pyrrolidone) graft copolymers in the presence of sodium dodecyl sulfate. *Colloids Surf. A Physicochem. Eng. Asp.* **2014**, *461*, 287–294. [CrossRef]
30. Fairley, N.; Hoang, B.; Allen, C. Morphological Control of Poly(ethylene glycol)-*block*-poly(ε-caprolactone) Copolymer Aggregates in Aqueous Solution. *Biomacromolecules* **2008**, *9*, 2283–2291. [CrossRef]

31. Qi, W.; Ghoroghchian, P.P.; Li, G.; Hammer, D.A.; Therien, M.J. Aqueous self-assembly of poly(ethylene oxide)-*block*-poly(ε-caprolactone) (PEO-*b*-PCL) copolymers: Disparate diblock copolymer compositions give rise to nano- and meso-scale bilayered vesicles. *Nanoscale* **2013**, *5*, 10908–10915. [CrossRef]
32. Danaei, M.; Dehghankhold, M.; Ataei, S.; Hasanzadeh Davarani, F.; Javanmard, R.; Dokhani, A.; Khorasani, S.; Mozafari, M.R. Impact of Particle Size and Polydispersity Index on the Clinical Applications of Lipidic Nanocarrier Systems. *Pharmaceutics* **2018**, *57*. [CrossRef]
33. Lallemand, F.; Schmitt, M.; Bourges, J.L.; Gurny, R.; Benita, S.; Garrigue, J.S. Cyclosporine A delivery to the eye: A comprehensive review of academic and industrial efforts. *Eur. J. Pharm. Biopharm.* **2017**, *117*, 14–28. [CrossRef] [PubMed]
34. Kennah, H.E., 2nd; Hignet, S.; Laux, P.E.; Dorko, J.D.; Barrow, C.S. An objective procedure for quantitating eye irritation based upon changes of corneal thickness. *Fundam. Appl. Toxicol.* **1989**, *12*, 258–268. [CrossRef]
35. Falahee, K.J.; Rose, C.S.; Olin, S.S. *Eye Irritation Testing: An Assessment of Methods and Guidelines for Testing Materials for Eye Irritancy*; Office of Pesticides and Toxic Substances, US Environmental Protection Agency: Washington, DC, USA, 1981.
36. Kay, J.H.; Calandra, J.C. Interpretation of eye irritation tests. *J. Soc. Cosmet. Chem.* **1962**, *13*, 281–289.
37. Di Tommaso, C.; Torriglia, A.; Furrer, P.; Behar-Cohen, F.; Gurny, R.; Moller, M. Ocular biocompatibility of novel Cyclosporin A formulations based on methoxy poly(ethylene glycol)-hexylsubstituted poly(lactide) micelle carriers. *Int. J. Pharm.* **2011**, *416*, 515–524. [CrossRef] [PubMed]
38. Civiale, C.; Licciardi, M.; Cavallaro, G.; Giammona, G.; Mazzone, M.G. Polyhydroxyethylaspartamide-based micelles for ocular drug delivery. *Int. J. Pharm.* **2009**, *378*, 177–186. [CrossRef] [PubMed]

Article

Assessment of Physicochemical and In Vivo Biological Properties of Polymeric Nanocapsules Based on Chitosan and Poly(*N*-vinyl pyrrolidone-*alt*-itaconic anhydride)

Kheira Zanoune Dellali [1], Mohammed Dellali [1], Delia Mihaela Rață [2,*], Anca Niculina Cadinoiu [2], Leonard Ionut Atanase [2,*], Marcel Popa [2,3,*], Mihaela-Claudia Spataru [4] and Carmen Solcan [4]

[1] Faculty of Technology, University Hassiba Benbouali, BP 151, Chlef 02000, Algeria; zanounekheira@yahoo.fr (K.Z.D.); m.dellali@univ-chlef.dz (M.D.)

[2] Faculty of Medical Dentistry, Apollonia University of Iasi, Pacurari Street, No. 11, 700511 Iasi, Romania; anca.n.cadinoiu@univapollonia.ro

[3] Academy of Romanian Scientists, Splaiul Independentei Street, No. 54, 050094 Bucharest, Romania

[4] Public Health Deparment, Faculty of Veterinary Medicine, Ion Ionescu de la Brad University of Life Sciences, Mihail Sadoveanu Alley, No. 8, 700489 Iasi, Romania; mspatarufmv@yahoo.com (M.-C.S.); carmensolcan@yahoo.com (C.S.)

* Correspondence: delia.rata@univapollonia.ro (D.M.R.); leonard.atanase@univapollonia.ro (L.I.A.); marpopa@ch.tuiasi.ro (M.P.)

Abstract: Drug delivery is an important field of nanomedicine, and its aim is to deliver specific active substances to a precise site of action in order to produce a desired pharmacological effect. In the present study nanocapsules were obtained by a process of interfacial condensation between chitosan (dissolved in the aqueous phase) and poly(*N*-vinyl pyrrolidone-*alt*-itaconic anhydride), a highly reactive copolymer capable of easily opening the anhydride ring under the action of amine groups of chitosan. The formed amide bonds led to the formation of a hydrogel membrane. The morphology of the obtained nanocapsules, their behavior in aqueous solution of physiological pH, and their ability to encapsulate and release a model drug can be modulated by the parameters of the synthesis process, such as the molar ratio between functional groups of polymers and the ratio of the phases in which the polymers are solubilized. Although a priori both polymers are biocompatible, this paper reports the results of a very detailed in vivo study conducted on experimental animals which have received the obtained nanocapsules by three administration routes—intraperitoneal, subcutaneous, and oral. The organs taken from the animals' kidney, liver, spleen, and lung and analyzed histologically demonstrated the ability of nanocapsules to stimulate the monocytic macrophage system without producing inflammatory changes. Moreover, their in vivo behavior has been shown to depend not only on the route of administration but also on the interaction with the cells of the organs with which they come into contact. The results clearly argue the biocompatibility of nanocapsules and hence the possibility of their safe use in biomedical applications.

Keywords: nanoparticles; natural and synthetic polymers; drug delivery systems; biocompatibility; in vivo tests

1. Introduction

Nanotechnology has gained considerable attention during last decades, and the development of different types of nanoparticles for biomedical applications is a growing field of research. Over the past twenty years, a significant number of nanoparticulate systems, composed of different materials including lipids, polymers, and inorganic materials, have been proposed in the biomedical field [1–3]. These nanocarriers, usually ranging from 1 to 1000 nm and suitable for the delivery of drugs, hormones, genes, nucleic acids, or imaging agents, have been designed in order to obtain improved specificity, drug targeting, and delivery efficiency, thus reaching a maximal therapeutic effect with minimal side effects [4].

Drug delivery is an important field of nanomedicine, and its aim is to deliver specific active substances to a precise site of action in order to produce a desired pharmacological effect. The development of a drug delivery system is influenced not only by the target site but also by the route of administration and the nature of the nanocarriers [5].

Polymer-based vehicles constitute the main branch of drug delivery systems and can be used for both diagnostic and therapeutic purposes. Polymer carriers with a spherical shape and smooth texture are considered ideal for delivering chemotherapeutic agents in order to easily transport them through the vascular system [6]. The most-investigated polymer carriers are micelles, nanospheres, nanocapsules, dendrimers, and polymersomes [7–10]. Generally, all these carriers can increase the local concentration of loaded active substances and improve their delivery, especially when they are poorly water soluble or when their bioavailability is low.

At the present, the majority of studies regarding polymeric drug delivery systems are focused on optimizing the nanocarrier physicochemical parameters, such as size, physical stability, and drug loading efficacy, but also on carrying on preliminary in vitro cytotoxicity tests in order to prove the effectiveness of the obtained formulations [11]. However, in vivo tests, which are useful in the investigation of the biological effects of these polymeric nanomaterials, are often not taken into account as they can be difficult to carry out. Despite the remarkable rapidity of development of nanomedicine, relatively little is known about the interaction at the nanoscale of polymeric carriers with living systems. The behavior of nanocarriers in the body depends not only on the administration route but also on their interactions with the cells with which they come into contact.

Thus, in the present study, a series of original polymeric nanocapsules (NCs) based on chitosan (CS) and poly(*N*-vinyl pyrrolidone-*alt*-itaconic anhydride) (NVPAI), were firstly investigated from a physicochemical point of view, and then their biological features were determined by in vivo testing in order to demonstrate their effectiveness as safe drug delivery systems. At this point it is worth mentioning that the nanocapsules (NCs), due to their morphology, have some practical advantages with respect to other nanocarriers, as they can be characterized by increased drug encapsulation efficiency and thus by an enhanced therapeutic effect. According to its polarity, the active substance is incorporated in the core, which then acts as a reservoir, or possibly adsorbed or covalently attached to the polymeric shell [12].

These hollow NCs were obtained by interfacial condensation method in absence of any kind of toxic crosslinking agents and in normal conditions (temperature and pressure). The polymer shell was formed by amide bridges at the contact between the NH_2 groups of CS and highly reactive anhydride cycles of poly(NVPAI). It has to be mentioned that a similar type of NCs was already prepared and characterized by our research group [13], but the novelty of the present system is that the exterior layer of the polymeric shell of the NCs is formed by the synthetic NVPAI copolymer.

CS was used in this study as it is one of the biopolymers that can form nanoparticles with unique properties and therefore is currently receiving great interest for drug delivery and tissue engineering applications in the medical and pharmaceutical field due to its interesting features, such as biocompatibility, biodegradability, and non-toxicity [14]. Compared with natural polymers, synthetic polymers such as NVPAI have high purity and good reproducibility and can control the release time of loaded active substances [15].

The obtained NCs have been analyzed structurally by Fourier transform–infrared spectroscopy (FT-IR), morphologically by TEM, and gravimetrically by thermo-gravimetric analysis (TGA). Moreover, the resulting NCs were characterized in terms of particle sizes, zeta potential, drug encapsulation efficiency, and in vitro drug release kinetics by using a hydrophilic model drug, 5-fluorouracil (5-FU). Furthermore, in vivo tests were performed on adult albino BALB/c line mice using different concentrations of NC suspensions and three types of administration routes, such as intraperitoneally, subcutaneously, and per o.s. for 21 days.

2. Experimental Part (Materials and Methods)

2.1. Materials

Chitosan (CS) (low molecular weight, degree of deacetylation 91%), acetone, dimethyl sulfoxide (DMSO), hexane, surfactants (Tween 80, Span 80), and drug (5-Fluorouracil), were purchased from Sigma Aldrich (St. Louis, MO, USA). Poly(N-vinylpyrrolidone-alt-itaconic anhydride) (NVPAI) is an alternant copolymer synthesized in laboratory by a radical copolymerization method [13]. Also, phosphate-buffered solution (PBS) with pH = 7.4 and double-distilled water was prepared in our laboratory.

2.2. Preparation Method of Nanocapsules

CS/poly(NVPAI)-based nanocapsules (NCs) were obtained by the interfacial condensation method. Initially, the polymer solutions were prepared in different phases. The aqueous phase was obtained by dissolving a specific amount of CS (Table 1) in 20 mL of 2% acetic acid solution (0.4 mL acetic acid was added to 20 mL distilled water) at a temperature of 65 °C under magnetic stirring. The solution was filtered before use and brought to room temperature. Then, an appropriate quantity of non-ionic surfactant Tween 80 (2% w/v) was added in the polymer solution and well homogenized. Separately, the organic phase was prepared by dissolving exactly 500 mg of poly(NVPAI) in 15 mL of DMSO under magnetic stirring. After complete dissolution of the copolymer, a specific volume of acetone was added, according to Table 1, under continuous stirring followed by the addition of the hydrophobic surfactant Span 80 (2% w/v). After this dissolution step, the aqueous solution of CS was slowly added drop-wise into the organic solution of poly(NVPAI) under vigorous magnetic stirring at room temperature. After 2 h, the formed NCs were separated from supernatant by centrifugation for 20 min at 7500 rpm. NCs were then purified by successive washes with distilled water (7 times) and acetone (5 times). After the last wash with acetone, the product was washed twice with hexane. Finally, the obtained product was dried from hexane at room temperature to constant weight.

Table 1. Experimental parameters used for the preparation of NCs.

Sample Code	CS/Poly(NVPAI) (mol/mol)	% CS (w/v)	Aqueous Phase/Organic Phase Ratio (v/v)
CN-1	0.2/1	0.50	
CN-2	0.3/1	0.75	1:2.0
CN-3	0.4/1	1.00	
CN-4	0.5/1	1.25	
CN-5			1:2.5
CN-6	0.3/1	0.75	1:3.0
CN-7			1:3.5

The variables taken into account in this study were the molar ratio between the functional groups involved in the condensation reaction, respectively NH$_2$/anhydride cycles, expressed in the Table 1 as the ratio between the two polymers [CS/poly (NVPAI)], and the volume ratio between the aqueous and organic phases.

The yield of the NCs (Table 2) was calculated according to the following equation:

$$\text{NCs yield (\%)} = \frac{amount\ of\ recovered\ nanocapsules}{total\ amount\ of\ polymers\ used} * 100 \qquad (1)$$

Table 2. Yield data for the obtained NCs.

Sample Code	CN-1	CN-2	CN-3	CN-4	CN-5	CN-6	CN-7
Yield (%)	28	39	45	49	62	77	83

2.3. Characterization Methods

2.3.1. Structural Characterization

The structural characterization of NCs was accomplished spectrally by Fourier transform infrared spectroscopy (Schimadzu Corporation, Kyoto, Japan) (FTIR) to confirm the formation of new amide groups. The structural characterization was performed with a IRSpirit FTIR Spectrometer spectrometer. All samples were prepared as KBr pellets and scanned over the wave number range of 400–4000 cm^{-1} at a resolution of 4.0 cm^{-1}. The relevant bands in the absorption spectrum have been attributed to corresponding functional groups.

2.3.2. Size, Morphology and Zeta Potential of NCs

Transmission electron microscopy (TEM) was used to determine the size, shape, and surface morphology of NCs. The samples for transmission electron microscopy (TEM) were prepared by slow evaporation of a suspension in acetone on a formvar-coated copper grid. The samples were analyzed with a Philips CM100 microscope equipped with an Olympus camera and transferred to a computer equipped with the Megaview system.

The mean diameter of NCs in dispersion and the size distribution were determined in triplicate at 25 °C at a suspension concentration of 1% (w/v) by dynamic light scattering (DLS) (Zeta Nanosizer Malvern). Anhydrous acetone was used as a dispersant to evaluate the average diameter of unswollen NCs. The particles' diameter in physiological saline solution was also evaluated, this medium being similar to the in vivo medium. This evaluation was performed as soon as possible after NCs came into contact with the aqueous environment. The zeta potential of NCs was determined by electrophoresis in phosphate buffer solution (PBS; pH = 7.4).

2.3.3. Thermal Properties

Thermogravimetric analyses (TGA) have allowed the determination of sample weight loss as a function of temperature. These analyses were accomplished with a TA Instrument Q600 analyzer in air atmosphere (100 mL/min) with a heating rate of 10 °C/min, in a temperature range from the room temperature to 700 °C. The nanocapsules samples with a weight between 8–10 mg were heated in a platinum crucible. The operation parameters were kept constant for all the tested samples. The thermal analysis results were processed with the Universal Analysis (V 2.0) software (TA instruments, New Castle, DE, USA).

2.3.4. Swelling Behaviour in Aqueous Solutions

In order to predict and understand the behaviour of NCs during the encapsulating and release process, and therefore to assess their behaviour as potential drug carriers, swelling studies were performed by gravimetric method. The swelling degree of the NCs samples was analyzed in slightly alkaline aqueous medium, pH = 7.4, that simulates physiological conditions. A specific amount (0.03 g) of dried NCs were weighted and immersed in Eppendorf's tube containing PBS. The obtained suspension was maintained at 37 ± 0.5 °C under magnetic stirring at 120 rpm. At pre-set times, the suspension was centrifuged, the supernatant was removed, and the swollen sample was weighed. All experiments were performed in triplicate. To ensure the swelling until equilibrium, the samples were allowed to swell for 24 h. The percentage of swelling ratio (Q%) was determined with Equation (2)

$$Q(\%) = \frac{W - W_0}{W_0} * 100 \qquad (2)$$

where W is the weight of swollen sample (mg) and W_0 is the initial weight of dry sample (mg).

In parallel with the gravimetric method, the change in the size of the NCs on swelling was also evaluated.

2.3.5. Drug Encapsulating Studies

The drug encapsulating process was carried out through diffusional mechanism. In this study, 5-Flourouracil (5-FU) was used as the model drug. Briefly, 20 mg NCs were dispersed in 1.5 mL aqueous drug solution with a concentration of 10 mg 5-FU/mL in ultrapure water. The suspension was maintained under magnetic stirring (120 rpm) and temperature (37 °C) for 24 h. The drug-loaded NCs were separated from supernatant by ultracentrifugation at 8000 rpm for 10 min. The drug-loaded NCs were lyophilized and stored as powder for further analyses. The amount of 5-FU encapsulated into NCs was calculated by the difference between the initial amount of 5-FU in solution and the amount of 5-FU from supernatant using a UV Spectrometer (Nanodrop One, Thermo Scientific, Waltham, MA, USA) at 266 nm [13,16]. The encapsulation efficiency (E_{ef}%) of 5-FU into NCs was calculated as follows:

$$m_l = m_i - m_s \tag{3}$$

$$E_{ef}\% = \frac{m_i - m_s}{m_i} \times 100 \tag{4}$$

where m_l—the amount of encapsulated 5-FU (mg); m_i—the initial amount of 5-FU (mg); m_s—the amount of 5-FU found in supernatant (mg).

The obtained drug-loaded NCs are designated as follows: CN-1-5FU, CN-2-5FU, CN-3-5FU, CN-4-5FU, CN-5-5FU, CN-6-5FU, and CN-7-5FU.

2.3.6. In Vitro Drug Release

The in vitro drug release studies were realized by the dialysis method. Each sample of 5-FU-encapsulated NCs was introduced into a dialysis membrane and, after that, was individually immersed in flasks with 13 mL PBS at pH = 7.4, a value which is similar to blood. This system was maintained at 37 ± 0.5 °C under continuous stirring at 120 rpm for all of the release period. At regular time intervals, 1 mL of solution was taken and replaced with fresh PBS. The 5-FU released and present in the medium was spectrophotometrically determined at 266 nm wavelength, using a Nanodrop One (Thermo Scientific). The release efficiency of 5-FU (R_{ef}%) was calculated using Equation (5):

$$Ref(\%) = \frac{m_r}{m_l} \times 100 \tag{5}$$

where m_r—the amount of drug released from NCs (mg); m_l—the amount of the drug encapsulated into the NCs (mg).

2.3.7. In Vivo Testing

The study was conducted on 42 adult albino BALB/c line mice, aged approximately 4 months, reared under conventional laboratory conditions (20–23 °C, 55% UR), fed standardized pelleted feed, fruits and vegetables, and water at discretion. Their bodyweight was monitored in the first, tenth, and last day of the experiment, and it was constant within the experimental error limits (Figure S1). Mice undergoing the experiment (equally female and male) were divided into 7 groups: one control and 6 experimental groups of 6 mice. To each group, CN-4 and CN-6 suspensions were administered daily by three different routes, such as intraperitoneally (0.01 mL and 0.02 mL), subcutaneously (0.1 and 0.2 mL), and per o.s. (0.2 and 0.3 mL) for 21 days. Suspensions of CN-4 and CN-6 were obtained by adding 5 mL of saline solution over 6.25 mg powder, the suspension being prepared approximately 2 h before use. The health status of the mice and body weight dynamics were followed and 2 days after the completion of the experiment the mice were euthanized by cervical dislocation and probes were harvested from internal organs (kidney, liver, spleen, and lung) and processed for histological examination.

2.3.8. Histological and Immunohistological Analysis

Organs samples were fixed in 10% formalin solution for 24 h. Approximately 0.5 cm-thick slices were dehydrated with a decreasing concentration of ethanol solution, then clarified in xylene and embedded in paraffin. After being cut with the microtometer, 10 microscope slides from each paraffin block were selected, specifically stained, and read on the Olympus CX41 microscope. They were initially stained with hematoxylin eosin (HE) then IHC using 4 antibodies: Cd147, p65, alpha SMA, HMC II, and Cox-2. Anti p65 (Nuclear factor-kB p65) antibodies, AA 143–158, antibodies, CD147 (ab188190), α-SMA (anti-alpha smooth muscle actin antibody), MA5-11547 (14A-asm-1), MHC II (Dako M0746), and Cox-2 (ab16701 SP-21), were used to perform immunohistochemical staining. After sections were deparaffinized in Xylen, hydrated in ethanol, and microwaved for 10 min at 95 °C in 10 mmol citrate acid buffer pH6, they were cooled for 20 min, then washed twice in PBS for 5 min. Slices were treated with 3% hydrogen peroxide and rinsed with PBS, after which they were incubated overnight at 4 °C in a humid atmosphere with primary antibodies in dilutions of 1:100 CD147, MHC II, Cox-2 and 1:500 p65, α-SMA. The following day, slides were washed 3 times in PBS for 5 min, being incubated with secondary antibodies. For CD147, MHC II, and Cox-2 activity of bone cells, goat anti-rabbit IgG secondary antibody was used, and for p65 and α-SMA, goat anti-mouse IgG secondary antibody was chosen. Microscope slides were developed in 3,3′-diaminobenzidine (DAB) and finally counterstained with hematoxylin. Images were interpreted using ImageJ IHC profile software.

DAB IHC profile scores were negative (−), low positive (±), positive (+), and over positive (++). Scoring of HE histological lesions was done by assessing changes and scoring as follows: no change (−), minor (+), medium (++), and major (+++).

2.4. Statistical Analysis

The statistical significance of cytotoxic activity was analyzed by Student's t-test. The values are expressed as mean ± SE of three parallel measurements, $p < 0.05$ being considered significant.

3. Results and Discussion

The objective of the present study was to assess the physicochemical and biological properties of a series of NCs based on CS and poly(NVPAI) which can be further used as drug delivery systems for the controlled and sustained release of different types of drugs.

A first result is that the yield of obtaining NCs increases with increasing the amount of CS in their composition (respectively of the CS/poly(NVPAI) molar ratio), as can be noticed in Table 2.

The increase in the initial amount of chitosan leads to the increase of the bonds formed between the poly(NVPAI)- and CS-reactive groups because a higher number of these reactive groups are available and can be involved in the interfacial polycondensation reaction. Consequently, a smaller amount of non-crosslinked polymer chains remains in the system and can be eliminated during the purification and washing process of the nanocapsules. In addition, enhancement of the aqueous phase/organic phase ratio, at the same CS/poly(NVPAI) ratio, has led to a visible increase in the nanocapsules final yield [17].

Different analysis techniques were used, and the obtained results are presented in the following.

3.1. FTIR Spectroscopy

The FT-IR spectra of CS, poly (NVPAI), and NCs presented in Figure 1 confirmed the reaction between the anhydride groups of the copolymer and chitosan amine groups.

Absorption bands at approximately 1778 cm^{-1} and 1858 cm^{-1} from the copolymer spectrum correspond to the stretching vibration of the –C=O anhydride groups. Another important peak is located at approximately 989 cm^{-1} and can be assigned to the C–O–C bond in the anhydride group [13].

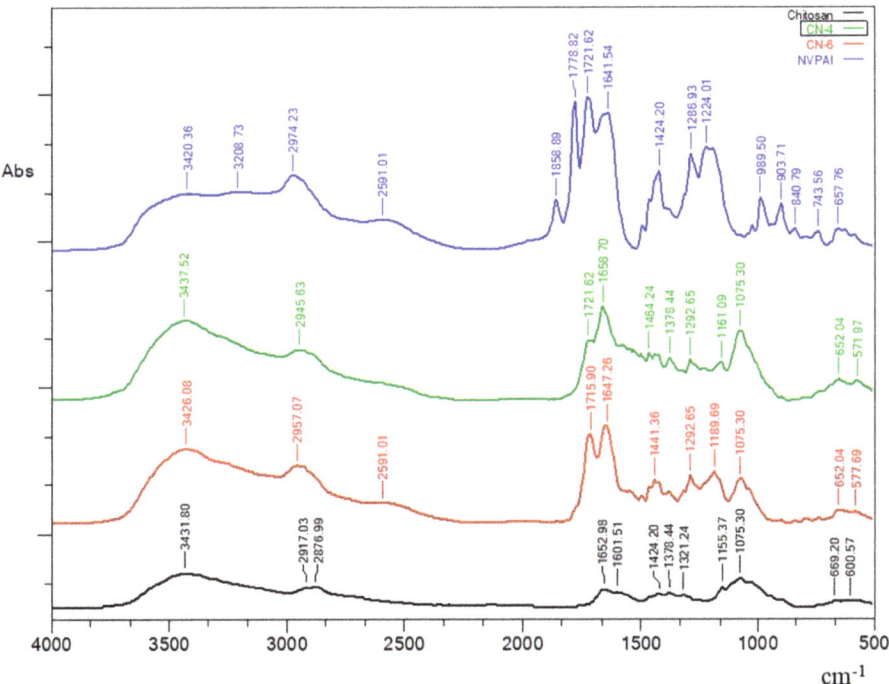

Figure 1. FTIR spectra of CS, NVPAI, CN-4 and CN-6 samples.

CS sample presents a peak at 1652 cm^{-1} that corresponds to carbonyl groups (C=O) and another peak at 1601 cm^{-1} attributed to free amino groups (–NH2).

The FT-IR spectra of NCs samples evidenced the disappearance of characteristic peaks for the anhydride groups and the appearance of the absorption bands at 1647 cm^{-1} and 1658 cm^{-1} corresponding to a carbonyl bond of the newly formed amide groups. The appearance of absorption bands at 1715 cm^{-1} (for sample CN-6) and at 1721 cm^{-1} (for sample CN-4) that are attributed to the –C=O of the carboxylic groups reveals that some of the anhydride groups have hydrolyzed. Finally, the disappearance of absorption bands at 989 cm^{-1} evidenced that all anhydride groups participated in the amidation or hydrolysis reaction.

Figure 2 shows the FTIR spectra for simple 5-FU, CN-6 without drug and drug-loaded NCs (CN-4-5FU and CN-6-5FU). The peaks characteristic of 5-FU spectrum (1429.92 cm^{-1}, 1246.89 cm^{-1}, 812.19 cm^{-1}, 755.00 cm^{-1}, and 640.6 cm^{-1}) are also found in the NC samples loaded with the model drug. In addition, these peaks are not visible in the spectrum of samples without drug (Figures 1 and 2).

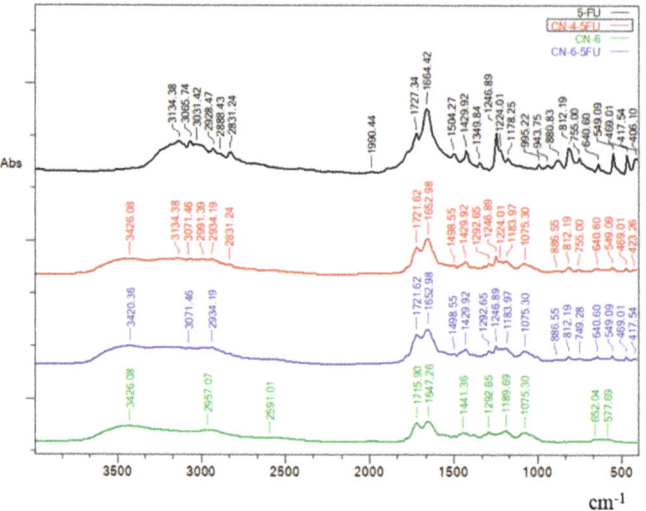

Figure 2. FTIR spectra of simple 5-FU, CN-6 without drug, and drug-loaded NCs (CN-4-5FU and CN-6-5FU).

3.2. Size, Morphology and Zeta Potential of NCs

The mean diameters for NCs in acetone varied between 107 and 250 nm (Figure 3 and Table 2). Size distribution curves evidenced the presence of two populations of nanocapsules in the case of CN-1, CN-2, and CN-3 samples. CN-4, CN-5, CN-6, and CN-7 samples presented a monomodal distribution. From the obtained results, it can be noticed that the NCs' diameter increases with the increasing of the quantity of CS. The explanation of this behavior is based on the fact that, by increasing the amount of CS, increasing amounts of polysaccharide are involved in the formation of the NCs' membrane, which becomes thicker, thus contributing to the increase of the diameter.

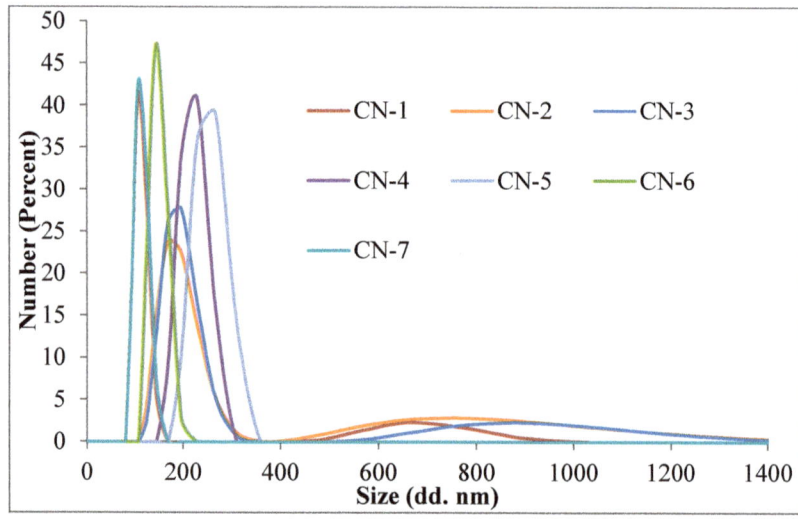

Figure 3. Size distribution curves of NCs in acetone at room temperature.

Increasing the volume of the organic phase, the number of macromolecules of copolymer that come into contact with a drop of the aqueous solution of CS is decreased gradually due to the dilution of the copolymer solution, and therefore the diameter of NCs decreases.

The NCs swell very quickly in aqueous environments. After 5 min in the physiological saline solution, diameters between 1089 and 2256 nm were recorded. The amount of CS and the volume of the organic phase had the same influence on the NCs size, even in an aqueous environment.

The zeta potential values evidenced the stability of the aqueous dispersion of NCs in a medium with slightly alkaline pH. In Table 3 the zeta potential values are presented which varied between -8 mV and -23.21 mV. Negative charge of NCs can be attributed to the presence of carboxylic groups formed as a result of the interfacial condensation reaction by amidation of the anhydride cycle in poly (NVPAI) or by their hydrolysis. By condensation, a carboxylic group also appears in the reaction of an amine cleavage with the copolymer. By hydrolysis, each anhydride cycle generates two carboxylic groups. This explains why by increasing the amount of CS, the zeta potential is reduced: the explanation lies in the overall formation of fewer carboxylic groups. From Table 3 it can be seen that, in the case of NCs without 5-FU, the increasing of the CS amount in the system leads to a decrease in the zeta potential values. On the other hand, NCs loaded with 5-FU show an increase in the zeta potential value as the amount of CS in the composition of the NCs increases.

Table 3. Diameter in volume, polydispersity index (PDI), and Zeta potential values of NCs samples.

Samples Code	Dv (nm) in Acetone	PDI	Dv (nm) in Physiological Saline Solution	PDI	ZP (mV) in PBS (pH = 7.4)
CN-1	107 ± 0.11	0.96 ± 0.04	1621 ± 84.42	0.43 ± 0.12	−19.8 ± 0.01
CN-2	188 ± 0.23	0.73 ± 0.02	1976 ± 34.51	0.42 ± 0.09	−18.4 ± 0.04
CN-3	192 ± 0.51	0.83 ± 0.04	1869 ± 57.84	0.42 ± 0.04	−18.0 ± 0.05
CN-4	220 ± 0.53	0.64 ± 0.03	2131 ± 51.61	0.41 ± 0.09	−16.8 ± 0.01
CN-5	250 ± 0.30	0.94 ± 0.03	2256 ± 43.82	0.31 ± 0.09	−18.4 ± 0.04
CN-6	150 ± 0.22	0.68 ± 0.06	1371 ± 30.91	0.23 ± 0.02	−8.5 ± 0.02
CN-7	114 ± 0.21	0.94 ± 0.04	1089 ± 53.84	0.46 ± 0.03	−11.06 ± 0.07
CN-1-5FU	169 ± 0.73	1.22 ± 0.07	2263 ± 157.62	1.30 ± 0.24	−20.07 ± 0.24
CN-2-5FU	226 ± 0.63	1.02 ± 0.10	2156 ± 203.41	1.27 ± 0.39	−20.57 ± 0.38
CN-3-5FU	254 ± 0.42	1.04 ± 0.04	2396 ± 107.04	1.07 ± 0.68	−21.1 ± 0.32
CN-4-5FU	240 ± 0.87	0.92 ± 0.08	2421 ± 124.95	1.12 ± 0.60	−21.52 ± 0.17
CN-5-5FU	297 ± 0.43	1.00 ± 0.12	2947 ± 191.39	1.10 ± 0.35	−22.22 ± 0.04
CN-6-5FU	227 ± 0.44	0.89 ± 0.05	1887 ± 54.22	0.54 ± 0.23	−23.21 ± 0.34
CN-7-5FU	181 ± 0.69	0.93 ± 0.24	1594 ± 79.73	0.81 ± 0.16	−22.8 ± 0.18

From the Figure 4 it appears that the NCs have a spherical morphology and a moderate polydispersity, but that their size is smaller than that in acetone obtained by DLS. This can probably be explained by the different principle of operation of the two techniques (in liquid media and in dry state). The solvent used for DLS, acetone, may diffuse to some extent inside the NCs causing a slight increase in its volume.

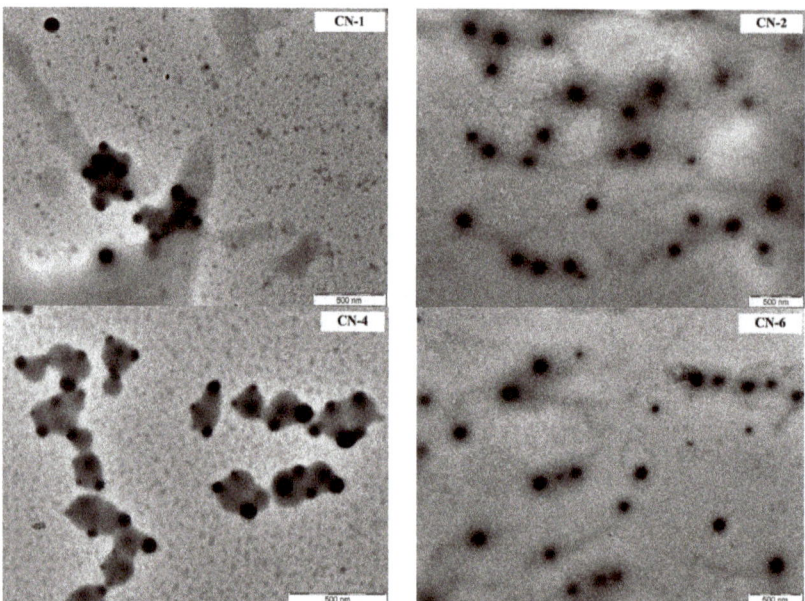

Figure 4. TEM images of CN-1, CN-2, CN-4 and CN-6 samples.

3.3. Thermal Behaviour

The evaluation of the temperature behavior of the NCs was performed, on one hand, in order to have additional evidence that their membrane is made of both polymers, and on the other hand, to determine whether it is possible to heat sterilize them before administration without suffering changes caused by possible thermal degradation. In Figure 5 are presented the TGA chromatograms of different NC samples.

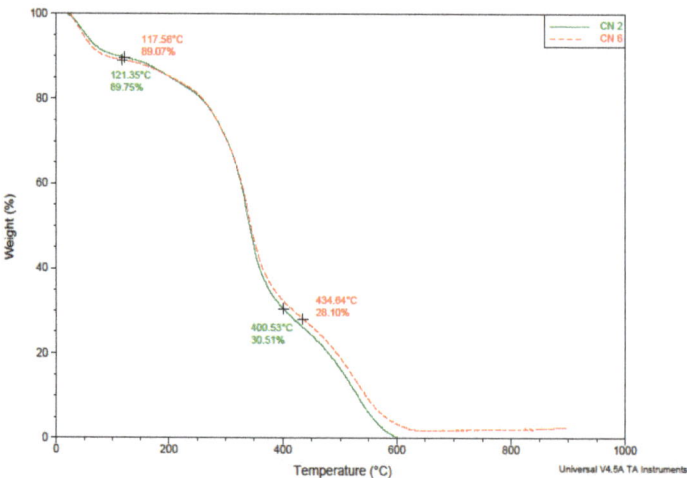

Figure 5. TGA chromatograms of NCs samples.

From the TGA results provided in Figure 5, it is obvious that the obtained NCs are stable until 400 °C, when a second degradation step occurs. This is a proof that these NCs are stable during the sterilization step. Moreover, analysis of the TGA chromatograms showed that the composition of the NCs shell does not qualitatively influence its thermal degradation behavior. It is found that indeed the CN-6 sample (with a higher yield, previously explained by the higher amount of reacted copolymer) has a slightly higher calcination residue content compared to the CN-2 sample.

3.4. Swelling Degree of NCs

The swelling behaviour of the obtained NCs was investigated in slightly alkaline medium (pH 7.4). This medium was chosen as these drug delivery systems are intended to be used in biological fluids, and the obtained results are illustrated in Figure 6.

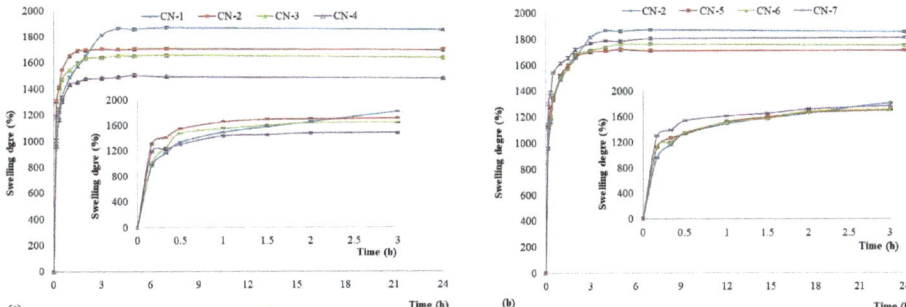

Figure 6. The swelling kinetics curves of the NCs in alkaline conditions (pH = 7.4) for samples (**a**) CN-1, CN-2, CN-3, and CN-4; (**b**) CN-2, CN-5, CN-6, and CN-7.

It is evident that all the NCs presented a high swelling degree which varies from 1476% to 1851%. The size, the composition, and the preparation parameters of NCs have an important influence on the swelling properties. In Figure 6a it can be observed that the swelling degree varied between 1851% and 1476% and decreased with the increase of the molar ratio of CS/NVPAI. The highest swelling degree was obtained for the CN-1 sample which is characterized by the smallest CS amount. The swelling of the NCs is caused by the water penetration within the empty core until complete filling as well as by the swelling of the polymer membrane, which has a hydrogel character [17].

The increase in the number of moles of -NH$_2$ groups from CS has as a consequence the increasing of the crosslinking density of the network of which the capsule membrane is formed, which reduces the diffusion of the water inside the capsule but also the amount of water that swells the membrane. In the other hand, it can be observed that the swelling degree of the NCs increases with the increase of the volume organic phase (Figure 6b) for an identical CS/poly(NVPAI) molar ratio of 0.3. This behaviour is explained as follows: as the volume of the organic phase increases (dilution of the copolymer solution), the number of macromolecules of poly(NVPAI) which come into contact with CS at the interface of the aqueous solution with organic phase is decreased and becomes smaller and smaller, so the crosslinking density of the NCs membrane will decrease, and therefore, the swelling rate will increase [18].

Since NCs can be used in physiological saline solutions for medical applications, it has also been necessary to evaluate their size in this environment (Figure 7). For the CN-4 and CN-6 samples, which were used for in vivo tests, the influence of the aqueous medium on the size over time was also evaluated (Figure 8).

After 24 h the diameter of the NCs had values between 5700 and 10,300 nm. Being capsules, it is easy to understand why they grow so much in size in the aqueous environment. Aqueous solution penetrates very easily inside the capsule and the network that creates the membrane of the capsule can relax freely.

Figure 8 shows that the diameter of the two samples increases quite rapidly after they are introduced into the saline solution. After only 2 min they go from nanometers to micrometers, and after half an hour they almost reach equilibrium.

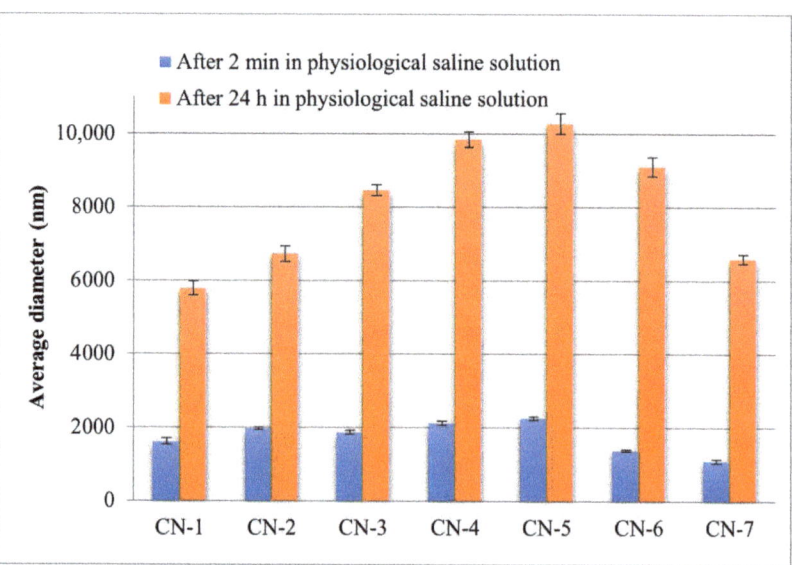

Figure 7. Evaluation of the average diameter of the NCs after 2 min and 24 h in physiological saline solutions.

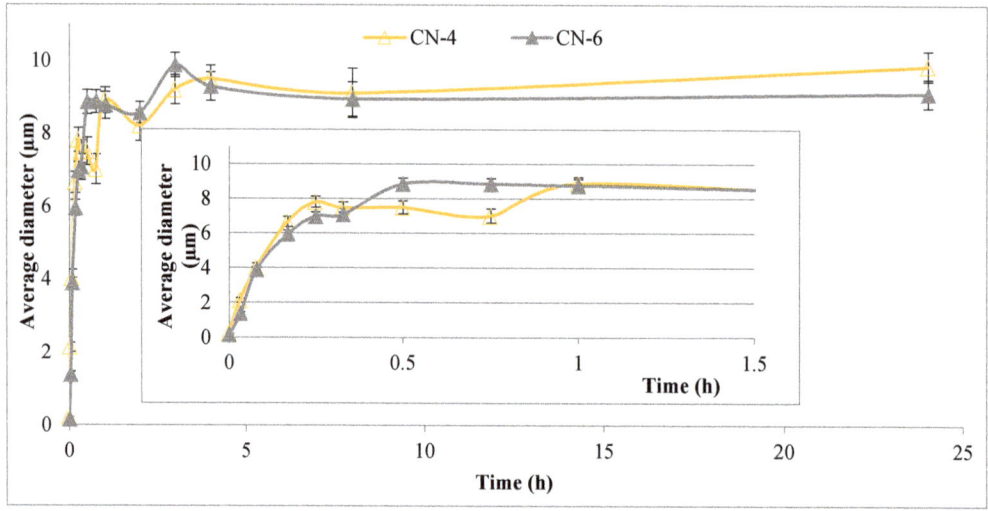

Figure 8. The influence of the aqueous environment on the size of CN-4 and CN-6 samples over time.

3.5. Encapsulation Efficiency of a Model Drug

This type of NC was developed in order to be used as efficient drug delivery systems for hydrophilic drugs. In this study, 5-FU was used as a model drug. The efficiency of encapsulating 5-FU into NCs and the amount of encapsulated drug (5-FU g/g NCs) are presented in Table 4.

Table 4. Drug encapsulation efficiency values.

Sample Code	Encapsulated Drug g/g NCs	The Encapsulation Efficiency (%)
CN-1	0.317 ± 0.001	42.3 ± 0.173
CN-2	0.244 ± 0.003	32.6 ± 0.333
CN-3	0.219 ± 0.001	29.3 ± 0.120
CN-4	0.198 ± 0.002	26.5 ± 0.289
CN-5	0.249 ± 0.001	33.2 ± 0.120
CN-6	0.258 ± 0.006	34.5 ± 0.866
CN-7	0.266 ± 0.004	36.0 ± 0.577

3.6. Drug Release Kinetics

The release studies of the model drug, 5-FU, were performed in slightly alkaline medium (pH = 7.4) and the results are reported in Figure 9.

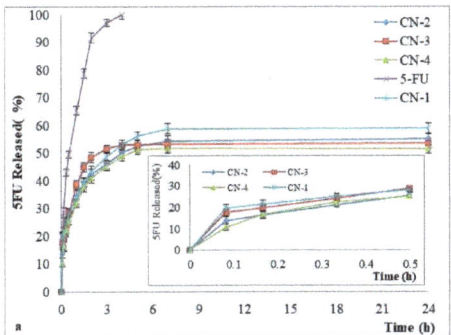
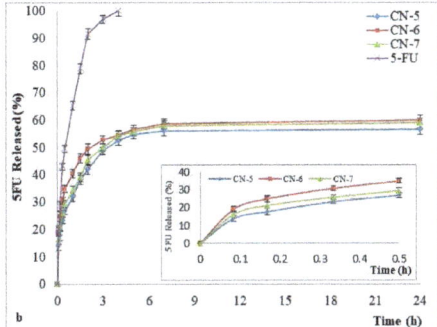

Figure 9. In vitro release kinetics of 5-FU from NCs in phosphate buffer solution (pH 7.4), with a zoom insert of the release kinetics between 0 and 30 min, for samples (**a**) CN-2, CN-3, and CN-4; (**b**) CN-5, CNM-6, and CN-7.

The results showed that 5-FU release from NCs was between 51% and 60%, and simple 5-FU was 100% released within 270 min of the start of the experiment. From Figure 9, it appears also that in the case of NCs loaded with 5-FU, a faster release was observed in the first hour which is due to the release of the drug adsorbed at the surface of NCs followed by a slow release of the encapsulated drug until the equilibrium was reached. Although the differences in release kinetics between the obtained NCs samples are not noticeable, there is still an influence on the amount of CS. If the CS amount increases, the reticulation density increases also, leading thus to a decrease of the drug release rate through the NCs shell. These results are in full accordance with the evolution of the previously discussed swelling process. These tests indicate that the obtained NCs can be used for the controlled and sustained release of hydrophilic drugs.

3.7. In Vivo Testing

For in vivo testing, mice were divided into 7 groups, one control and 6 experimental groups that received "CN" which received the NCs. At this point, different administration routes of the NCs were investigated.

Under light microscopy, the liver in the control group is normal; the hepatocytes are large and cuboidal with a prominent round nucleus and eosinophilic cytoplasm. The cords are radially arranged from the centrilobular venule to the periphery of the lobule. Hepatic sinusoidal capillaries are arranged between the hepatic plates with a sparse arrangement of Kupffer cells. In the control group, the kidney has a normal appearance of both the cortex and medulla. Malpighian corpuscles and urinary tubules show no changes (Figure 10). Hematoxylin- and eosin (HE)-stained sections from the control group (group I) showed normal histological architecture of the lungs: thin-walled alveoli, alveolar sacs, clear alveolar spaces, and thin-walled blood vessels. The epithelial lining of the alveoli was composed of squamous alveolar cells with dense nuclei (type I pneumocytes) and large alveolar cells with large, rounded nuclei (type II pneumocytes). The spleen in the control group is covered by a thin capsule of connective tissue and shows white pulp and red pulp without changes. In the liver, kidney, and lung, congestion of blood vessels was observed.

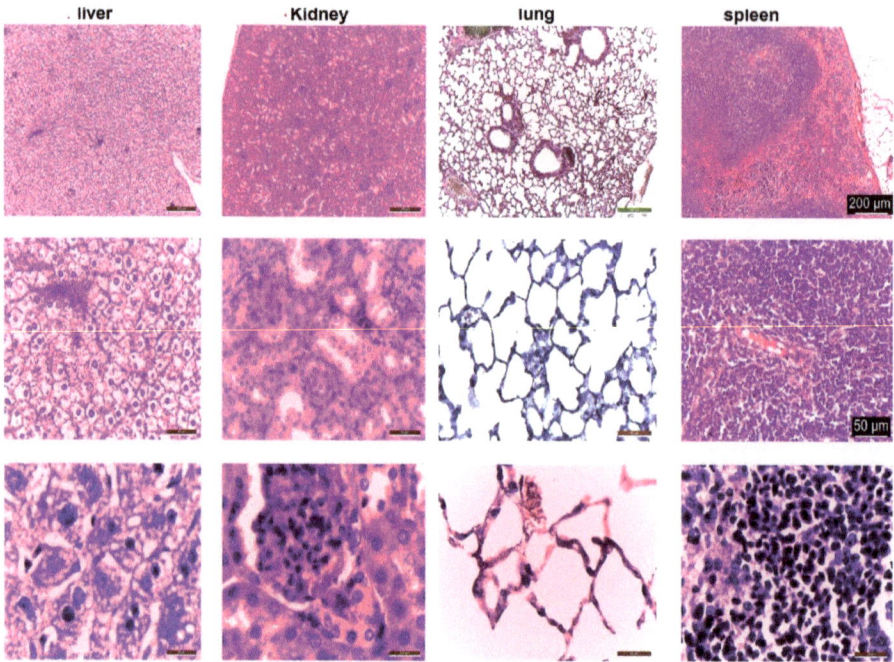

Figure 10. Histological structure of liver, kidney, lung, and spleen in control mice. HE staining.

In the case of the experimental groups, by HE staining, no major changes in the histological structure of the organs studied were observed (Table 5).

CN-4 and CN-6 produced reduced histological changes in the organelles studied depending on the route of administration and dose. Administration p.o. and s.c., irrespective of dose, caused ectasia of a small number of veins in the liver (Figure 11), kidney (Figure 12), lung (Figure 13), and spleen (Figure 14). In the liver the hepatocytes show microvesicles and an increase in the number of Kupffer cells (6–7.5/12,879.8 mm^2). In the lung there is a high number of alveolar macrophages residing in the septal wall (14–24/12,879.8 mm^2). The alveoli have an intact, thin septal wall, but in some places it is thickened by macrophages. In the kidneys, the proximal convoluted tubules show rare brush border changes, and a small number of Malpighian corpuscles have a condensed glomerulus and increased capsular space volume (one in each preparation examined). In the interstitial space of the urinary tubules in the cortical area macrophages are observed (2–5/12,879.8 mm^2). In the spleen the lymphoid follicles show a slight increase, where pigments and megakaryocytes, some of them giant, appear.

Administration of i.p. CN-4 0.01 mL and 0.02 mL in the liver, in hepatocytes with micro- and macrovesicles, resulted in 7–9.3 Kupffer cells /12,878.2 mm^2, alveolar macrophages (9–13/12,879.8 mm^2), and interstitial macrophages between the urinary tubules 4–6.6/12,879.8 mm^2. A small number (1–3 out of 10 fields examined of each individual) of Malpighian corpuscles were found with sclerotic changes of the glomerulus, with dilated convoluted tubules and with damaged brush border. Administration of i.p. CN-6

0.01 mL and 0.02 mL caused ectasia of some veins in the liver, kidney, and lung. The 0.02 mL dose produced changes in hepatocytes represented by macrovesicles, small focal necrosis, frequent Kupffer cells (9–11/12,879.8 mm^2), rare Malpighian corpuscles with sclerosing glomerular lesions (3 in each preparation examined), a reduced frequency of dilated proximal convoluted tubules with secretion in the lumen, macrophages in the interstitial space between the urinary tubules (5–7.5/12,879.8 mm^2), and alveolar macrophages in the septal wall of the pulmonary alveoli (9–18.5/12,879.8 mm^2). In the lung, clusters of 15–20 macrophages were observed in the lamina propria of the bronchioles and perivascular.

Table 5. Assessment of cellular and vascular changes produced in the organs under study.

Organ/Lesions	Control	PO 0.3 mL	SC 0.2 mL	Ip 0.01 mL CN-4	Ip 0.02 mL CN-4	Ip 0.01 mL CN-6	Ip 0.02 mL CN-6
Liver							
-cords are radially arranged from the centrilobular venule to the periphery of the lobule	+	+	+	+	+	+	+
No of Kupffer cells/12,879.8 mm^2	3.5	6–7	6–7.5	7–8	8–9.3	9–10.5	9–11
Vascular congestion	−	−	−	−	++	++	++
Kidney							
Malpighian corpuscles and urinary tubules show changes	−	1	1	1	1	3	3
No of. Interstitial macrophages/12,879.8 mm^2	−	2–4	2–5	4–6	4–6.6	5–6	5–7.5
Increased in volume of nephrocytes of the proximal convoluted tubules	−	−	−	+	+	++	++
Vascular congestion	−	−	−	+	+	+	++
Lungs							
thin-walled alveoli	+	+	+	+	+	+	+
No of alveolar macrophages/12,879.8 mm^2	3–4	14–22	14–24	9–11	9–13	9.5–17	9–18.5
Vascular congestion	−	−	−	+	+	++	++
Spleen							
Lymphoid follicles and lymphatic cords	−	−	++	++	++	++	++
Vascular congestion	−	−	+	+	+	++	++
Megakaryocytes and pigment cells	−	−	+	+	+	++	++

Scoring of HE histological lesions was done by assessing changes and scoring as follows: no change (−), minor (+), medium (++).

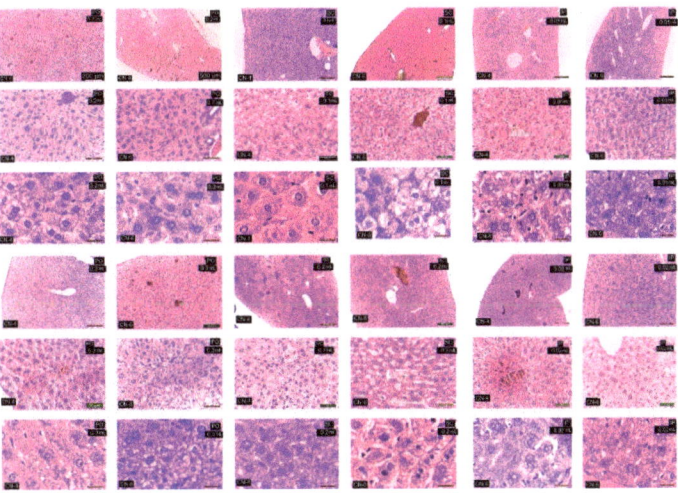

Figure 11. Histological structure of the liver in experimental groups. HE staining.

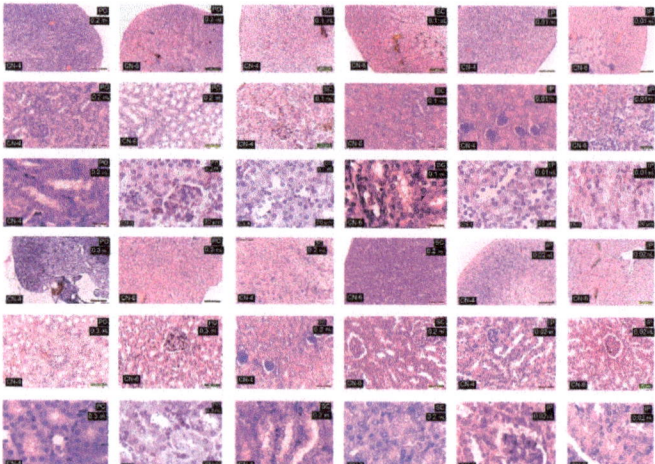

Figure 12. Kidney in experimental groups. HE stain.

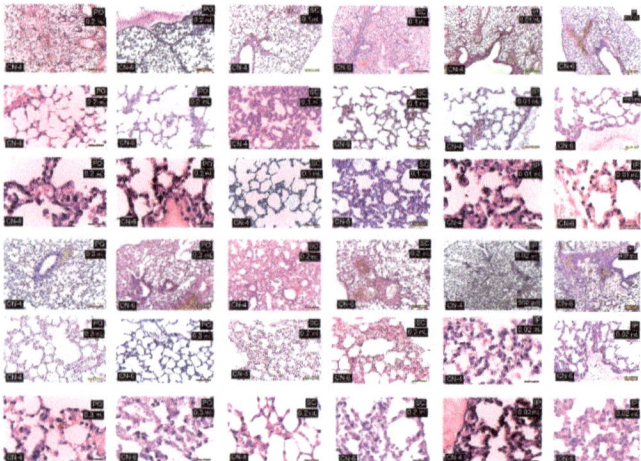

Figure 13. Lungs in the experimental group. HE staining.

From this figure it can be seen that in all experimental groups, the main change is in congestive blood vessels and a higher frequency of Kupffer cells. The arrangement of hepatocytes in radial cords is preserved. Cell shape and nuclei are similar to the cells in the control group. Hepatocytes from experimental groups receiving CN-4 and CN-6 p.o. have hepatocytes with intracytoplasmic microvesicles. In the groups administered s.c. and i.p., cytoplasmic micro and macrovesicles and focal necrosis are observed.

From this figure it can be seen that in all experimental groups the main change is in the vascular congestion. In the experimental groups that received CN-4 and CN-6 p.o., a reduced number of interstitial macrophages appear among the urinary tubules in the cortical area. In the groups given s.c. and i.p., the Malpighian corpuscles with glomerular compaction, modified brush border in some proximal convoluted tubules, and small hemorrhages in the cortical area are observed. Some tubules with increased nephrocytes in volume were apparent.

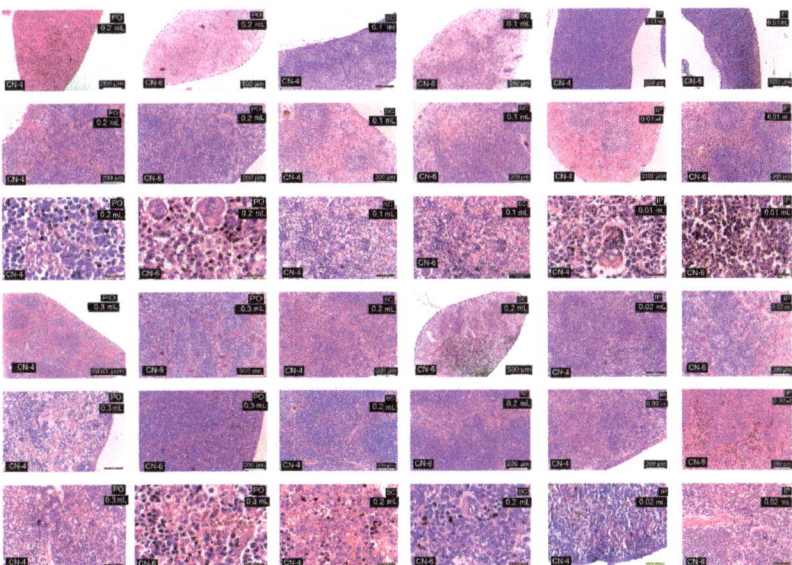

Figure 14. Spleen in the experimental group. HE staining.

Sections from the experimental groups showed preserved lung architecture. CN administration produced at some sites a change in alveolar septal diameter due to the accumulation of alveolar macrophages. Alveoli in the vicinity of congested vessels were observed in collapse. The number of alveolar macrophages located in the angular septal wall between alveoli was higher upon p.o. and i.p. administration. In the i.p. administration groups, macrophage clusters were observed in the lamina propria of the bronchi or perivascular. No alveolar macrophages were observed in the alveolar lumen.

In the experimental groups, the number of lymphoid follicles is increased, especially when administered i.p.; lymphatic cords have a slightly increased volume, and the number and frequency of megakaryocytes and pigment cells is higher in the experimental groups compared to the control group.

Following the evaluation of the IHC images, a low positive CD-4 label was observed for all markers regardless of the route of administration. Positive labeling was recorded for CN-6 only at intraperitoneal administration.

The CD147 marker was positive in the liver on Kupffer, endothelial, and stellate cells; in the kidney on endothelial and interstitial macrophages and mesangial and nephrocyte base cells; in the lung on alveolar macrophages and endothelial cells; in the spleen on endothelial lymphocytes and activated macrophages, platelets, and megakaryocytes. In the groups with intraperitoneal administration of CN-6, this CD147 marker was also expressed on hepatocytes around the centrilobular venule, on the nephrocyte basement, on a higher number of lymphocytes in the spleen, and on endothelial cells and type I pneumocytes in the lung. Although the IHC profile is positive in these organs, no pathological changes were observed (Figures 15–18).

In experimental groups receiving the CN-4 sample by various administration routes, P65 is expressed on endothelial cells in the liver, very little on endothelial cells in the lung, on nephrocyte bases and endothelial cells in the kidney, and on some lymphocytes and lymph nodes. Positive labeling was recorded in organelles from CN-6 with intraperitoneal administration (Figures 15–18).

α-SMA positively labelled blood vessel smooth muscle fibers in all organelles examined, some lymphocytes, endothelial cells, and red blood cells in capillaries. No excess pericapillary extracellular matrix was found. Blood vessels were positively marked for red blood cells (Figures 15–18).

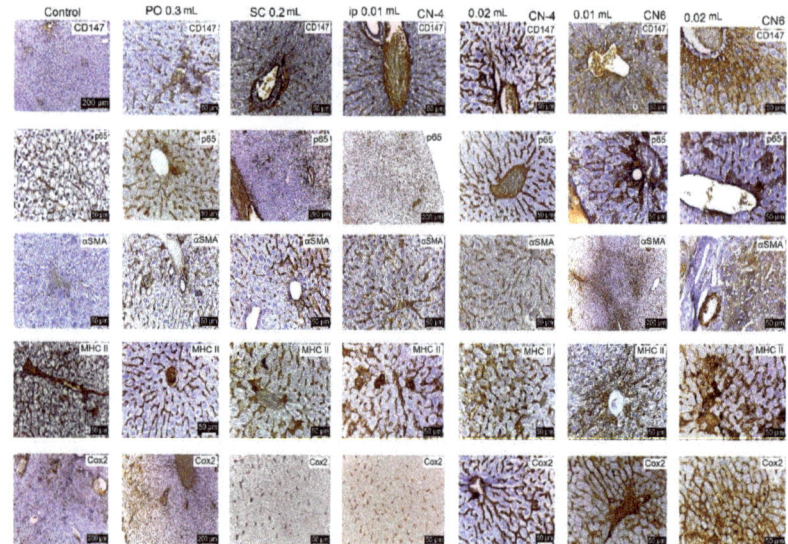

Figure 15. IHC marking in the liver.

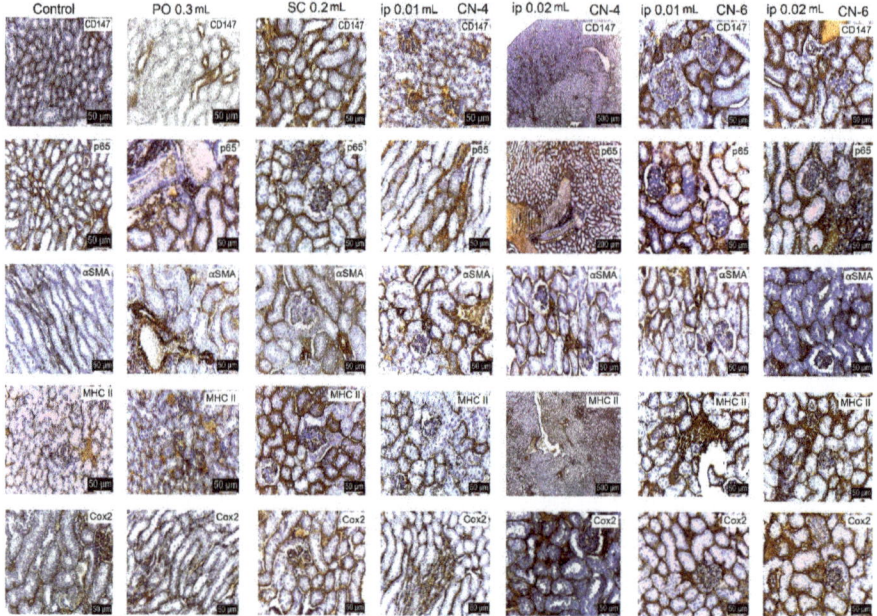

Figure 16. IHC labeling in kidneys.

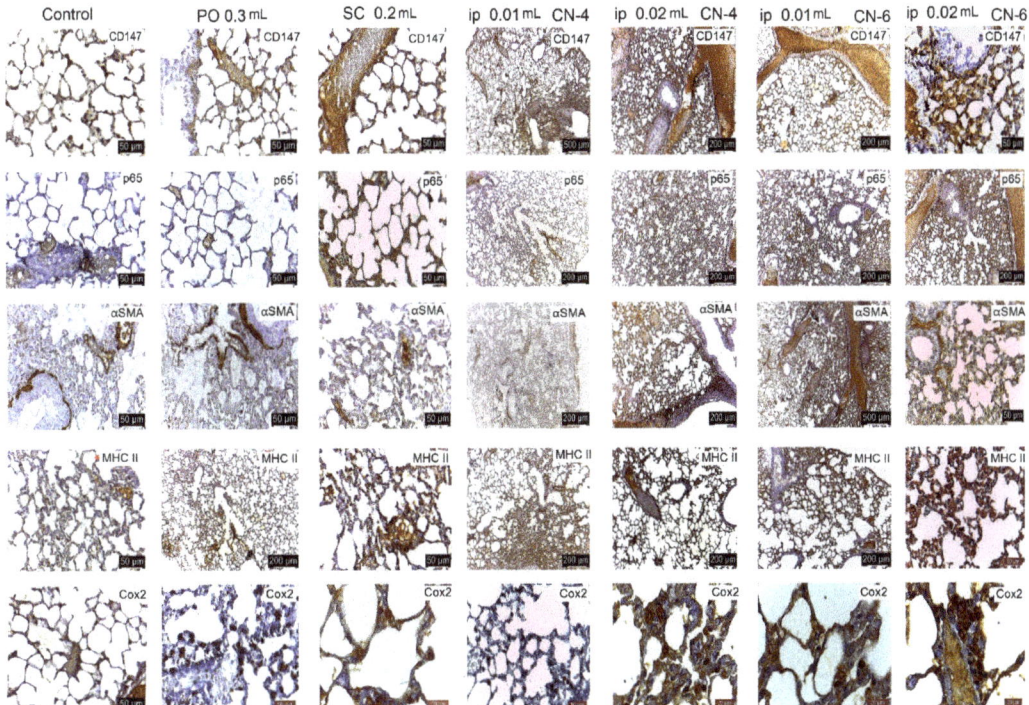

Figure 17. IHC labelling in the lung.

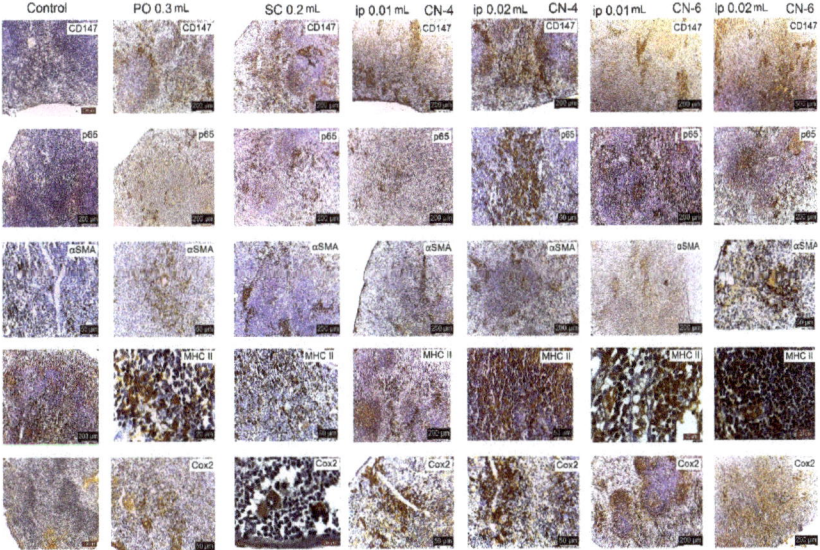

Figure 18. IHC labeling in the spleen.

MHC-II antigens were expressed on some lymphocytes (LB); monocytic lineage cells; a small population of T helper cells; activated T cells; arteriolar, sinusoidal, and venous endothelium; Kupffer cells; spindle cells in the connective tissue of the portal tract; large hepatic veins; and liver capsule with the exception of the liver where duct cells were not labelled with MHCII (Figures 15–18). MHC II is expressed on bronchial epithelium, type II pneumocytes, and ciliated epithelial cells.

Cox-2 labeling was noted on cells of the monocytic macrophage system, namely Kupffer cells, splenic macrophages, alveolar macrophages, mesangial cells in kidneys, and at the base of nephrocytes (Figures 15–18).

All antibodies taken in the study label endothelial, stellate, and Kupffer cells in the CN-4 exposed batches. The batches exposed to CN-6, intraperitoneal administration also have hepatocytes positively labelled to CD147, p65, αSMA, MHC II, and COX-2. Positive labelling was also recorded in the erythrocytes in capillaries and venules.

The antibodies taken in the study mark endothelial cells, the membrane of the basal pole of nephrocytes, which has multiple invaginations which confers more intense positivity. Cox-2 and MHCII can also distinguish positively labelled mesangial cells. Higher positivity is recorded in the experimental batch CN6 intraperitoneal with both doses.

In all experimental groups, the number of alveolar macrophages detected by Cox-2 increased. Positive and low-positive labelling was also influenced by the positivity of hematomas in septal capillaries and blood vessels.

Low positive labeling CD147, p65, MHC II, and Cox2 was recorded on activated lymphocytes and macrophages in the CN-4 group. Positive MHCII and Cox-2 labelling was recorded in all experimental groups.

NPC has been successfully used for mucosal administration such as oral, nasal, ocular, and pulmonary, due to its mucoadhesive and mucosal permeability properties [19,20]. NPCs have been shown to bind efficiently to intestinal epithelial cells and effectively cross the epithelial barrier to penetrate tissues. This suggests that it is possible that NPCs are taken up by antigen-presenting cells of the mucosal immune system which subsequently transport the delivered antigens to immune initiation sites, such as Peyer's patches, to induce an immune response [21].

The presence of mononuclear phagocyte system macrophages in the liver, lung, kidney and spleen upon per o.s. administration in our experiment underlines this property of stimulation and distribution by macrophages in the body of chitosan nanoparticles. Following per o.s. and s.c. administration, a significant reactivity was achieved in the lung, followed by the liver, spleen, and kidney. In addition, the i.p. administration induced increased numbers of splenic lymphoid follicles and a slight increase in the number of macrophages in the interstitial space of the urinary tubules in the cortical area of the kidney.

Kupffer cells are specialized in the internalization of foreign nanoparticles, playing an essential role in their uptake, trafficking, and destination in the body (He et al., 2010). After intravenous administration, chitosan nanoparticles are opsonized in the blood before being phagocytosed by macrophages and accumulating in cells of the mononuclear phagocyte system. This passive targeting promotes the accumulation of NPs in the liver [22]. Transport from Kupffer cells to hepatocytes was also studied in mice, and it was observed that the majority of CsNps were localized in hepatocytes after intravenous injection. Micro and macrovesicles in hepatocytes could be given by the accumulation of chitosan nanoparticles in these cells in our experiment.

Kupffer cells internalize NP via several toll-like receptors, mannose, and Fc (Gustafson et al., 2015). The mechanisms involved are macropinocytosis, clathrin-mediated endocytosis, caveolin-mediated endocytosis, and the additional endocytotic pathway [23–25]. Clathrin-mediated endocytosis has been shown to be responsible for the internalization of approximately 100–350 nm size, while caveolin-mediated endocytosis is responsible for endocytosis of 20–100 nm size [26–28]. Macropinocytosis enables internalization of 0.5–5 μm nanosystems [29].

The liver is a major site of accumulation of CS nanoparticles after intravenous administration [30–32]. Administration of CNP by various routes did not produce major histological changes in the organs studied, but there was a proliferation of liver Kupffer cells and alveolar macrophages in the septal space of the pulmonary alveoli without changing its diameter. Intravenous administration of NCs does not cause significant hemodynamic changes, and 30 min after administration of NCs, they accumulate mainly in the liver and lung without causing hemolysis and leukocytosis [33]. The toxicity of CS nanoparticles was manifested by a short-term delay in weight gain as reported in the literature in rats. We did not encounter granulomas in the liver as reported by other researchers [33]. Granulomas found in the lung and liver indicate slow biodegradation of chitosan nanoparticles. Overall, the results obtained indicate good tolerability of intravenous administration of an unmodified chitosan suspension in the dose range studied [33]. The effect of chitosan on the viability of hepatocytes has been investigated in a series of studies on the creation of artificial livers in which chitosan served as a framework (nanofibre scaffold) for hepatocyte cultures. The presence of CS fibres in the intercellular space, performing a supportive function, improved the function of hepatocytes [34,35].

The biocompatibility of NCs is also explained by the fact that, under in vivo conditions, negatively charged plasma proteins are adsorbed on the surfaces of nanoparticles, thus preventing erythrocyte aggregation and hemolysis. The process of thrombosis during systemic NC incorporation is directly dependent on the magnitude of positive NCs' charge. Thus, pronounced agglutination, hemolysis, and intravascular thrombosis develop in the case of NCs with high charge and high concentration [36], whereas NCs with low charge adsorb coagulation factors, mainly fibrinogen, upon contact with blood and cause only weak inhibition of platelet aggregation [37]. In addition, the charge of NCs and their derivatives depends on the pH of the medium and is determined by the concentration of amino groups in the polymer molecule [38,39]. Resuspension in saline reduces NCs load, thus allowing intravenous administration of those without the risk of hemolysis [38] at particles doses of 4–6 mg/kg [40–42]. Including s.c. and i.p. administration was possible in the present experiment without thrombosis or hemolysis phenomena. Lee et al. [43] suggest that the lungs may be the primary barrier organ for intravenously administered CPN. This phenomenon has already been described and has, in fact, been proposed as a targeted delivery strategy for the lungs [43]. Proliferation of alveolar macrophages was also observed in our experiment regardless of the route of administration.

After entering the body, NMs (>6 nm) predominantly accumulate in organs of the mononuclear macrophage system (MPS), such as the liver, spleen, lungs, etc. [40], due to subsequent recognition and internalization by the macrophages of this system [44]. Some particles, such as gold NPs (40 nm) and [45] dextran-coated magnetite NPs (core diameter of 8–10 nm) [46,47] are gradually degraded in cells over several months or even longer periods and sequestered in MPS for a long period of time. Indeed, it has been observed that long-term sequestration in MPS induces some potential side effects, especially immunotoxicity [48]. Accordingly, removal of NCs from MPS organs, especially the liver, sequesters 30–99% of administered NCs from the bloodstream [40] being essential for the clinical safety of NCs. Therefore, it is necessary that all injected NCs are completely removed within a reasonable period of time [49]. Following distribution in hepatocytes, the hepatobiliary–fecal excretion route has been shown to be the primary elimination pathway for CsNps in addition to the reno-urinary excretion pathway. Elimination of CsNps in mice was a lengthy process with a half-life of approximately 2 months.

CD147, a transmembrane glycoprotein with two immunoglobulin-like glycoproteins of the immunoglobulin G (IgG) superfamily, is widely expressed on the surface of various cells, activated lymphocytes, and epithelial cells including cancer cells. As a metalloproteinase inducer, CD147 has been shown to be involved in multiple biological processes such as immune response, tumour progression, and tissue repair. CD147 is recognized as a regulator of lipid metabolism in a variety of cell types and autophagy [50]. In experimental groups, this marker was captured on various cells, but no histopathological changes were reported. Modified CS nanoparticles activate macrophages and are phagocytosed by them [51], processes that lead to aseptic inflammation [38,52]. CD147 is expressed by platelets and is overexpressed following platelet activation [53]. The CS nanoparticles used in this experiment have a small cytotoxic effect and have a weak antiplatelet and anticoagulant effect as mentioned by Sonin et al. [33]. CD147-mediated cell–cell interaction on the cell surface induces platelet activation and via NF-B-factor-mediated monocyte activation will occur [54].

Nuclear factor-κB (NF-κB/p65) is a family of transcription factors that plays a critical role in inflammation, immunity, cell proliferation, differentiation, and survival. Inducible NF-κB activation depends on proteosomal degradation induced by phosphorylation of NF-κB inhibitory proteins (IκB), which retain inactive NF-κB dimers in the cytosol in unstimulated cells. Most of the signalling pathways leading to NF-κB activation converge on the IκB kinase complex (IKK) which is responsible for IκB phosphorylation and is essential for signal transduction to NF-κB. Further regulation of NF-κB activity is achieved by various posttranslational modifications of the basic components of NF-κB signalling pathways. In addition to cytosolic modifications of IKK and IκB proteins, as well as other pathway-specific mediators, transcription factors are themselves extensively modified [55]. The low positive labeling after CN-4 administration demonstrates the presence of modified transcription factors not involved in transcriptional processes. In animals exposed to CN-6, positive labeling may indicate involvement of this factor in transcription. Numerous labelled cells were observed in the spleen of mice from experimental groups. Macrophages are able to integrate an impressive amount of information on the identity and virulence of pathogens, as well as endogenous cues present in their microenvironment, to modulate the immune response for optimal host protection. Central to this capacity are the numerous ways in which NF-κB signaling is modulated based on shifting activation thresholds, integration of information from different classes of pattern recognition receptors, and tight regulation of transcription through rigorous positive and negative feedback loops [54]. The presence of this p65 marker was low positive in the studied organelles with the exception of the CN-6 intraperitoneal exposed batches, denoting a reduced inflammatory response.

Circulating monocytes are recruited into tissues, where they differentiate into macrophages and take part in the process of inflammation or tissue remodeling. According to the traditional concept, macrophages are classified into pro-inflammatory (M1), non-activated (M0), or anti-inflammatory (M2) subsets that play distinct roles in the initiation and resolution of inflammation. More recent experimental findings have led to a substantial update of the monocyte–macrophage nomenclature to include the nature of the polarization signal. In response to proinflammatory stimuli, monocytes can be polarized directly into three subsets of macrophages with M1-type proinflammatory phenotype; interferon-γ-induced macrophages have the strongest proinflammatory properties. When exposed to various anti-inflammatory stimuli, monocytes can differentiate into at least five subsets of M2-type macrophages. Of these, a subset generated upon exposure to IL-4 (IL-13) has the most typical M2-type characteristics. In both humans and mice, differentiation of monocytes into macrophages involves global transcriptome changes that are tightly controlled by various transcriptional regulators and signaling mechanisms [56]. There are many key checkpoints in the transcriptional control and signaling network that trigger either pro-inflammatory or anti-inflammatory polarization [57]. Regulation of NF-κB family transcriptional activity plays a central role in M1–M2 switching and macrophage polarization towards an anti-inflammatory or pro-inflammatory phenotype. The absence of intense p65 labeling also implies insignificant macrophage activation in this experiment.

MHC-II or HLA-DR is a molecule involved in antigen presentation and is the main signal in immunological cooperation [58]. MHC-II antigens are normally expressed on B lymphocytes, monocytic lineage cells, a small population of T helper cells, activated T cells, thymic epithelium, and vascular endothelium [59]. Normal arteriolar, sinusoidal, and central venous endothelium often express MHC II. Kupffer cells have always expressed these antigens. MHC II-positive spindle cell fibroblasts were identified in the connective tissue of the portal tract, large hepatic veins, and liver capsule: most shared antigens common to all leukocytes and reacted with MHC II. Bile duct epithelium expressed MHC II in primary biliary cirrhosis, large duct obstruction, and drug-induced cholestasis, indicating that MHC II positive spindle cells are phenotypically similar to histiocytes. In our experiment liver duct cells were not labeled with MHCII. Although Kupffer cells express lower levels of MHC class II molecules than classical dendritic cells, they are able to interact with T cells. However, unlike dendritic cells, Kupffer cells favor the development of regulatory T cells, thereby promoting immune tolerance [60] which explains the lack of liver injury in this experiment following NC administration. Hepatic macrophages may be highly targeted by nanoparticle drug carriers due to their efficient phagocytosis function in the liver [61].

MHCII is expressed on both bronchial and alveolar epithelium, especially on type II pneumocytes and ciliated cells, and in this experiment as reported in the literature [58].

Various studies have shown that CS and its derivatives can effectively activate antigen-presenting cells and induce cytokine stimulation to produce an effective immune response and promote Th1/Th2 response balance [62]. Catalytically active Cox-2 (and Cox-1) is located in the nuclear envelope (NE) and endoplasmic reticulum (ER), where it mediates PGE2 biosynthesis. Cox-2 dissociated from the nuclear envelope is catalytically inactive [63]. Activated macrophages overexpress these enzymes (Cox-2), which would lead to the production of large amounts of PGs. In addition, NF-κB is a transcription factor that induces copying of proinflammatory genes to produce large amounts of proinflammatory mediators, such as Cox-2, in activated macrophages [55]. The presence of a positive low label at this marker indicates reduced transcription of proinflammatory mediators. Although a large number of alveolar macrophages can be observed in the lung, low positive for alpha SMA indicates a lack of synthesis of collagen precursors. This is also observed in other organs.

Cyclooxygenases are responsible for the synthesis of prostaglandins from arachidonic acid and are present in two isoforms in the kidney, Cox-1 and Cox-2. Cox-1 is involved in the regulation of basic cellular functions, while Cox-2 is a proinflammatory enzyme and is induced by inflammatory stimuli [64]. Cox-2 labeling has been noted in cells of the monocytic macrophage system, namely Kupffer cells, splenic macrophages, alveolar macrophages, and mesangial cells in the kidney. The absence of an inflammatory process denotes that Cox-2 is in an inactive phase.

The adjuvant effect of CS is mainly evaluated from several points of view such as biocompatibility, biodegradability, and cell permeability [65]. Size and surface charge are key characteristics that define how cells interact with and internalize NCs [66]. The affinity of NCs to the cell membrane is related to the cationic component, a characteristic of CS [67].

Upon oral administration an important role in nanoparticle uptake is played by M cells in the intestinal mucosa. The key point for the initiation of the mucosal immune response is antigen uptake, in the case of the NCs based on CS. Many experiments have shown that M cells can carry various macromolecular substances and microorganisms. After M cells in Peyer plates adsorb NCs, they are actively transported to the underlying immune cells, dendritic cells, to stimulate the local immune system or mucosal immunity [68]. Previous research has shown that CS nanoparticles accumulate predominantly in the macrophages of the monocytic macrophage system [69]. Administered systemically, nanoparticles are taken up by Peyer's plaque cells, pass into the lymph and then into the general blood circulation, and can subsequently be taken up by the liver, kidney, spleen, heart, and other vital organs [70]. CS nanoparticles are also known as immunomodulators. An important role in this case is played by the size and physicochemical characteristics of the particles that can influence the interaction with immune cells to induce the desired therapeutic benefit [65]. Bioactive CS nanoparticles are internalized by macropinocytosis, clathrin-mediated endocytosis, and phagocytosis. They are then transported intracellularly by endosomes, multivesicular bodies, and lysosomes. Proteomics elucidated that chitosan nanoparticles induced an increase in proteins involved in immunoregulatory functions and antioxidant activities. They also promoted the production of anti-inflammatory/pro-regenerative mediators but suppressed pro-inflammatory ones. Therefore, CS nanoparticles could prevent persistent post-treatment inflammation [71].

Most research that has addressed the toxicity of chitosan-based nanoparticles has typically conducted biocompatibility studies 2–4 days after intravenous administration in animal models [72–75]. However, the toxicity determined from these types of studies, which use short observation periods for long-circulating dispersed solutions [72,74], may not be fully representative because neither biodistribution nor biodegradation processes are completed within these short periods. However, studies over longer observation periods after intravenous administration in subacute or chronic experiments are rare [33], even though long-term observations of biological effects of dispersed systems are no less important than acute toxicity analysis. Based on this consideration, we chose chronic exposure with per o.s., s.c., and i.p. administration.

4. Conclusions

Interfacial condensation of CS with poly(NVPAI) is an effective method of obtaining NCs capable of encapsulating, transporting, and delivering drugs. The morphological characteristics of NCs as well as their physicochemical properties can be modeled by varying some parameters of the preparation process (the ratio between the functional groups of the polymers involved in the reaction, the ratio of the phases in which the polymers are solubilized). Consistent with these properties also varies the ability of NCs to encapsulate and release the model drug in a simulated physiological environment (pH = 7.4).

Low positive labeling at all markers taken in the study in experimental groups subjected to CN-4 sample denotes biocompatibility and does not result in histological changes. These NCs have demonstrated the ability to stimulate cells of the mononuclear phagocyte system in the liver, lung, spleen, and kidney without producing inflammatory changes. When administered per o.s., proliferation of alveolar macrophages was observed, followed by Kupffer cells. Administration of CN4 and CN6 i.p. at a dose of 0.02 mL induced micro- and macrovacuoles of hepatocytes. Overexpression of CD147, p65, MHCII, and Cox-2 markers was observed in the macrophages of the organs studied in all experimental groups. The absence of alpha SMA labelling denotes the absence of an inflammatory process.

From this study, it can be concluded that the obtained NCs have the advantage of being biocompatible, non-toxic, and safe for in vivo administration as drug delivery systems.

Supplementary Materials: The following supporting information can be downloaded at: https://www.mdpi.com/article/10.3390/polym14091811/s1, Figure S1: Evolution of mice bodyweight through the experiment (the first measurement was made in the first day of the experiment, the second one at 10 day and the last one at the final of the experiment); Table S1: p-values of Student's t-test for simple 5-FU compared with 5-FU loaded NCs.

Author Contributions: Conceptualization, K.Z.D., D.M.R., A.N.C., M.P. and L.I.A.; methodology, K.Z.D., M.D., D.M.R., A.N.C., L.I.A., M.-C.S. and C.S.; software, D.M.R., A.N.C., M.-C.S. and C.S.; investigation, K.Z.D., D.M.R., A.N.C., M.-C.S. and C.S.; writing—original draft preparation, D.M.R., M.P., L.I.A. and C.S.; visualization, D.M.R., A.N.C., L.I.A., M.P. and C.S.; supervision, M.P., L.I.A. and C.S.; project administration, L.I.A. and M.P. All authors have read and agreed to the published version of the manuscript.

Funding: The research leading to these results has received funding from the NO Grants 2014–2021, under project contract no. 15/2020.

Institutional Review Board Statement: Not applicable.

Informed Consent Statement: Not applicable.

Data Availability Statement: The data presented in this study are available on request from the corresponding author.

Conflicts of Interest: The author declares no conflict of interest.

References

1. Jeevanandam, J.; Barhoum, A.; Chan, Y.S.; Dufresne, A.; Danquah, M.K. Review on nanoparticles and nanostructured materials: History, sources, toxicity and regulations. *Beilstein J. Nanotechnol.* **2018**, *9*, 1050–1074. [CrossRef] [PubMed]
2. Khan, I.; Saeed, K.; Khan, I. Nanoparticles: Properties, applications and toxicities. *Arab. J. Chem.* **2019**, *12*, 908–931. [CrossRef]

3. Hano, C.; Abbasi, B.H. Plant-based green synthesis of nanoparticles: Production, characterization and applications. *Biomolecules* **2022**, *12*, 31. [CrossRef] [PubMed]
4. Zhang, P.; Li, Y.; Tang, W.; Zhao, J.; Jing, L.; McHugh, K.J. Theranostic nanoparticles with disease-specific administration strategies. *Nano Today* **2022**, *42*, 101335. [CrossRef]
5. Vasant, V.R.; Mannfred, A.H. *Drug Delivery Systems*, 2nd ed.; Lewis Publisher: Boca Raton, FL, USA, 2003.
6. Babu, A.; Templeton, A.K.; Munshi, A.; Ramesh, R. Nanoparticle-based drug delivery for therapy of lung cancer: Progress and challenges. *J. Nanomater.* **2013**, *2013*, 863951. [CrossRef]
7. Daraba, O.M.; Cadinoiu, A.N.; Rata, D.M.; Atanase, L.I.; Vochita, G. Antitumoral drug-loaded biocompatible polymeric nanoparticles obtained by non-aqueous emulsion polymerization. *Polymers* **2020**, *12*, 1018. [CrossRef]
8. Iurciuc-Tincu, C.E.; Cretan, M.S.; Purcar, V.; Popa, M.; Daraba, O.M.; Atanase, L.I.; Ochiuz, L. Drug delivery system based on pH-sensitive biocompatible poly(2-vinyl pyridine)-b-poly(ethylene oxide) nanomicelles loaded with curcumin and 5-Fluorouracil. *Polymers* **2020**, *12*, 1450. [CrossRef]
9. Sung, Y.K.; Kim, S.W. Recent advances in polymeric drug delivery systems. *Biomater. Res.* **2020**, *24*, 12. [CrossRef]
10. Iurciuc-Tincu, C.E.; Atanase, L.I.; Jerome, C.; Sol, V.; Martin, P.; Popa, M.; Ochiuz, L. Polysaccharides-based complex particles' protective role on the stability and bioactivity of immobilized Curcumin. *Int. J. Mol. Sci.* **2021**, *22*, 3075. [CrossRef]
11. Jesus, S.; Schmutz, M.; Som, C.; Borchard, G.; Wick, P.; Borges, O. Hazard assessment of polymeric nanobiomaterials for drug delivery: What can we learn from literature so far. *Front. Bioeng. Biotechnol.* **2019**, *7*, 261. [CrossRef]
12. Couvreur, P.; Barratt, G.; Fattal, E.; Legrand, P.; Vauthier, C. Nanocapsule technology: A review. *Crit. Rev. Ther. Drug Carrier Syst.* **2002**, *19*, 99–134. [CrossRef] [PubMed]
13. Rata, D.M.; Chailan, J.F.; Peptu, C.A.; Costuleanu, M.; Popa, M. Chitosan:poly(N-vinylpyrrolidone-alt-itaconic anhydride) nanocapsules—A promising alternative for the lung cancer treatment. *J. Nanopart. Res.* **2015**, *17*, 316–327. [CrossRef]
14. Jou, S.; Peters, L.; Mucalo, M. Chitosan: A review of molecular structure, bioactivities and interactions with the human body and micro-organisms. *Carbohydr. Polym.* **2022**, *282*, 119132.
15. Rață, D.M.; Popa, M.; Chailan, J.F.; Zamfir, C.L.; Peptu, C.A. Biomaterial properties evaluation of poly(vinyl acetate-alt-maleic anhydride)/chitosan nanocapsules. *J. Nanopart. Res.* **2014**, *16*, 2569. [CrossRef]
16. Alupei, L.; Lisa, G.; Butnariu, A.; Desbrieres, J.; Cadinoiu, A.N.; Peptu, C.A.; Calin, G.; Popa, M. New Folic Acid-Chitosan Derivative Based Nanoparticles—Potential Applications in Cancer Therapy. *Cellulose Chem. Technol.* **2017**, *51*, 631–648.
17. Iurea, D.M.; Peptu, C.A.; Chailan, J.F.; Carriere, P.; Popa, M. Sub-micronic capsules based on gelatin and poly(maleic anhydride-alt-vinylacetate) obtained by interfacial condensation with potential biomedical applications. *J. Nanosci. Nanotechnol.* **2013**, *13*, 3841–3850. [CrossRef]
18. Dellali, K.Z.; Rata, D.M.; Popa, M.; Djennad, M.; Ouagued, A.; Gherghel, D. Antitumoral drug: Loaded hybrid nanocapsules based on chitosan with potential effects in breast cancer therapy. *Int. J. Mol. Sci.* **2020**, *21*, 5659. [CrossRef]
19. Zhou, C.; Hao, G.; Thomas, P.; Liu, J.; Yu, M.; Sun, S.; Öz, O.K.; Sun, X.; Zheng, J. Near-infrared emitting radioactive gold nanoparticles with molecular pharmacokinetics. *Angew. Chem.* **2012**, *51*, 10118–10122. [CrossRef]
20. Burns, A.A.; Vider, J.; Ow, H.; Herz, E.; Penate-Medina, O.; Baumgart, M.; Larson, S.M.; Wiesner, U.; Bradbury, M. Fluorescent silica nanoparticles with efficient urinary excretion for nanomedicine. *Nano Lett.* **2009**, *9*, 442–448. [CrossRef]
21. Hajam, I.A.; Senevirathne, A.; Hewawaduge, C.; Kim, J.; Lee, J.H. Intranasally administered protein coated chitosan nanoparticles encapsulating influenza H9N2 HA2 and M2e mRNA molecules elicit protective immunity against avian influenza viruses in chickens. *Vet. Res.* **2020**, *51*, 37. [CrossRef]
22. Li, L.; Wang, H.; Ong, Z.Y.; Xu, K.; Ee, P.L.R.; Zheng, S.; Hedrick, J.L.; Yang, Y.Y. Polymer- and lipidbased nanoparticle therapeutics for the treatment of liver diseases. *Nano Today* **2010**, *5*, 296–312. [CrossRef]
23. Ma, Y.; Yang, M.; He, Z.; Wei, Q.; Li, J. The biological function of kupffer cells in liver disease. In *Biology of Myelomonocytic Cells*; Ghosh, A., Ed.; IntechOpen: London, UK, 2017.
24. Dobrovolskaia, M.A.; McNeil, S.E. Immunological properties of engineered nanomaterials. *Nat. Nanotechnol.* **2007**, *2*, 469–478. [CrossRef] [PubMed]
25. Xiao, K.; Li, Y.; Luo, J.; Lee, J.S.; Xiao, W.; Gonik, A.M.; Agarwal, R.; Lama, K.S. The effect of surface charge on in vivo biodistribution of PEGoligocholic acid based micellar nanoparticles. *Biomaterials* **2011**, *32*, 3435–3446. [CrossRef] [PubMed]
26. Parton, R.G. Caveolae meet endosomes: A stable relationship? *Dev. Cell.* **2004**, *7*, 458–460. [CrossRef] [PubMed]
27. Pelkmans, L.; Burli, T.; Zerial, M.; Helenius, A. Caveolin-stabilized membrane domains as multifunctional transport and sorting devices in endocytic membrane traffic. *Cell* **2004**, *118*, 767–780. [CrossRef]
28. Gratton, S.E.; Ropp, P.A.; Pohlhaus, P.D.; Luft, J.C.; Madden, V.J.; Napier, M.E.; DeSimone, J.M. The effect of particle design on cellular internalization pathways. *Proc. Natl. Acad. Sci. USA* **2008**, *105*, 11613–11618. [CrossRef]
29. Gustafson, H.H.; Holt-Casper, D.; Grainger, D.W.; Ghandehari, H. Nanoparticle uptake: The phagocyte problem. *Nano Today* **2015**, *10*, 487–510. [CrossRef]
30. Yan, C.; Gu, J.; Guo, Y.; Chen, D. In vivo biodistribution for tumor targeting of 5-fluorouracil (5-FU) loaded N-succinyl-chitosan (Suc-Chi) nanoparticles. *Yakugaku Zasshi* **2010**, *130*, 801–804. [CrossRef]
31. Zhang, C.; Qu, G.; Sun, Y.; Wu, X.; Yao, Z.; Guo, Q.; Ding, Q.; Yuan, S.; Shen, Z.; Ping, Q.; et al. Pharmacokinetics, biodistribution, efficacy and safety of N-octyl-O-sulfate chitosan micelles loaded with paclitaxel. *Biomaterials* **2008**, *29*, 1233–1241. [CrossRef]
32. Kean, T.; Thanou, M. Biodegradation, biodistribution and toxicity of chitosan. *Adv. Drug Deliv. Rev.* **2010**, *62*, 3–11. [CrossRef]

33. Sonin, D.; Pochkaeva, E.; Zhuravskii, S.; Postnov, V.; Korolev, D.; Vasina, L.; Kostina, D.; Mukhametdinova, D.; Zelinskaya, I.; Skorik, Y.; et al. Biological Safety and Biodistribution of Chitosan Nanoparticles. *Nanomaterials* **2020**, *10*, 810. [CrossRef] [PubMed]
34. Lee, J.H.; Lee, D.H.; Son, J.H.; Park, J.K.; Kim, S.K. Optimization of chitosan-alginate encapsulation process using pig hepatocytes for development of bioartificial liver. *J. Microbiol. Biotechnol.* **2005**, *15*, 7–13.
35. Chu, X.H.; Shi, X.L.; Feng, Z.Q.; Gu, Z.Z.; Ding, Y.T. Chitosan nanofiber scaffold enhances hepatocyte adhesion and function. *Biotechnol. Lett.* **2009**, *31*, 347–352. [CrossRef] [PubMed]
36. Carreno-Gómez, B.; Duncan, R. Valuation of the biological properties of soluble chitosan and chitosan microspheres. *Int. J. Pharm.* **1997**, *148*, 231–240. [CrossRef]
37. Li, X.; Radomski, A.; Corrigan, O.I.; Tajber, L.; De Sousa Menezes, F.; Endter, S.; Medina, C.; Radomski, M.W. Platelet compatibility of PLGA, chitosan and PLGA-chitosan nanoparticles. *Nanomedicine* **2009**, *4*, 735–746. [CrossRef]
38. Nadesh, R.; Narayanan, D.; Sreerekha, P.R.; Vadakumpully, S.; Mony, U.; Koyakkutty, M.; Nair, S.V.; Menon, D. Hematotoxicological analysis of surface-modified and -unmodified chitosan nanoparticles. *J. Biomed. Mater. Res. A* **2013**, *101*, 2957–2966. [CrossRef]
39. Zhou, X.; Zhang, X.; Zhou, J.; Li, L. An investigation of chitosan and its derivatives on red blood cell agglutination. *RSC Adv.* **2017**, *7*, 12247–12254. [CrossRef]
40. Zhang, Y.N.; Poon, W.; Tavares, A.J.; Mcgilvray, I.D.; Chan, W.C.W. Nanoparticle-liver interactions: Cellular uptake and hepatobiliary elimination. *J. Control. Release* **2016**, *240*, 332–348. [CrossRef]
41. Hirano, S.; Seino, H.; Akiyama, Y.; Nonaka, I. Chitosan: A biocompatible material for oral and intravenous administrations. In *Progress in Biomedical Polymers*; Gebelein, C.G., Dunn, R.L., Eds.; Springer: Boston, MA, USA, 1990; pp. 283–290.
42. Banerjee, T.; Singh, A.K.; Sharma, R.K.; Maitra, A.N. Labeling efficiency and biodistribution of Technetium-99m labeled nanoparticles: Interference by colloidal tin oxide particles. *Int. J. Pharm.* **2005**, *289*, 189–195. [CrossRef]
43. Lee, S.Y.; Jung, E.; Park, J.H.; Park, J.W.; Shim, C.K.; Kim, D.D.; Yoon, I.S.; Cho, H.J. Transient aggregation of chitosan-modified poly(D,L-lactic-co-glycolic) acid nanoparticles in the blood stream and improved lung targeting efficiency. *J. Colloid. Interface Sci.* **2016**, *480*, 102–108. [CrossRef]
44. Walkey, C.D.; Olsen, J.B.; Guo, H.; Emili, A.; Chan, W.C. Nanoparticle size and surface chemistry determine serum protein adsorption and macrophage uptake. *J. Am. Chem. Soc.* **2012**, *134*, 2139–2147. [CrossRef] [PubMed]
45. Sadauskas, E.; Danscher, G.; Stoltenberg, M.; Vogel, U.; Larsen, A.; Wallin, H. Protracted elimination of gold nanoparticles from mouse liver. *Nanomedicine* **2009**, *5*, 162–169. [CrossRef] [PubMed]
46. Lacava, L.M.; Garcia, V.A.P.; Kückelhaus, S.; Azevedo, R.B.; Sadeghiani, N.; Buske, N.; Morais, P.C.; Lacava, Z.G.M. Long-term retention of dextran-coated magnetite nanoparticles in the liver and spleen. *J. Magn. Magn. Mater.* **2004**, *272–276*, 2434–2435. [CrossRef]
47. Levy, M.; Luciani, N.; Alloyeau, D.; Elgrabli, D.; Deveaux, V.; Pechoux, C.; Chat, S.; Wang, G.; Vats, N.; Gendron, F.; et al. Long term in vivo biotransformation of iron oxide nanoparticles. *Biomaterials* **2011**, *32*, 3988–3999. [CrossRef] [PubMed]
48. Lee, S.; Kim, M.S.; Lee, D.; Kwon, T.K.; Khang, D.; Yun, H.S.; Kim, S.H. The comparative immunotoxicity of mesoporous silica nanoparticles and colloidal silica nanoparticles in mice. *Int. J. Nanomed.* **2013**, *8*, 147–158.
49. Choi, H.S.; Liu, W.; Misra, P.; Tanaka, E.; Zimmer, J.P.; Ipe, B.I.; Bawendi, M.G.; Frangioni, J.V. Renal clearance of quantum dots. *Nat. Biotechnol.* **2007**, *25*, 1165–1170. [CrossRef]
50. Muramatsu, T. Basigin (CD147), a multifunctional transmembrane glycoprotein with various binding partners. *J. Biochem.* **2016**, *159*, 481–490. [CrossRef]
51. He, C.; Hu, Y.; Yin, L.; Tang, C.; Yin, C. Effects of particle size and surface charge on cellular uptake and biodistribution of polymeric nanoparticles. *Biomaterials* **2010**, *31*, 3657–3666. [CrossRef]
52. Stefan, J.; Lorkowska-Zawicka, B.; Kaminski, K.; Szczubialka, K.; Nowakowska, M.; Korbut, R. The current view on biological potency of cationically modified chitosan. *J. Physiol. Pharmacol.* **2014**, *65*, 341–347.
53. Lusis, A.J. Atherosclerosis. *Nature* **2000**, *407*, 233–241. [CrossRef]
54. Dorrington, M.G.; Fraser, I.D.C. NF-κB signaling in macrophages: Dynamics, crosstalk, and signal integration. *Front. Immunol.* **2019**, *10*, 705. [CrossRef] [PubMed]
55. Liu, T.; Zhang, L.; Joo, D.; Sun, S.-C. NF-κB signaling in inflammation. *Sig. Transduct. Target Ther.* **2017**, *2*, 17023. [CrossRef] [PubMed]
56. Atri, C.; Guerfali, F.Z.; Laouini, D. Role of human macrophage polarization in inflammation during infectious diseases. *Int. J. Mol. Sci.* **2018**, *19*, 1801. [CrossRef]
57. Orekhov, A.N.; Orekhova, V.A.; Nikiforov, N.G.; Myasoedova, V.A.; Grechko, A.V.; Romanenko, E.B.; Zhang, D.; Chistiakov, D.A. Monocyte differentiation and macrophage polarization. *Vessel Plus* **2019**, *3*, 10. [CrossRef]
58. Wosen, J.E.; Mukhopadhyay, D.; Macaubas, C.; Mellins, E.D. Epithelial MHC class II expression and its role in antigen presentation in the gastrointestinal and respiratory tracts. *Front. Immunol.* **2018**, *25*, 2144. [CrossRef] [PubMed]
59. Barbatis, C.; Kelly, P.; Greveson, J.; Heryet, A.; McGee, J.O. Immunocytochemical analysis of HLA class II (DR) antigens in liver disease in man. *J. Clin. Pathol.* **1987**, *40*, 879–884. [CrossRef] [PubMed]
60. Heymann, F.; Peusquens, J.; Ludwig-Portugall, I.; Kohlhepp, M.; Ergen, C.; Niemietz, P.; Martin, C.; van Rooijen, N.; Ochando, J.C.; Randolph, G.J.; et al. Liver inflammation abrogates immunological tolerance induced by Kupffer cells. *Hepatology* **2015**, *62*, 279–291. [CrossRef]

61. Bartneck, M.; Warzecha, K.T.; Tacke, F. Therapeutic targeting of liver inflammation and fibrosis by nanomedicine. *Hepatobiliary Surg. Nutr.* **2014**, *3*, 364–376.
62. Franco-Molina, M.A.; Coronado-Cerda, E.E.; López-Pacheco, E.; Zarate-Triviño, D.G.; Galindo-Rodríguez, S.A.; Salazar-Rodríguez, M.C.; Ramos-Zayas, Y.; Tamez-Guerra, R.; Rodríguez-Padilla, C. Chitosan nanoparticles plus KLH adjuvant as an alternative for human dendritic cell differentiation. *Curr. Nanosci.* **2019**, *15*, 532–540. [CrossRef]
63. Harris, R.C. An update on cyclooxygenase-2 expression and metabolites in the kidney. *Curr. Opin. Nephrol. Hypertens.* **2008**, *17*, 64–69. [CrossRef]
64. Shinohara, T.; Pantuso, T.; Shinohara, S.; Kogiso, M.; Myrvik, Q.N.; Henriksen, R.A.; Shibata, Y. Persistent inactivation of macrophage cyclooxygenase-2 in mycobacterial pulmonary inflammation. *Am. J. Respir. Cell Mol. Biol.* **2008**, *41*, 146–154. [CrossRef] [PubMed]
65. Nag, M.; Lahiri, D.; Mukherjee, D.; Banerjee, R.; Garai, S.; Sarkar, T.; Ghosh, S.; Dey, A.; Ghosh, S.; Pattnaik, S.; et al. Functionalized chitosan nanomaterials: A jammer for quorum sensing. *Polymers* **2021**, *13*, 2533. [CrossRef] [PubMed]
66. Bannunah, M.; Vllasaliu, D.; Lord, J.; Stolnik, S. Mechanisms of nanoparticle internalization and transport across an intestinal epithelial cell model: Effect of size and surface charge. *Mol. Pharm.* **2014**, *11*, 4363–4373. [CrossRef] [PubMed]
67. Oh, N.; Park, J.H. Endocytosis and exocytosis of nanoparticles in mammalian cells. *Int. J. Nanomed.* **2014**, *9*, 51–63.
68. Saraf, S.; Jain, S.; Sahoo, R.N.; Mallick, S. Present scenario of m-cell targeting ligands for oral mucosal immunization. *Curr. Drug Targets* **2020**, *21*, 1276–1284. [CrossRef] [PubMed]
69. Nawroth, I.; Alsner, J.; Deleuran, B.W.; Dagnaes-Hansen, F.; Yang, C.; Horsman, M.R.; Overgaard, J.; Howard, K.A.; Kjems, J.; Gao, S. Peritoneal macrophages mediated delivery of chitosan/siRNA nanoparticle to the lesion site in a murine radiation-induced fibrosis model. *Acta Oncol.* **2013**, *52*, 1730–1738. [CrossRef]
70. Lewinski, N.; Colvin, V.; Drezek, R. Cytotoxicity of nanoparticles. *Small* **2004**, *4*, 26–49. [CrossRef]
71. Hussein, H.; Kishen, A. Engineered chitosan-based nanoparticles modulate macrophage–periodontal ligament fibroblast interactions in biofilm-mediated inflammation. *J. Endod.* **2021**, *47*, 1435–1444. [CrossRef]
72. Yan, C.; Chen, D.; Gu, J.; Hu, H.; Zhao, X.; Qiao, M. Preparation of N-succinyl-chitosan and its physical-chemical properties as a novel excipient. *Yakugaku Zasshi* **2006**, *126*, 789–793. [CrossRef]
73. Kato, Y.; Onishi, H.; Machida, Y. Biological fate of highly-succinylated N-succinyl-chitosan and antitumor characteristics of its water-soluble conjugate with mitomycin C at i.v. and i.p. administration into tumor-bearing mice. *Biol. Pharm. Bull.* **2000**, *23*, 1497–1503. [CrossRef]
74. Wilson, B.; Samanta, M.K.; Muthu, M.S.; Vinothapooshan, G. Design and evaluation of chitosan nanoparticles as novel drug carrier for the delivery of rivastigmine to treat. *Ther. Deliv.* **2011**, *2*, 599–609. [CrossRef] [PubMed]
75. Zhu, H.; Liu, F.; Guo, J.; Xue, J.; Qian, Z.; Gu, Y. Folate-modified chitosan micelles with enhanced tumor targeting evaluated by near infrared imaging system. *Carbohydr. Polym.* **2011**, *86*, 1118–1129. [CrossRef]

Review

Paclitaxel Drug Delivery Systems: Focus on Nanocrystals' Surface Modifications

Razan Haddad [1,*], Nasr Alrabadi [2,*], Bashar Altaani [1] and Tonglei Li [3]

[1] Department of Pharmaceutical Technology, Faculty of Pharmacy, Jordan University of Science and Technology, Irbid 22110, Jordan; altaani@just.edu.jo
[2] Department of Pharmacology, Faculty of Medicine, Jordan University of Science and Technology, Irbid 22110, Jordan
[3] Department of Industrial and Physical Pharmacy, Purdue University, West Lafayette, IN 47907, USA; tonglei@purdue.edu
* Correspondence: rhhaddad17@ph.just.edu.jo (R.H.); nnalrabadi@just.edu.jo (N.A.)

Abstract: Paclitaxel (PTX) is a chemotherapeutic agent that belongs to the taxane family and which was approved to treat various kinds of cancers including breast cancer, ovarian cancer, advanced non-small-cell lung cancer, and acquired immunodeficiency syndrome (AIDS)-related Kaposi's sarcoma. Several delivery systems for PTX have been developed to enhance its solubility and pharmacological properties involving liposomes, nanoparticles, microparticles, micelles, cosolvent methods, and the complexation with cyclodextrins and other materials that are summarized in this article. Specifically, this review discusses deeply the developed paclitaxel nanocrystal formulations. As PTX is a hydrophobic drug with inferior water solubility properties, which are improved a lot by nanocrystal formulation. Based on that, many studies employed nano-crystallization techniques not only to improve the oral delivery of PTX, but IV, intraperitoneal (IP), and local and intertumoral delivery systems were also developed. Additionally, superior and interesting properties of PTX NCs were achieved by performing additional modifications to the NCs, such as stabilization with surfactants and coating with polymers. This review summarizes these delivery systems by shedding light on their route of administration, the methods used in the preparation and modifications, the in vitro or in vivo models used, and the advantages obtained based on the developed formulations.

Keywords: paclitaxel; nanocrystals; surface modification; chemotherapy; cancer; drug delivery; nanotechnology

1. Introduction

Currently, cancer is considered a serious disease that is globally widespread, and it is one of the most life-threatening illnesses [1], accounting for about 10 million deaths in 2020 [2]. Additionally, the economic burden of this disease is enormous, and it is anticipated to increase in the future [3,4]. On the other hand, chemotherapeutic agents are considered effective at fighting cancer and preventing its development and progress [5]. However, there is still an urgent need for more therapeutic options or strategies to improve the currently available treatments in terms of safety and efficacy.

The improvements of chemotherapeutic agents mainly depend on two research lines [5]. The first one is related to explaining cancer-specific mechanisms and molecular targets, such as signal transduction inhibitors concerning essential processes of cells such as growth, survival, and differentiation. These substances may have the ability to prevent the injuries caused by cancer cells, including proliferation and tissue invasion [6]. The second line is considering the enhancement of the available cytotoxic drugs which act on abundant targets (e.g., DNA or tubulin) [5]. These cytotoxic drugs are either natural products or their derivatives obtained from plants, marine species, and microorganisms, but unfortunately, these drugs are still toxic to normal cells [7]. Therefore, the improvement of their efficacy and safety is always warranted.

Eventually, many anticancer agents were obtained. but most of them are inefficient and cause severe side effects. Therefore, there is an emerging need to develop new therapeutic agents or delivery approaches. Several drug delivery systems based on nanotechnology modalities have been obtained for different anticancer drugs such as solid lipid nanoparticles, liposomes, micelles, polymeric nanoparticles, nano-emulsions, implants, and nanocrystals [8–10]. All these approaches are aimed at either enhancing the efficacy or reducing the side effects of the currently available chemotherapeutic agents. Finding novel and appropriate drug delivery systems is crucial, especially for chemotherapies where intravenous delivery remains the main route used for drug administration [8]. This returns to the fact that most anticancer drugs have low solubility or gastrointestinal tract (GIT) toxic side effects over oral administration, which in turn can reduce their oral absorption below the therapeutic effective levels [10].

2. Paclitaxel

Paclitaxel (PTX) is an important chemotherapeutic agent that belongs to the taxane family. Taxanes were initially obtained from plants of the genus *Taxus*. PTX was first derived from the bark of the Pacific yew (*Taxus brevifolia*), which is an evergreen tree and small to medium in size and also known as western yew, native to the Pacific Northwest of North America [11,12]. PTX was approved by the United States (US) Food and Drug Administration (FDA) to treat various kinds of cancers including breast cancer, ovarian cancer, advanced non-small-cell lung cancer, and acquired immunodeficiency syndrome (AIDS)-related Kaposi's sarcoma [13]. In general, PTX is not well tolerated and related to serious adverse drug effects such as hypersensitivity reactions, hematological toxicity, peripheral sensory neuropathy, and myalgia or arthralgia [13], even though PTX has been used for two decades either as a single drug or in combination with other chemotherapeutics.

The antitumor activity of paclitaxel comes from its high binding affinity to microtubules, stabilizing and improving the polymerization of tubulin and destruction of the dynamics of the spindle microtubule [14,15]. Such activities provide effective inhibition of cell mitosis, intracellular transport, and motility, which end up with cell death by apoptosis. However, the clinical developments of the natural form of paclitaxel have been restricted due to its physicochemical properties, particularly its very low solubility [16]. Additionally, the absence of modifiable functional moieties in its structure makes the chemical alteration of the natural paclitaxel very complicated when attempting to enhance its solubility [17]. Considering that, the selection of a proper delivery system to paclitaxel is considered very crucial to improving its clinical development, safety, and efficiency.

Regarding the chemical structure of PTX (Figure 1), its 20-carbon compound (C_{20}) belongs to the diterpene class of natural compounds [18]. The anticancer activity is mainly recognized for ring A, ring D (the oxetane ring), the C2 benzoyl group, and some components such as the C3′ amide-acyl group and the OH group at C2′, which attaches on the side chain to C13 [19]. On the other hand, other groups slightly affect the therapeutic activity of PTX such as the carbonyl group on C9 and the acetyl group on C10. Moreover, the specified conformation of the paclitaxel molecule is provided by the acetyl group [19].

PTX has poor aqueous solubility, low permeability, and as it is a P-gp substrate, it also has limited capabilities for being delivered via the oral route [20–23]. The poor permeability of PTX is related to the following molecular factors: its molecular weight is more than 500, the hydrogen bond acceptor (HBA) is greater than 10, and the polar surface area (PSA) is more than 140 $Å^2$, which results in a permeability coefficient value in the range of 10^{-6} cm/s [24,25]. Therefore, PTX is administered parentally via the intravenous (IV) route with a suitable cosolvent (cremophor EL and ethanol), which unfortunately ends up with several direct, problematic, adverse effects such as acro-anesthesia and neurovirulence, causing pain and high cost [26–29].

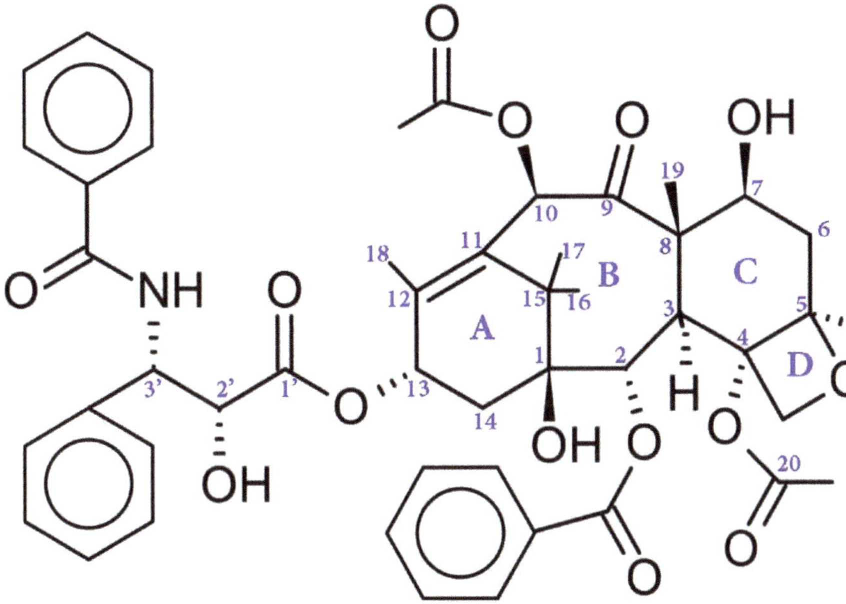

Figure 1. The chemical structure of a PTX drug.

3. PTX Formulations

To improve the benefit and delivery of PTX, several formulations have been developed. The most commonly used delivery system is a cosolvent strategy based on a 50:50 mixture of ethanol and Cremophor EL™ (a polyoxyethylated castor oil). Taxol® is the first generic product of paclitaxel, and it consists of this cosolvent mixture. Although this method overcomes the problem of solubility, Cremophor EL has been associated with non-linear pharmacokinetics and serious and dose-limiting toxicities, such as hypersensitivity, neurotoxicity, and nephrotoxicity [11]. Due to these adverse effects, Taxol® is given slowly in 135- or 175-mg/m^2 doses by infusion over 3–24 h every 3 weeks [30,31].

Abraxane™ is another marketed drug of PTX, which was produced by Abraxis BioScience (later obtained by the Celgene company) and approved by the FDA in 2005 [32]. The formulation of PTX in this product is performed with human serum albumin (HSA) [33]. HSA is the most abundant plasma protein in the blood, with a large half-life that reaches up to 19 h and which can bind hydrophobic substances irreversibly, transport them through the body, and deliver them to the cell surface [34]. Additionally, HSA plays a significant role in cellular uptake and transcytosis, as it is bound to gp60 and other proteins which are highly expressed in malignant cells, such as secreted proteins acidic and rich in cysteine (SPARC). Nevertheless, it is still ambiguous how exactly HAS improved the biological response of PTX. However, it is significantly clear that the removal of Cremphor EL contributes to the ability to administer a higher dose of PTX with an analogous toxicity [35]. Moreover, Abraxane™ has a linear pharmacokinetic profile and a higher intratumoral concentration by 33% in comparison with Taxol®, based on results obtained by the Abraxis BioSciences company [36].

Another marketed drug of PTX is Lipusu™, which was formulated by Luye Pharmaceutical Co. Ltd. and approved in China in 2003. It is a liposome composed of PTX, lecithin, and cholesterol. In comparison with Taxol®, Lipusu™ has similar activities toward breast cancer, non-small cell lung, and gastric cancer but with considerably lesser side effects [37,38].

Finally, Genexol-PM™ is marketed by Samyang Corporation and was approved in South Korea in 2007. It is composed of PTX and poly (ethylene glycol)-b-poly (lactic acid) (PEG-b-PLA) block copolymers. Clinical studies showed that Genexol-PM™ has dose-dependent pharmacokinetics and good tolerance, especially for patients with advanced pancreatic cancer or metastatic breast cancer [39–41].

4. Drug Delivery of PTX

As previously mentioned, the physicochemical properties and the nature of PTX complicated its formulations. Consequently, several delivery systems for PTX have been developed to enhance its solubility and pharmacological properties involving micelles, liposomes, nanoparticles, the prodrug approach, emulsions, implants, and nanocrystals [42–45]. Figure 2 summarizes the most common strategies utilized for PTX delivery systems.

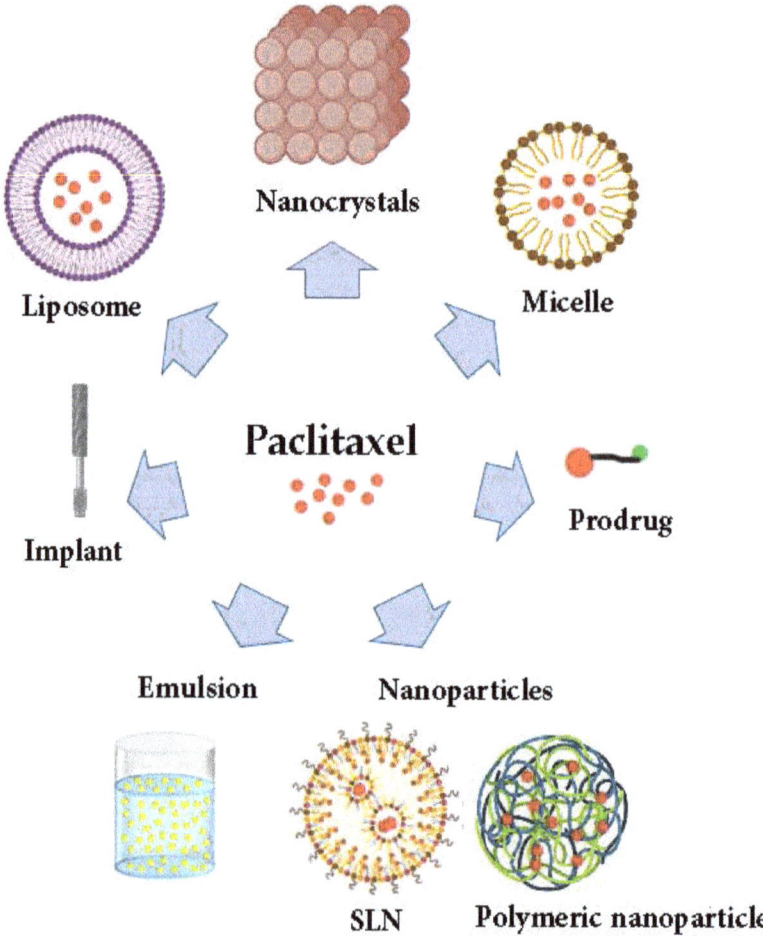

Figure 2. The most common developed strategies to improve the delivery of paclitaxel drugs. SLN: solid lipid nanoparticles.

4.1. Micelles

Generally, micelles consist of polar heads that encounter the outside aqueous environment and non-polar tails, which form the interior hydrophobic core. Above the critical micelle concentration, micelles spontaneously form, and drugs with low solubility encapsulate efficiently in the lipidic core [46,47]. The properties of micelles and the hydrophobic regions can be modified and tailored by using various polymer structures [48]. The targeted delivery of PTX micelles developed using an Asn-Gly-Arg (NGR) peptide, which covalently bonds to PEG chains to deliver PTX through a brain tumor [49]. Moreover, the oral delivery of PTX was obtained using multi-functional chitosan polymeric micelles [50]. The development of redox-sensitive PEG2000-S-S-PTX micelles resulted in a reduction of PTX cytotoxicity in ovarian and breast cancer cells [51].

4.2. Liposomes

The liposome is a spherical structure with a membrane composed of a single or multiple phospholipid bilayers. It has an aqueous core to encapsulate hydrophilic drugs, while the hydrophobic drug can be loaded in the region of the bilayer membrane. Liposomes have been used to deliver PTX because they showed that they can enhance solubility and efficacy by modulating its pharmacokinetic properties. Additionally, the used excipients are clinically approved [52]. Lipusu^{TM} is the first injected PTX liposome, and it has been used in China to treat non-small cell lung cancer, breast cancer, and other cancers [19]. It maintains the original activity of PTX but with a significant reduction in the side effects. Moreover, LEP-ETU is another liposome loaded with PTX. Based on the phase I study, LEP-ETU showed little difference in the pharmacokinetic properties in comparison to Taxol® while being safer at higher doses [53].

Long-term instability is the main obstacle related to liposomes. Despite liposomes having the ability to deliver cytotoxic compounds to certain tissues, they can be eliminated by the mononuclear phagocytic system (MPS) in the spleen and liver [54]. Interestingly, the average circulation time of liposomes can be enhanced by 10 folders with PEGylation, which results in an improvement of the half-life of PTX and its antitumor properties [55–57]. The PEGylated liposome can be also modified by active targeting strategies to improve its efficacy [58,59]. This can be obtained by covalently binding species to the surface of the liposomes, such as the peptides [60], proteins [61], and tissue-specific antibodies [62]. Specifically, a multifunctional peptide was incorporated to the surface of the liposomes loaded with PTX, which improved its targeting activity and also its efficacy [59]. Moreover, triphenylphosphonium (TPP) was incorporated into the surface of the PEGylated PTX liposomes, which consequently enhanced their cytotoxicity and antitumor efficacy and provided efficient mitochondrial targeting in cancer cells [58]. Moreover, PTX loaded to a pH-sensitive lipid that was incorporated into a liposomal membrane prevented liposome degradation by lysozymes and consequently caused more suppression of tumors by providing more PTX accumulation at pH of 7.4 instead of 5.5 [63,64].

4.3. Nanoparticles

4.3.1. Solid Lipid Nanoparticles

Generally, solid lipid nanoparticles (SLN) are obtained from solid lipids, such as complex glyceride, highly purified triglycerides, and waxes [65]. Several kinds of lipids and surfactants can be used for SLN production and engineering. More specifically, lipids such as phospholipids and glycerides and surfactants such as tween 80, sodium glycolate, lecithin, and poloxamer 188 are considered suitable for IV injection [66]. There are several advantages related to the SLN, such as the simplicity of the preparation method and the scaling up, biocompatibility, stability, low cost, low toxicity, controlled drug release, and versatile chemistry [52].

To obtain high drug loading and the slow release of PTX, SLN should obtain high drug solubility and miscibility [67]. The cellular uptake and cytotoxicity properties of PTX-loaded SLN can be vary based on the lipid materials used. For instance, studies showed that the cellular uptake of SLN was concentration- and time-dependent and related to the melting point of the lipidic materials, the length of its hydrocarbon chain, and the particle size [68–70]. PTX-loaded PEGylated steric acid SLN proved to have a high cellular uptake and up to 10-fold greater cytotoxicity in comparison with PTX. Moreover, SLN showed an ability to affect P-gp-mediated multidrug resistance (MDR), as PTX loaded SLN provided an inhibition of P-gp activity and a rapid depletion of ATP [71,72].

As with the other noncompaction systems, surface modification of the particles by different chemical moieties is useful for obtaining prolonged SLN circulation by avoiding the clearance with the reticuloendothelial system (RES) [52].

4.3.2. Polymeric Nanoparticles

Polymers have been used in nanoparticle preparation to provide them with suitable properties and characteristics. Examples of some of the polymers that are commonly used in developing paclitaxel nanoparticles are poly (lactic-co-glycolic acid) (PLGA) and chitosan, which will be discussed in the following sections.

Poly Lactic-co-Glycolic Acid (PLGA)

Poly (lactic-co-glycolic acid) (PLGA) is a biocompatible, biodegradable, nontoxic synthetic polymer derived from poly (lactic acid) (PLA) and poly (glycolic acid) (PGA) [73,74]. It has been approved by the US Food and Drug Administration (FDA) for drug delivery, as it has superior properties in the delivery of many therapeutic agents. PLGA is a very useful and successful polymer in nanomedicine and the nano-delivery of drugs. In addition, it has a favorable ability to target tumors and DNA [73,75–77]. PLGA is available commercially with various molecular weights and copolymer ratios. Based on that, the duration of the degradation can vary, as can the release time. Glycolic acid is more hydrophilic than lactic acid, and thus PLGA with higher glycolic acid is more hydrophilic and can adsorb water more and degrade faster [78,79]. The loading of PTX to PLGA nanoparticles has been obtained by various methods such as emulsion solvent evaporation [77], interfacial deposition methods [80], and the nanoprecipitation method [81]. Studies showed that PTX loaded to the PLGA nanoparticles had superior antitumor properties and efficacy in comparison with Taxol® [81,82]. Moreover, surface modification of the nanoparticles has a crucial impact on their properties, such as efficacy and targeting. The delivery of PTX was improved by surface modification of PLGA nanoparticles with albumin, as the circulation time of these nanoparticles in the blood was increased even as it became more toxic in the in vitro study [83]. The targeted delivery of PTX to breast cancer cells was developed by loading it into PLGA nanoparticles coated with hyaluronic acid (HA), and the results showed that the cellular uptake was increased using this system [84]. Moreover, PTX has been loaded to lipid PLGA hybrid nanoparticles, and the results showed that the release profile was affected with this lipid coat. Also, these nanoparticles provided a prolongation in the circulation time in the blood [85].

Chitosan

Chitosan is a natural polysaccharide polymer produced by the diacylation of chitin. It has many attractive properties such as non-toxicity, biocompatibility, biodegradability, and bio-adhesivity, which necessitates its use in drug delivery [86,87]. The solubility of chitosan in acidic solutions and its limited solubility in biological solutions (pH 7.4) are considered the main drawbacks of its application in drug delivery. Lately, many chitosan derivatives have been prepared by adding various hydrophobic or hydrophilic groups to the chitosan structure [88,89]. Moreover, studies showed that chitosan has antitumor properties, and it can affect the cancer cells by interfering with its metabolism, inhibiting its growth, or inducing its apoptosis [90].

Chitosan has been introduced to many PTX delivery systems, and it improved various aspects (e.g., decreasing the toxicity and enhancing the efficiency and targeting capabilities) [91–93]. A study showed that the PTX-loaded micelle based on N-octyl-O-sulfate chitosan (OSC), which is a novel derivative of water-soluble chitosan used for the delivery of PTX, has superior toxic properties, as lower side effects were observed, and the AUC was about 3.5 lower than the marketed drug Taxol® while preserving the antitumor efficacy at equivalent doses [94]. Additionally, other studies showed that the targeted delivery of PTX chitosan nanoparticles had been achieved in combination with other polymers such as PEGylated chitosan nanoparticles grafted with Arg-Gly-Asp (RGD) [92], poly NIPAAm [95], transferrin [96], and biotinylated N-palmitoyl chitosan [97].

4.4. Prodrug Approach

Prodrugs are derivatives of a drug molecule that can be transformed chemically or enzymatically in the body to release the active ingredient that possesses pharmacological effects [98]. Differing from other delivery systems or formulations, prodrugs are usually formulated by chemical linkage with proper quality control and less variation from batch to batch. Generally, prodrugs are developed to overcome problems related to the parent drug itself, such as poor aqueous solubility, limited permeability, inadequate oral absorption and delivery, non-targeting, and toxic side effects [99,100].

The PTX prodrug is usually fabricated at the carbon no. 123; 2′ or 7-OH group [100]. PTX prodrugs are constructed using various strategies, such as polymer-based prodrugs, which are formulated using polymers such as PEG [101], PLA [102–104], poly(amidoamine) [PAMAM] [105], N-(2-hydroxypropyl) methacry'lamide (HPMA) [106], and poly(L-glutamic acid) (PGA) [107]. Moreover, a protein-based prodrug of PTX has been developed using different proteins, such as the marketed product Abraxane™, which is tumor-targeted and formulated using CREKA and LyP-1 [108]. Additionally, PTX prodrugs were obtained using transferrin (Tf) and Fmoc-L-glutamic acid 5-tert-butyl ester (linker) to specifically target tumor tissues and cells [109]. Similarly, peptide-based prodrugs of PTX were also formulated such as the Tat-based self-assembling peptide, which is used to deliver PTX intracellularly [110], the tumor-homing cell-penetrating peptide (CPP) [111], and recombinant chimeric polypeptides (CPs) [112]. Additionally, PTX prodrugs can be obtained using small molecules such as docosahexaenoic acid (DHA) [113], conjugated linoleic acids (CLAs) [114], and oligo(lactic acid)$_8$ [115]. Finally, hybrid prodrugs for PTX also exist, which are a combination of two drugs or more that is capable of producing synergistic effects, reducing the adverse effects related to a high dose of a single drug, and overcoming the multidrug resistance mechanism of cancer cells during treatment [116]. PTX hybrid prodrugs were delivered using other anticancer drugs such as doxorubicin (DOX) [117], camptothecin (CPT) [118], and the nucleic acid oligonucleotide [119].

4.5. Emulsions

Generally, macroemulsions are defined as the dispersion of one liquid in another liquid and it is considered a two-phase system [120]. They are turbid or opaque, viscus, and thermodynamically unstable, and their preparation is complicated as sheer is needed. On the other hand, microemulsions are translucent, thermodynamically stable, have a lower viscosity, and form spontaneously [121]. Based on the name, nano-emulsions should have a droplet size lower than microemulsions. As a matter of fact, nano-emulsions have a droplet size of 20–200 nm and a narrow particle size distribution [122–124].

A Tocosol™ nano-emulsion was established early in 2000. It was formulated using an a-tocopherol isomer of vitamin E as a solubilizing agent for PTX and vitamin E TPGS as an emulsifier. Unfortunately, studies in phase III showed that the overall response rate was only 37%, while it was 45% with Taxol®. Based on that, the Tocosol™ nano-emulsion was terminated [125]. Recently, Shakhwar et al. tried to reform a Tocosol™ nano-emulsion using the c-tocotrienol (c-T3) isomer instead of a-tocopherol and the PEGylated c-T3 surfactant instead of vitamin E TPGS. Their results showed that the reformulated PTX was more active toward pancreatic tumor cell lines than the previous formulation [126].

Moreover, self-emulsifying drug delivery systems (SEDDSs) and self-microemulsifying drug delivery systems (SMEDDSs) are combinations of the non-aqueous components of emulsions and microemulsions, respectively [127], such as oils, surfactants, and if present, cosurfactant or cosolvents. These mixtures can be readily dispersed when diluted with an aqueous phase (gastric fluids) in the body and then spontaneously emulsified to form fine oil-in-water (O/W) microemulsions. This process can be sped up by slight mechanical agitation, and in vivo, this can be obtained by gastrointestinal motility [18,122,128]. A novel SMEDDS was developed for oral delivery of PTX, and it was administered to patients with advanced cancer and compared with orally administered Taxol®. The SMEDDS was co-administered with cyclosporin A to inhibit P-gp and CYP3A4. This formula was safe and well-tolerated by patients and had comparable bioavailability to oral Taxol®. In addition, the T-max of the SMEDDS was lower than the orally delivered Taxol®. This means that the absorption was higher in the novel formula, and this may be related to the added excipients [129]. In another study, the oral delivery of PTX was designed as an SEDDS. In this study, tocopheryl polyethylene glycol succinate was used to assist the emulsification. The results indicated that this system had higher G2M cell cycle arrest, apoptosis, mitochondrial membrane potential disruption, and ROS production in comparison with Taxol®. Moreover, the oral bioavailability of the SEDDS was about fourfold greater than Taxol®. Considerable reductions in the volumes and weights of the tumors were detected in syngeneic mammary tumors in SD rats. Additionally, this system was safe, stable, and caused low lung metastasis [130].

4.6. Implants

Drug-loaded polymeric implants are considered a pioneering approach in drug delivery. Active ingredients can be delivered to malignant cells using biodegradable polymers in continuous, sustained, and predictable patterns. Owing to their nature, biodegradable polymers do not need to be removed surgically after their application and thus eliminate complications associated with the long-term safety of implanted devices with non-biodegradable polymers. Additionally, the postsurgical local insertion of a biodegradable implant device loaded with an anticancer drug can avoid the further spread of cancer cells while avoiding toxic chemotherapy adverse effects in the patient [131]. Recently, an in situ depot-forming implant (ISFI) has been developed which can be injected as a liquid and then subsequently solidified [132,133]. In this way, an effective dosage form can be delivered with the avoidance of surgical insertion [134]. Moreover, ISFIs have relatively simpler preparation conditions and fewer complications than solid implants [135,136]. The PTX ISFI was formulated using PLGA to improve its efficiency and toxicity. This formula provided an in vitro sustained release of PTX for 28 days [137].

4.7. Nanocrystals

Nanocrystal formulations have become more attractive for the delivery of chemotherapies due to their superior properties in comparison with other nano-delivery approaches [138–140]. Nanocrystals eliminate the need for chemical carriers, therefore eradicating any toxic side effects induced by the excipients used for solubilization or coating and also providing about 100% drug loading, which ensures suitable concentrations of the drug even at low doses [141]. Additionally, due to the stable and uniform physical properties of crystalline particles, the enhancement of the pharmacokinetics and biodistribution properties of the anticancer drugs are anticipated [142–145].

Nanocrystals can be produced either by top-down or bottom-up methods. The top-down technique involves utilizing a high mechanical energy force to produce nanocrystals from large crystals by media milling or high-pressure homogenization [142,143]. These techniques are generally used to formulate insoluble drugs, especially those used for oral drug delivery [146,147]. In the high-pressure homogenization method, large drug crystals are forced across fluidic pressure and an impact valve, which leads the drug crystals to break down into tinier particles. The control of the particle size is achieved through the pressure and space among the impact valves. On the other hand, in media milling, the grinding of large crystals of the drug is obtained using solid particles like yttrium-stabilized zirconia, cerium, highly crosslinked polystyrene resin-coated beads, and stainless steel [142].

In the bottom-up approach, which involves the antisolvent perception method, nanocrystals can be produced directly from the drug solution. When the drug solution is mixed with an antisolvent with poor drug solubility, in such a case, the decrease in solubility leads to nucleation and crystal growth of the drug, and these are the two critical steps of this method [148].

As more nuclei form during the nucleation stage, then the growth of each nucleus is lower, and based on that, the nucleation step needs to be monitored carefully. The ultrasonic waves produced by sonication can help reduce the size of the nanocrystals by decreasing the particle agglomeration, achieved by breaking down the contact between particles. Consequently, perception and ultrasonication (PU) are commonly used in the bottom-up method [149,150].

The anti-solvent method produced nanocrystals with a smaller size that were cost-effective, simpler, and easy to scale up in comparison with other methods of the top-down approach [151,152]. However, various factors during nanocrystal preparation can be controlled to influence the size and morphology of nanocrystals obtained by the antisolvent method, such as the drug concentration, drug solution flow rate, temperature, solvent-to-antisolvent volume ratio, stirring speed, and the ultrasound wave characteristics [152–154]. In addition, the addition of surfactants and polymers during the crystallization process has an impact on the size or the shape of the drug's nanocrystals [155]. This shows that engineering the modifications of nanocrystals according to our preference and usefulness is possible. Moreover, a combination of both approaches—the top-down and bottom-up methods—can also possibly obtain NCs with a smaller size (<100 nm), narrow distribution, and less production time [156]. In addition, the shape of nanocrystals is also considered important in controlling the activity and toxicity of anticancer drugs. For instance, the rod shape of some drug nanocrystals has superior anticancer activity and toxicity in comparison with the spherical shape [157]. Another study showed that the needle shape of some drug nanocrystals provides better accumulation in some cancers, which may be referred to as an increasing ability of these nanocrystals to be entrapped [144]. Moreover, the size of the nanocrystals is very critical for the in vivo performance of drugs. For instance, smaller nanocrystals have more dissolution rates than larger ones. Conversely, larger nanocrystals may provide sustained release behavior, which results in greater drug accumulation in tumors similar to drug depots. On the other hand, the smaller one is more stable because of the lower accumulation. Finally, the surface treatment or coating of nanocrystals with a polymer or surfactant can further improve the anticancer properties of the nanocrystals and the stability [10,158].

Manipulation during the preparation of the nanocrystals is possible and might end up in unexpected favorable outcomes. Therefore, this indicates the significance of controlling nanocrystals' properties based on the efficiency, effectiveness, and safety of the anticancer drug, as these can be improved and manipulated indirectly during nanocrystal preparation, especially in the case of the nanocoating.

5. PTX Nanocrystals

As PTX is a hydrophobic drug with inferior water solubility properties, it is improved greatly by nanocrystal formulation [159]. Based on that, many studies employed the nano-crystallization techniques not only to improve the oral delivery of PTX, but IV, intraperitoneal (IP), and local and intertumoral delivery systems were also developed. Additionally, superior and interesting properties of the PTX NCs were achieved by performing additional modifications to the NCs, such as stabilizing them with surfactants and polymers or coating them with polymers. Tables 1–3 summarize these modified delivery systems by classifying them into three main categories, according to their route of administration: either IV (Table 1), oral (Table 2), or local and intraperitoneal (Table 3) delivery systems. Additionally, the summary tables shed light on the methods used in the preparation of these modified NCs, the in vitro or in vivo models used, and the advantages obtained based on the developed formulations. Nearly the majority of these NCs had a rode-like shape with drug-loading capabilities (>50%), and their size was between <50 and 500 nm (mainly 100–300 nm). The most common route of administration for these novel formulas was the intravenous (IV) route (Table 1), and the most common method of preparation was the antisolvent or precipitation method. The most common cancer cell lines or types of cancer tested were breast cancer (MCF-7 cell lines), followed by ovarian and then lung cancer. Finally, the aims for modifications were mainly focused on providing more solubility and tumor and cancer cell targeting, less elimination and side effects, and more anti-cancer effects with a smaller dose. In addition, it appears that the NCs' formulation provided a suitable method for multiple drug combinations.

Table 1. Modified PTX NC formulations for intravenous (IV) drug delivery.

PTX NC	Method of Preparation	The Models Used and the Reference or Control Formula	Benefits, Aims, and Other Notes	Refs.
Albumin-coated PTX-NC (Alb-PTX NCs)	NC crystallized in the medium containing Pluronic F-127 and then coated with albumin "Cim-F-alb"	The new formula was compared to Abraxane and solvent-dissolved PTX In vitro models including Biolayer interferometry analysisCell culture models: J774A.1 macrophages and SPARC+ B16F10 melanoma cells In vivo model: mouse model of B16F10 melanoma	High drug loading (90%) and serum stability Equivalent cytotoxicity. More stability in undiluted serum. Less interaction with serum proteins. In cell culture studies, demonstrated suitable cell interaction profiles (depressed uptake by macrophages and great uptake by melanoma cells). In the in vivo studies, exhibited prolonged plasma $t_{1/2}$ and superior accumulation in tumors by about 1.5 and 4.6 times, respectively. Exhibited superior antitumor efficacy.	[160,161]
Surface modified PTX-NCs with apo-transferrin (Tf) or hyaluronic acid (HA)	PTX NCs were prepared by the nanoprecipitation Method, and then the surface was modified by grafting with Tf or HA	The new formula was compared to PTX-NC and pure PTX drug In vitro models: drug release in PBS with or without tween 80 Cell culture models: HaCaT normal cells and MCF-7 cancer cells	PTX release was faster. Improve the cellular uptake, permeability, and cell growth inhibition (60%) against the cancer cells. The effect on the normal cells was inferior. Provide targeted delivery to cancer cells.	[162]
Hyaluronic acid (HA) coated PTX NCs	The NCs were prepared by the top-down method using homogenization	The new formula was compared to Taxol® and heparin-coated PTX NCs In vitro models: 2D monolayer and 3D spheroids Cell culture models: MDA-MB 231 cells In vivo model: LA-7 tumor-bearing rat model	Exhibited superior in vitro efficacy. HA-PTX NCs incur receptor-mediated endocytosis by binding to CD44 receptors. The in vivo studies indicated significantly prolonged blood circulation time of PTX. Exhibited superior efficacy with reduced lung metastasis and toxicity.	[163]

Table 1. Cont.

PTX NC	Method of Preparation	The Models Used and the Reference or Control Formula	Benefits, Aims, and Other Notes	Refs.
PEGylated PTX NCs	The NCs were prepared by the antisolvent precipitation method combined with probe sonication	The new formula was compared to PTX NCs and Taxol®. In vivo model: breast cancer xenografted mice model and a model of lung tumor metastasis quantified by the luciferase activity	Superior stability under both storage and physiological conditions. In vivo studies showed significant improvement of the antitumor activity in facing in situ or metastatic tumors.	[164]
PEGylated polyelectrolyte multilayer-coated PTX NCs	The layer-by-layer method was used to coat PTX NCs with alternating layers of oppositely charged polyelectrolytes, utilizing a PEGylated copolymer as the upper layer, and PTX NCs were prepared by a wet milling approach	The new formula was compared to Abraxane and PTX NCs. In vitro models: physiologically relevant media and human RBC hemolysis. Cell culture models: HT-29 cells. In vivo model: NMRI-nu mice bearing HT-29 subcutaneous xenografts	Slowed down the dissolution. Offered colloidal stability in physiologically simulated media. Showed no innate effect on cell viability using HT-29 cells. No hemolytic activity was detected. Quickly eliminated from the bloodstream and accumulated in the liver and spleen (mononuclear phagocyte organs). Poor tumor accumulation.	[165]
PTX NCs modified with PEG and folic acid (FA)(PTX NCs-PEG-FA)	PTX NCs were prepared by thin-film hydration method, which is a bottom-up method, and then modified with both PEG and FA derivatives using thin-film hydration technique	The new formula was compared to Taxol®, PTX NCs, and PTX NCs-PEG. In vitro models: plasma. Cell culture models: 4T1 breast cancer cells. In vivo model: PK rat model and 4T1 orthotopic breast cancer-bearing nude mice	More size stability in plasma. Improved cellular uptake and growth inhibition in cells. An in vivo pharmacokinetic study showed a significant increase in the circulation of PTX. In vivo cancer model showed that it significantly enhanced the accumulation of PTX in the tumor and effectively inhibited tumor growth.	[166]
Surface hybridization of PTX NCs by DSPE-PEG 2000	PTX NCs were prepared by anti-solvent method, and DSPE-PEG 2000 was incorporated by hybridization	The new formula was compared to PTX solution and PTX NCs. In vitro models: in vitro release study. In vivo model: PK rats' model	Similar size with an increased negative charge. The in vitro study showed that the release of PTX was significantly slower. The pharmacokinetics studies showed a greater area under the curve (AUC) and a lower clearance rate.	[167]

Table 1. Cont.

PTX NC	Method of Preparation	The Models Used and the Reference or Control Formula	Benefits, Aims, and Other Notes	Refs.
Cube-shaped PTX NC prodrug with surface functionalization of SPC and MPEG-DSPE	PTX was labeled with fluorophore conjugate 4-chloro-7-nitro-1, 2, 3-benzoxadiazole (NBD-Cl) (PTX-NBD), which was synthesized by a nucleophilic substitution reaction of PTX with NBD-Cl in high yield. The PTX-NBD NCs were prepared by the anti-solvent method followed by surface functionalization of SPC and MPEG-DSPE.	The new formula was compared to free PTX-NBD and the sphere-shaped PTX-NBD nanocrystals with surface functionalization of SPC and MPEG-DSPE (PTX-NBD@PC-PEG NSs) Cell culture models: HeLa cells	The cube-shaped PTX-NBD@PC-PEG NCs exhibited better drug loading and stability properties. It showed a remarkable decrease in burst release, efficiently enhanced cellular uptake, and had a better ability to kill cancer cells in vitro using HeLa cells. These NCs can be useful for cell imaging and chemotherapy.	[168]
Surface-modified PTX with positively charged poly(allylamine hydrochloride) (PAH)	Nano-precipitation method (bottom-up approach) was employed to prepare PTX NCs, and the surface-modified NCs were obtained by an absorption method with the positively charged polymer	The new formula was compared to pure PTX, PTX NCs, and negatively charged poly (sodium 4-styrene sulfonate) PSS PTX NCs In vitro models: PBS (pH 7.4) containing 0.5% (w/v) tween 80 and bovine serum albumin (BSA) Cell culture models: A549 cells	Higher drug release. Stronger interaction with bovine serum albumin. Greater cellular internalization, uptake, and cytotoxicity.	[169]
A non-covalent transferrin-stabilized PTX NCs	The NCs were prepared by the antisolvent precipitation method augmented by sonication	The new formula was compared to PTX solution, PTX NCs, and Taxol® Cell culture models: human KB epidermal carcinoma cells and SKOV-3 ovarian cancer cells In vivo model: mice inoculated with KB cells	The in vivo efficacy studies on KB-bearing mice showed a significantly superior tumor inhibition rate compared with PTX NCs and less efficacy than Taxol, but with a better toxicity profile. However, in cellular models, it showed similar efficacy 72 h after treatment.	[158]
PTX NCs stabilized by D-α-tocopheryl polyethylene glycol 1000 succinate (TPGS)	The NCs were prepared by three-phase nanoparticle engineering technology (3PNET)	The new formula was compared to Taxol® and PTX/Pluronic F127 (F127) NCs Cell culture models: P-glycoprotein-overexpressing PTX-resistant (H460/TaxR) cancer cells In vivo model: PK using CD-1 mice	The greater the amount of TPGS in the formula, the greater cytotoxicity and cellular internalization. TPGS PTX NCs demonstrated a significantly sustained and prolonged in vitro release pattern. PK studies indicated more rapid clearance. However, they were more effective in promoting the accumulation of PTX in drug-resistant tumors.	[170]

Table 1. Cont.

PTX NC	Method of Preparation	The Models Used and the Reference or Control Formula	Benefits, Aims, and Other Notes	Refs.
Herceptin (HCT)-functionalized PTX NCs	PTX NCs were prepared by sono-precipitation approach, and then HCT was coated, applying a facile non-covalent technique	The new formula was compared to PTX NCs and PTX powder In vitro models: release study Cell culture models: HER2-positive breast cancer cell lines	Exhibited a sustained release pattern comparable to PTX NCs. Demonstrated a higher binding affinity, greater cell-specific internalization, and inhibition of growth to HER2-positive breast cancer cell lines.	[171]
PTX-NCs coated with Pluronic® F68 (PEG-PPG-PEG block polymer)	The NCs were prepared by the anti-solvent method	The new formula was compared to Taxol® and PTX NCs In vivo model: tumor-bearing (HT-29 and KB cells) mice and female nude outbred mice	These NCs exhibited similar or better antitumor efficacy and lower toxicity in comparison with Taxol. The in vivo study showed a significant enhancement in the blood circulation of PTX and accumulation in tumor tissue. However, the definite amount that reached the tumor was still minimal for the administered dose. The maximum amount of the coated NCs was significantly obtained in the liver compared with the other organs relative to the uncoated PTX NCs.	[172]
Triphenylphosphonium (TPP$^+$)-stabilized PTX NCs (TPP$^+$ PTX NCs)	Precipitation-resuspending method	The new formula was compared to free PTX and unmodified PTX NCs In vitro cell culture models: 2D monolayer and 3D multicellular spheroids (MCs) of MCF-7 cells and MCF-7/ADR cells	A mitochondria-targeted system was developed. Showed the strongest cytotoxicity that was associated with a reduction in mitochondrial membrane potential. Showed greater penetration and superior growth inhibition.	[173]
Platelet membrane-coated or cloaked PEG-PTX NCs (PPNCs)	The modified emulsion-lyophilized crystallization method	The new formula was compared to PTX NCs Platelet aggregation was examined using a spectrophotometric method In vitro drug releasee Cell culture models: 4T1 breast cancer cells In vivo model: BALB/c mice injected with 4T1 cells model	Minor risk of thrombus formation after injection was observed. Higher cellular uptake and greater cytotoxicity. In vivo studies showed the ability to deliver a higher dose of the drug and target the site of the coagulation (surgery or vascular disrupting), which improved the antitumor efficacy and decreased toxicities.	[174]

Table 1. *Cont.*

PTX NC	Method of Preparation	The Models Used and the Reference or Control Formula	Benefits, Aims, and Other Notes	Refs.
RGD peptide -PEGylated PTX NCs coated by polydopamine (PDA) (NC@PDA-PEG-RGD)	The NCs were prepared using modified antisolvent–sonication method	The new formula was compared to free PTX, PTX NCs, PTX NCs-PEG, and PTX NCs-PDA-PEG In vitro models: plasma for size stability Cell culture models: A549 lung cancer cell line In vivo model: nude mice A549 bearing cancer model	More size stability in plasma. Showed superior cellular uptake, growth inhibition, and cytotoxicity on A549 lung cancer cell line. In vivo demonstrated significantly greater accumulation in the tumor and slower tumor growth.	[175]
PTX and lapatinib (LAPA) composite nanocrystals with PDA and PEG modification (cNC@PDA-PEG)	PEG coat was introduced into the cNC via PDA) coat to get PEGylated composite NCs (cNC@PDA-PEG). The NCs were prepared using the bottom-up method or precipitation-resuspending method.	The new formula was compared to free PTX and unmodified PTX NCs In vitro models: plasma and blood Cell culture models: MCF-7/ADR cancer cells	cNC@PDA-PEG had optimum size and stability. The in vitro release study showed that both PTX and LAPA were released completely from cNC@PDA-PEG in 3 days, while only 30% of the drug was released from bulk drugs or unmodified NCs. Showed negligible hemocytolysis and improved therapeutic effect on MCF-7/ADR through endocytosis of whole NCs.	[176]

Table 2. Modified PTX NC formulations for oral drug delivery.

PTX NC	Method of Preparation	The Models Used and the Reference or Control Formula	Benefits, Aims, and Other Notes	Ref.
Pluronic-grafted chitosan as a stabilizer for PTX NC (Pl-g-CH PTX NCs)	A novel Pluronic-grafted chitosan copolymer was established and then utilized as a functional stabilizer for PTX NCs. Generally, the NCs were prepared using a high-pressure homogenizer.	The new formula was compared to Taxol® Cell culture models: Caco-2 cells and B16 F10 murine melanoma cells In vivo model for oral PK evaluation: Wistar rats In vivo model for efficacy study: healthy Balb/C mice injected with B16 F10 murine melanoma model	Improving intra-cellular accumulation. Improving the absorption by the transcellular and paracellular routes. Showed a P-gp inhibitory property. The in vivo model demonstrated more anti-tumor efficacy and growth reduction after oral delivery, and this was related to the enhancement in the systemic circulation as both the absorption and bioavailability were improved significantly.	[177]

Table 2. Cont.

PTX NC	Method of Preparation	The Models Used and the Reference or Control Formula	Benefits, Aims, and Other Notes	Ref.
PTX NCs stabilized by tween 80 or low molecular weight synthetic polymer sodium polystyrene sulfonate (PSS)	The top-down method was performed using a microfluidizer as a high-pressure homogenizer that was used to prepare the NCs without using any organic solvent	The new formula was compared to formulas stabilized with high molecular weight polymers glycol chitosan (GC) and sodium alginate (SA), as well as with PTX solution and PTX-NCs Cell culture models: MCF7 and MDA-MB breast cancer cell lines In vivo model: PK in male Wistar rat model	The prepared NCs were more suitable, efficient, and exhibited a considerable increase in the dissolution rate, which indicated an enhancement in its bioavailability. The in vitro cell culture study showed more efficiency and potency in killing and inhibiting the growth of the cancer cells. In vivo pharmacokinetic studies demonstrated a considerable increase in AUC_{0-t}, C_{max}, and MRT and a decrease in T_{max}.	[178]
Transferrin (TF)-modified PTX NCs	PTX NCs were prepared using the precipitation–resuspension method	The new formula was compared to Taxol® and unmodified PTX NCs In vitro models: in situ intestinal perfusion study Cell culture models: Caco-2 cells and MCF-7 cancer cells In vivo model: PK Sprague Dawley rat model	Showed an enhancement of cellular monolayer penetration. Had superior suppression in MCF-7 cell growth. Showed an enhancement of intestinal absorption. The pharmacokinetic studies also demonstrated greater C_{max} and AUC than both PTX NCs and Taxol® while having the lowest t_{max}.	[179]
Poly(sodium pstyrenesulfonate) (PSS)-modified PTX NCs	Not mentioned	In vitro models: interactions with biomolecules in oral delivery pathways Cell culture models: Caco-2 cell lines	Suitable mono-dispersion and stability in the gastrointestinal tract (GIT) environments for at least 24 h. No substantial interactions with pepsin or trypsin enzymes were detected in the GIT environments. PSS-modified PTX NCs passed through the mimical intestinal epithelial cell (Caco-2 cell lines) with about 25% transmittance. However, the concentration of the NCs should be controlled to avoid toxic effects on the cells.	[180]

Table 3. Modified PTX NC formulations for local and intraperitoneal drug delivery.

PTX NC	Route of Administration	Method of Preparation	The Models Used and the Reference or Control Formula	Benefits, Aims, and Other Notes	Ref.
PTX NC-loaded PECT hydrogels	Local delivery and peritumoral administration	PTX NCs were prepared by three-phase nanoparticle engineering technology (3PNET), while PTX-NC-based PECT (PTX-NC-PECT) gel was prepared based on the "cold" method	The new formula was compared to a nanoparticle-based system (PTX-NP-PECT) and controlled hydrogel of Pluronic® F127 In vitro models: release study In vivo model: MCF-7 tumor-bearing mouse models	High loading capacity of the drug. In vitro release was more effective and homogeneous. In vivo near-infrared fluorescence (NIRF) imaging indicated the ability to maintain the payloads of 1,1-dioctadecyltetramethyl indotricarbocyanine iodide (DiR) at a peri-tumoral site for about 21 days. Exhibited the most complete release system with the greatest anti-tumor efficacy and apoptosis effect.	[181]
Silica-coated PTX NCs Si	Intra-peritoneal (IP)	Precipitation–resuspending method	The new formula was compared to uncoated PTX-NC or Abraxane Cell culture model: neural stem cells and OVCAR-8 cells In vivo model: athymic nude mice which inoculated with 2 M OVCAR-8.eGFP.ffluc human ovarian cancer cells	More effective in loading neural stem cells (NSCs). In vivo studies showed that loaded NSCs preserved their migratory ability and, for low PTX dose, were more effective against ovarian tumors.	[182]
Poly-tannic acid-coated PTX NCs (PTA-PTX NCs)	Intertumoral injection	The NCs were prepared using the thin-film hydration method followed by probe sonication	The new formula was compared with or without laser irradiation to PTX Cell culture models: 4T1, A549, and HepG2 cells In vivo model: 4T1 tumor-bearing mice	PTX NCs were prepared to act as a chemo-therapeutic agent and poly-tannic acid (pTA)-coated PTX NCs in the presence of Fe^{3+} acting as a potential agent for photothermal therapy (PTT). The cellular uptake was significantly improved. A synergistic effect with laser irradiation was observed. Demonstrated mild photothermal effect in vivo and the greatest effect in tumor inhibition upon laser irradiation.	[183]
PTX NC with F127 hydrogel	Intertumoral injection	Precipitation–resuspending method. The cold method was used for hydrogel preparation.	The new formula was compared to PTX or PTX microcrystal-based hydrogels In vitro erosion of the hydrogels and drug release In vivo model: 4T1 tumor-bearing BALB/c mice	PTX NCs gel offered optimum properties with high drug loading combined with moderate drug release and erosion profiles. Superior anti-tumor efficacy in 4T1 tumor-bearing BALB/c mice.	[184]

Table 3. *Cont.*

PTX NC	Route of Administration	Method of Preparation	The Models Used and the Reference or Control Formula	Benefits, Aims, and Other Notes	Ref.
In situ cross-linkable hydrogel depot containing PTX NCs	Intraperitoneal (IP)	Anti-solvent and temperature-induced crystallization method	The new formula was compared to Taxol® and microparticulate PTX precipitates (PPT). Cell culture models: SKOV3 cells. In vivo model: healthy Balb/c mice for toxicity studies and Balb/c mice (SKOV3-Luc) cell-bearing mice for the efficacy study	Superior killing efficiency and more toxicity in SKOV3 cell line. The in vivo study indicated improved dissolution, cellular uptake, and lower maximum tolerated dose. It also showed that a single IP dose was sufficient in extending the survival of tumor-bearing mice.	[185]
PTX-NCs combined with niclosamide (NLM) NLM-NCs co-loaded PLGA-PEG-PLGA thermosensitive hydrogel (PN-NCs-Ts)	Intratumoral injection	PTX-NCs were prepared by the "3PNET" method	The new formula was compared to PTX-NCs, PTX-NCs-Ts Gel, NLM-NCs, NLM-NCs-Ts gel, and PN–NCs-Ts gel. In vitro drug release. Cell culture models: MDA-MB-231 cells. In vivo model: BALB/c nude mice inoculated with MDA-MB-231 cells	Sustained and significantly delayed drug release both in vitro and in vivo. The combination with NLM improved PTX cellular uptake, apoptosis, and provided inhibition of cell migration. The in vivo studies showed significant inhibition of tumor growth with acceptable safety and effectively overcoming it. Triple-negative breast cancer (TNBC) progress and drastically prevented breast cancer stem cells (BCSCs).	[186]

6. Future Aspects

It is worth mentioning that the nanocrystals are formed by weak, non-covalent interactions. This leads drug nanocrystals to continue to dissolve, albeit slowly, when in contact with water. As such, any surface-coated materials on drug nanocrystals will eventually be detached during the dissolution process. This not only makes it a challenging task to develop surface-treated nanocrystals but also results in transient target-homing effects.

In this regard, the concept of hybrid nanocrystals may overcome this limitation by physically integrating guest molecules among the crystal lattices of nanocrystals. Small molecules such as fluorescent dyes have been demonstrated in vitro and in vivo of paclitaxel nanocrystals. It is thus possible to utilize larger molecules as a guest in making hybrid nanocrystals.

Finally, it is pertinent to understand and eventually predict drug release and dissolution kinetics of paclitaxel nanocrystals in a biological environment. This may be aided by in vitro experimentation and physics-based simulation. One ultimate goal in developing paclitaxel nanocrystals is precision medicine for cancer treatment, which can only be enabled by a thorough understanding of the interactions and the pharmacokinetic characteristics of drug nanocrystals in tissues and cells.

7. Conclusions and Remarks

Several delivery systems for paclitaxel drugs have been developed to enhance their solubility and pharmacological properties. Of these delivery systems, nanocrystal formulations are considered a promising modality that can also have the advantage of providing a suitable platform for surface modifications. Based on that, many studies employed nano-crystallization techniques not only to improve the oral delivery of PTX but also to improve the IV, intraperitoneal (IP), and local and intertumoral delivery systems, where the applications of surface modifications can be of greater value in terms of targeted delivery.

Moreover, these systems can provide 100% loading and releasing capacities for the drugs as well as gain the advantages of being formulated as particles that have different circulation patterns, fates, cellular uptake mechanisms, and sometimes preferable efficacy and safety profiles compared with free drugs. Finally, more studies are needed to understand the molecular basis for the formation and interaction of these nanocrystals with biological systems, and consequently providing better platforms for useful modifications in the future.

Author Contributions: Conceptualization, R.H. and N.A.; methodology, R.H. and N.A.; software, R.H.; validation, R.H., N.A., B.A. and T.L.; formal analysis, R.H. and N.A.; investigation, R.H.; resources, R.H.; data curation, R.H., N.A., B.A. and T.L.; writing—original draft preparation, R.H.; writing—review and editing, N.A. and T.L.; visualization, N.A.; supervision, N.A., B.A. and T.L.; project administration, N.A. and B.A.; funding acquisition, N.A. and B.A. All authors have read and agreed to the published version of the manuscript.

Funding: This research was funded by Jordan University of Science and Technology, grant number [298/2021].

Informed Consent Statement: Not applicable.

Data Availability Statement: Not applicable.

Conflicts of Interest: The authors declare no conflict of interest.

References

1. WHO. Cancer-Fact Fheets. Available online: www.who.int/mediacentre/factsheets/fs297/en/12/9/20182/3/2020 (accessed on 2 March 2020).
2. World Health Organization. Available online: https://www.who.int/news-room/fact-sheets/detail/cancer (accessed on 5 March 2021).
3. Patterson, R.; Fischman, V.G.; Wasserman, I.; Siu, J.; Shrime, M.G.; Fagan, J.J.; Koch, W.; Alkire, B.C. Global Burden of Head and Neck Cancer: Economic Consequences, Health, and the Role of Surgery. *Otolaryngol. Neck Surg.* **2020**, *162*, 296–303. [CrossRef] [PubMed]
4. Smith, G.L.; Lopez-Olivo, M.; Advani, P.G.; Ning, M.S.; Geng, Y.; Giordano, S.H.; Volk, R.J. Financial Burdens of Cancer Treatment: A Systematic Review of Risk Factors and Outcomes. *J. Natl. Compr. Cancer Netw.* **2019**, *17*, 1184–1192. [CrossRef] [PubMed]
5. Altmann, K.-H.; Gertsch, J. Anticancer drugs from nature—Natural products as a unique source of new microtubule-stabilizing agents. *Nat. Prod. Rep.* **2007**, *24*, 327–357. [CrossRef] [PubMed]
6. Rowinsky, E.K. Signal events: Cell signal transduction and its inhibition in cancer. *Oncologist* **2003**, *8*, 5–17. [CrossRef]
7. Yue, Q.-X.; Liu, X.; Guo, D.-A. Microtubule-Binding Natural Products for Cancer Therapy. *Planta Med.* **2010**, *76*, 1037–1043. [CrossRef]
8. Hollis, C.P.; Li, T. Hybrid Nanocrystal as a Versatile Platform for Cancer Theranostics. In *Biomaterials for Cancer Thera-peutics: Diagnosis, Prevention and Therapy*; Woodhead Publishinged: Cambridge, UK, 2013.
9. Chaturvedi, V.K.; Singh, A.; Singh, V.K.; Singh, M.P. Cancer Nanotechnology: A New Revolution for Cancer Diagnosis and Therapy. *Curr. Drug Metab.* **2019**, *20*, 416–429. [CrossRef]
10. Lu, Y.; Chen, Y.; A Gemeinhart, R.; Wu, W.; Li, T. Developing nanocrystals for cancer treatment. *Nanomedicine* **2015**, *10*, 2537–2552. [CrossRef]
11. Miele, E.; Spinelli, G.P.; Miele, E.; Tomao, F.; Tomao, S. Albumin-bound formulation of paclitaxel (Abraxane®ABI-007) in the treatment of breast cancer. *Int. J. Nanomed.* **2009**, *4*, 99.
12. Heinig, U.; Scholz, S.; Jennewein, S. Getting to the bottom of Taxol biosynthesis by fungi. *Fungal Divers.* **2013**, *60*, 161–170. [CrossRef]
13. Markman, M. Managing taxane toxicities. *Support. Care Cancer* **2003**, *11*, 144–147. [CrossRef]
14. Rowinsky, E.K.; Cazenave, L.A.; Donehower, R.C. Taxol: A Novel Investigational Antimicrotubule Agent. *JNCI J. Natl. Cancer Inst.* **1990**, *82*, 1247–1259. [CrossRef] [PubMed]
15. Schiff, P.; Horwitz, S.B. Taxol stabilizes microtubules in mouse fibroblast cells. *Proc. Natl. Acad. Sci. USA* **1980**, *77*, 1561–1565. [CrossRef] [PubMed]
16. Yoncheva, K.; Calleja, P.; Agüeros, M.; Petrov, P.; Miladinova, I.; Tsvetanov, C.; Irache, J.M. Stabilized micelles as delivery vehicles for paclitaxel. *Int. J. Pharm.* **2012**, *436*, 258–264. [CrossRef]
17. Deepa, G.; Ashwanikumar, N.; Pillai, J.J.; Kumar, G.S.V. Polymer nanoparticles-a novel strategy for administration of paclitaxel in cancer chemotherapy. *Curr. Med. Chem.* **2012**, *19*, 6207–6213. [CrossRef]
18. Ezrahi, S.; Aserin, A.; Garti, N. Basic principles of drug delivery systems—The case of paclitaxel. *Adv. Colloid Interface Sci.* **2019**, *263*, 95–130. [CrossRef]
19. Zhang, Z.; Mei, L.; Feng, S.-S. Paclitaxel drug delivery systems. *Expert Opin. Drug Deliv.* **2013**, *10*, 325–340. [CrossRef]

20. Schiff, P.; Fant, J.; Horwitz, S.B. Promotion of microtubule assembly in vitro by taxol. *Nature* **1979**, *277*, 665–667. [CrossRef]
21. Stinchcombe, T.E. Nanoparticle albumin-bound paclitaxel: A novel Cremphor-EL®-free formulation of paclitaxel. *Nanomedicine* **2007**, *2*, 415–423. [CrossRef]
22. Ghadi, R.; Dand, N. BCS class IV drugs: Highly notorious candidates for formulation development. *J. Control. Release* **2017**, *248*, 71–95. [CrossRef]
23. Malingré, M.M.; Beijnen, J.H.; Schellens, J.H. Oral delivery of taxanes. *Investig. New Drugs* **2001**, *19*, 155–162. [CrossRef]
24. Lee, J.; Lee, S.C.; Acharya, G.; Chang, C.; Park, K. Hydrotropic Solubilization of Paclitaxel: Analysis of Chemical Structures for Hydrotropic Property. *Pharm. Res.* **2003**, *20*, 1022–1030. [CrossRef] [PubMed]
25. Thomas, V.H.; Bhattachar, S.; Hitchingham, L.; Zocharski, P.; Naath, M.; Surendran, N.; Stoner, C.L.; El-Kattan, A. The road map to oral bioavailability: An industrial perspective. *Expert Opin. Drug Metab. Toxicol.* **2006**, *2*, 591–608. [CrossRef] [PubMed]
26. Bradley, J.D.; Paulus, R.; Komaki, R.; Masters, G.; Blumenschein, G.; Schild, S.; Bogart, J.; Hu, C.; Forster, K.; Magliocco, A.; et al. Standard-dose versus high-dose conformal radiotherapy with concurrent and consolidation carboplatin plus paclitaxel with or without cetuximab for patients with stage IIIA or IIIB non-small-cell lung cancer (RTOG 0617): A randomised, two-by-two factorial phase 3 study. *Lancet Oncol.* **2015**, *16*, 187–199. [CrossRef] [PubMed]
27. Song, W.; Tang, Z.; Li, M.; Lv, S.; Sun, H.; Deng, M.; Liu, H.; Chen, X. Polypeptide-based combination of paclitaxel and cisplatin for enhanced chemotherapy efficacy and reduced side-effects. *Acta Biomater.* **2014**, *10*, 1392–1402. [CrossRef] [PubMed]
28. Brotto, L.; Brundage, M.; Hoskins, P.; Vergote, I.; Cervantes, A.; Casado, H.A.; Poveda, A.; Eisenhauer, E.; Tu, N. Randomized study of sequential cisplatin-topotecan/carboplatin-paclitaxel versus carboplatin-paclitaxel: Effects on quality of life. *Support Care Cancer* **2016**, *24*, 1241–1249. [CrossRef] [PubMed]
29. Zhang, R.; Yang, J.; Sima, M.; Zhou, Y.; Kopeček, J. Sequential combination therapy of ovarian cancer with degradable N-(2-hydroxypropyl)methacrylamide copolymer paclitaxel and gemcitabine conjugates. *Proc. Natl. Acad. Sci. USA* **2014**, *111*, 12181–12186. [CrossRef]
30. Panchagnula, R. Pharmaceutical aspects of paclitaxel. *Int. J. Pharm.* **1998**, *172*, 1–15. [CrossRef]
31. Marupudi, N.; E Han, J.; Li, K.W.; Renard, V.M.; Tyler, B.M.; Brem, H. Paclitaxel: A review of adverse toxicities and novel delivery strategies. *Expert Opin. Drug Saf.* **2007**, *6*, 609–621. [CrossRef]
32. Green, R.M.; Manikhas, M.G.; Orlov, S.; Afanasyev, B.; Makhson, M.A.; Bhar, P.; Hawkins, J.M. Abraxane®, a novel Cremophor®-free, albumin-bound particle form of paclitaxel for the treatment of advanced non-small-cell lung cancer. *Ann. Oncol.* **2006**, *17*, 1263–1268. [CrossRef]
33. Gradishar, W.J. Albumin-bound paclitaxel: A next-generation taxane. *Expert Opin. Pharmacother.* **2006**, *7*, 1041–1053. [CrossRef]
34. Paál, K.; Müller, J.; Hegedûs, L. High affinity binding of paclitaxel to human serum albumin. *JBIC J. Biol. Inorg. Chem.* **2001**, *268*, 2187–2191. [CrossRef] [PubMed]
35. Singh, S.; Dash, A.K. Paclitaxel in cancer treatment: Perspectives and prospects of its delivery challenges. *Crit. Rev. Ther. Drug Carr. Syst.* **2009**, *26*, 333–372. [CrossRef] [PubMed]
36. Gradishar, W.J.; Tjulandin, S.; Davidson, N.; Shaw, H.; Desai, N.; Bhar, P.; Hawkins, M.; O'Shaughnessy, J. Phase III Trial of Nanoparticle Albumin-Bound Paclitaxel Compared with Polyethylated Castor Oil–Based Paclitaxel in Women with Breast Cancer. *J. Clin. Oncol.* **2005**, *23*, 7794–7803. [CrossRef] [PubMed]
37. Chen, Q.; Zhang, Q.-Z.; Liu, J.; Li, L.-Q.; Zhao, W.-H.; Wang, Y.-J.; Zhou, Q.-H.; Li, L. Multi-center prospective randomized trial on paclitaxel liposome and traditional taxol in the treatment of breast cancer and non-small-cell lung cancer. *Zhonghua Zhong Liu Za Zhi Chinese J. Oncol.* **2003**, *25*, 190–192.
38. Xu, X.; Wang, L.; Xu, Q.H.; Huang, E.X.; Qian, D.Y.; Xiang, J. Clinical comparison between paclitaxel liposome (Lipusu®) and paclitaxel for treatment of patients with meta-static gastric cancer. *Asian Pac. J. Cancer Prev.* **2013**, *14*, 2591–2594. [CrossRef]
39. Kim, T.-Y.; Kim, D.-W.; Chung, J.-Y.; Shin, S.G.; Kim, S.-C.; Heo, D.S.; Kim, N.K.; Bang, Y.-J. Phase I and Pharmacokinetic Study of Genexol PM, a Cremophor-Free, Polymeric Micelle-Formulated Paclitaxel, in Patients with Advanced Malignancies. *Clin. Cancer Res.* **2004**, *10*, 3708–3716. [CrossRef]
40. Lim, W.T.; Tan, E.H.; Toh, C.K.; Hee, S.W.; Leong, S.S.; Ang, P.C.S.; Wong, N.S.; Chowbay, B. Phase I pharmacokinetic study of a weekly liposomal paclitaxel formulation (Genexol®-PM) in patients with solid tumors. *Ann. Oncol.* **2009**, *21*, 382–388. [CrossRef]
41. Saif, M.W.; Podoltsev, N.A.; Rubin, M.S.; Figueroa, J.A.; Lee, M.Y.; Kwon, J.; Rowen, E.; Yu, J.; Kerr, R.O. Phase II Clinical Trial of Paclitaxel Loaded Polymeric Micelle in Patients with Advanced Pancreatic Cancer. *Cancer Investig.* **2010**, *28*, 186–194. [CrossRef]
42. Sartori, S.; Caporale, A.; Rechichi, A.; Cufari, D.; Cristallini, C.; Barbani, N.; Giusti, P.; Ciardelli, G. Biodegradable paclitaxel-loaded microparticles prepared from novel block copolymers: Influence of polymer composition on drug encapsulation and release. *J. Pept. Sci.* **2013**, *19*, 205–213. [CrossRef]
43. He, H.; Chen, S.; Zhou, J.; Dou, Y.; Song, L.; Che, L.; Zhou, X.; Chen, X.; Jia, Y.; Zhang, J.; et al. Cyclodextrin-derived pH-responsive nanoparticles for delivery of paclitaxel. *Biomaterials* **2013**, *34*, 5344–5358. [CrossRef]
44. Wang, H.; Cheng, G.; Du, Y.; Ye, L.; Chen, W.; Zhang, L.; Wang, T.; Tian, J.; Fu, F. Hypersensitivity reaction studies of a polyethoxylated castor oil-free, liposome-based alternative paclitaxel formulation. *Mol. Med. Rep.* **2013**, *7*, 947–952. [CrossRef] [PubMed]
45. Xia, X.-J.; Guo, R.-F.; Liu, Y.-L.; Zhang, P.-X.; Zhou, C.-P.; Jin, D.-J.; Wang, R.-Y. Formulation, Characterization and Hypersensitivity Evaluation of an Intravenous Emulsion Loaded with a Paclitaxel-Cholesterol Complex. *Chem. Pharm. Bull.* **2011**, *59*, 321–326. [CrossRef]

46. Torchilin, V.P. Micellar Nanocarriers: Pharmaceutical Perspectives. *Pharm. Res.* **2007**, *24*, 1. [CrossRef] [PubMed]
47. May, S.; Ben-Shaul, A. Molecular Theory of Lipid-Protein Interaction and the Lα-HII Transition. *Biophys. J.* **1999**, *76*, 751–767. [CrossRef]
48. Lukyanov, A.N.; Torchilin, V.P. Micelles from lipid derivatives of water-soluble polymers as delivery systems for poorly soluble drugs. *Adv. Drug Deliv. Rev.* **2004**, *56*, 1273–1289. [CrossRef] [PubMed]
49. Zhao, B.J.; Ke, X.Y.; Huang, Y.; Chen, X.M.; Zhao, X.; Zhao, B.X.; Lu, W.L.; Lou, J.N.; Zhang, X.; Zhang, Q. The antiangiogenic efficacy of NGR-modified PEG–DSPE micelles containing paclitaxel (NGR-M-PTX) for the treatment of glioma in rats. *J. Drug Target.* **2011**, *19*, 382–390. [CrossRef]
50. Chen, T.; Tu, L.; Wang, G.; Qi, N.; Wu, W.; Zhang, W.; Feng, J. Multi-functional chitosan polymeric micelles as oral paclitaxel delivery systems for enhanced bioavailability and anti-tumor efficacy. *Int. J. Pharm.* **2020**, *578*, 119105. [CrossRef]
51. Mutlu-Agardan, N.B.; Sarisozen, C.; Torchilin, V. Cytotoxicity of Novel Redox Sensitive PEG 2000-SS-PTX Micelles against Drug-Resistant Ovarian and Breast Cancer Cells. *Pharm. Res.* **2020**, *37*, 65. [CrossRef]
52. Feng, L.; Mumper, R.J. A critical review of lipid-based nanoparticles for taxane delivery. *Cancer Lett.* **2013**, *334*, 157–175. [CrossRef]
53. Fetterly, G.J.; Grasela, T.H.; Sherman, J.W.; Dul, J.L.; Grahn, A.; LeComte, D.; Fiedler-Kelly, J.; Damjanov, N.; Fishman, M.; Kane, M.P.; et al. Pharmacokinetic/Pharmacodynamic Modeling and Simulation of Neutropenia during Phase I Development of Liposome-Entrapped Paclitaxel. *Clin. Cancer Res.* **2008**, *14*, 5856–5863. [CrossRef]
54. Crosasso, P.; Ceruti, M.; Brusa, P.; Arpicco, S.; Dosio, F.; Cattel, L. Preparation, characterization and properties of sterically stabilized paclitaxel-containing liposomes. *J. Control. Release* **2000**, *63*, 19–30. [CrossRef]
55. Klibanov, A.L.; Maruyama, K.; Torchilin, V.P.; Huang, L. Amphipathic polyethyleneglycols effectively prolong the circulation time of liposomes. *FEBS Lett.* **1990**, *268*, 235–237. [CrossRef]
56. Yoshizawa, Y.; Kono, Y.; Ogawara, K.-I.; Kimura, T.; Higaki, K. PEG liposomalization of paclitaxel improved its in vivo disposition and anti-tumor efficacy. *Int. J. Pharm.* **2011**, *412*, 132–141. [CrossRef] [PubMed]
57. Abu Lila, A.; Kiwada, H.; Ishida, T. The accelerated blood clearance (ABC) phenomenon: Clinical challenge and approaches to manage. *J. Control. Release* **2013**, *172*, 38–47. [CrossRef]
58. Biswas, S.; Dodwadkar, N.S.; Deshpande, P.; Torchilin, V.P. Liposomes loaded with paclitaxel and modified with novel triphenylphosphonium-PEG-PE conjugate possess low toxicity, target mitochondria and demonstrate enhanced antitumor effects in vitro and in vivo. *J. Control. Release* **2012**, *159*, 393–402. [CrossRef] [PubMed]
59. Liu, Y.; Ran, R.; Chen, J.; Kuang, Q.; Tang, J.; Mei, L.; Zhang, Q.; Gao, H.; Zhang, Z.; He, Q. Paclitaxel loaded liposomes decorated with a multifunctional tandem peptide for glioma targeting. *Biomaterials* **2014**, *35*, 4835–4847. [CrossRef]
60. Luo, L.-M.; Huang, Y.; Zhao, B.-X.; Zhao, X.; Duan, Y.; Du, R.; Yu, K.-F.; Song, P.; Zhao, Y.; Zhang, X.; et al. Anti-tumor and anti-angiogenic effect of metronomic cyclic NGR-modified liposomes containing paclitaxel. *Biomaterials* **2013**, *34*, 1102–1114. [CrossRef]
61. Qin, L.I.; Wang, C.Z.; Fan, H.J.; Zhang, C.J.; Zhang, H.W.; Lv, M.H.; Cui, S.D. A dual-targeting liposome conjugated with transferrin and arginine-glycine-aspartic acid peptide for glio-ma-targeting therapy. *Oncol. Lett.* **2014**, *8*, 2000–2006. [CrossRef]
62. Büyükköroğlu, G.; Şenel, B.; Başaran, E.; Gezgin, S. Development of paclitaxel-loaded liposomal systems with anti-her2 antibody for targeted therapy. *Trop. J. Pharm. Res.* **2016**, *15*, 895. [CrossRef]
63. Chen, D.; Jiang, X.; Liu, J.; Jin, X.; Zhang, C.; Ping, Q. In vivo evaluation of novel pH-sensitive mPEG-Hz-Chol conjugate in liposomes: Pharmacokinetics, tissue distribution, efficacy assessment. *Artif. Cells Blood Substit. Biotechnol.* **2010**, *38*, 136–142. [CrossRef]
64. Monteiro, L.O.; Malachias, A.; Pound-Lana, G.; Magalhaes-Paniago, R.; Mosqueira, V.C.; Oliveira, M.C.; de Barros, A.L.B.; Leite, E.A. Paclitaxel-loaded pH-sensitive liposome: New insights on structural and physicochemical characterization. *Langmuir* **2018**, *34*, 5728–5737. [CrossRef]
65. Qi, J.; Lu, Y.; Wu, W. Absorption, Disposition and Pharmacokinetics of Solid Lipid Nanoparticles. *Curr. Drug Metab.* **2012**, *13*, 418–428. [CrossRef]
66. Shahgaldian, P.; Da Silva, E.; Coleman, A.W.; Rather, B.; Zaworotko, M.J. Para-acyl-calix-arene based solid lipid nanoparticles (SLNs): A detailed study of preparation and stability parameters. *Int. J. Pharm.* **2003**, *253*, 23–38. [CrossRef]
67. Yegin, A.B.; Benoît, J.-P.; Lamprecht, A. Paclitaxel-loaded lipid nanoparticles prepared by solvent injection or ultra-sound emulsification. *Drug Dev. Ind. Pharm.* **2006**, *32*, 1089–1094. [CrossRef] [PubMed]
68. Yuan, H.; Miao, J.; Du, Y.-Z.; You, J.; Hu, F.-Q.; Zeng, S. Cellular uptake of solid lipid nanoparticles and cytotoxicity of encapsulated paclitaxel in A549 cancer cells. *Int. J. Pharm.* **2008**, *348*, 137–145. [CrossRef] [PubMed]
69. Xu, W.; Bae, E.J.; Lee, M.-K. Enhanced anticancer activity and intracellular uptake of paclitaxel-containing solid lipid nanoparticles in multidrug-resistant breast cancer cells. *Int. J. Nanomed.* **2018**, *13*, 7549–7563. [CrossRef] [PubMed]
70. Valsalakumari, R.; Yadava, S.K.; Szwed, M.; Pandya, A.D.; Mælandsmo, G.M.; Torgersen, M.L.; Iversen, T.-G.; Skotland, T.; Sandvig, K.; Giri, J. Mechanism of cellular uptake and cytotoxicity of paclitaxel loaded lipid nanocapsules in breast cancer cells. *Int. J. Pharm.* **2021**, *597*, 120217. [CrossRef]
71. Dong, X.; Mattingly, C.A.; Tseng, M.T.; Cho, M.J.; Liu, Y.; Adams, V.R.; Mumper, R.J. Doxorubicin and Paclitaxel-Loaded Lipid-Based Nanoparticles Overcome Multidrug Resistance by Inhibiting P-Glycoprotein and Depleting ATP. *Cancer Res.* **2009**, *69*, 3918–3926. [CrossRef]

72. Tammam, S.N. Lipid Based Nanoparticles as Inherent Reversing Agents of Multidrug Resistance in Cancer. *Curr. Pharm. Des.* **2017**, *23*, 6714–6729. [CrossRef]
73. Pandey, A.; Jain, D.S.; Chakraborty, S. Poly Lactic-Co-Glycolic Acid (PLGA) Copolymer and Its Pharmaceutical Application. *Handb. Polym. Pharm. Technol.* **2015**, *2*, 151–172.
74. Astete, C.E.; Sabliov, C.M. Synthesis and characterization of PLGA nanoparticles. *J. Biomater. Sci. Polym. Ed.* **2006**, *17*, 247–289. [CrossRef] [PubMed]
75. Berthet, M.; Gauthier, Y.; Lacroix, C.; Verrier, B.; Monge, C. Nanoparticle-based dressing: The future of wound treatment? *Trends Biotechnol.* **2017**, *35*, 770–784. [CrossRef] [PubMed]
76. Rezvantalab, S.; Drude, N.; Moraveji, M.K.; Güvener, N.; Koons, E.K.; Shi, Y.; Lammers, T.; Kiessling, F. PLGA-Based Nanoparticles in Cancer Treatment. *Front. Pharmacol.* **2018**, *9*, 1260. [CrossRef]
77. Jin, C.; Wu, H.; Liu, J.; Bai, L.; Guo, G. The effect of paclitaxel-loaded nanoparticles with radiation on hypoxic MCF-7 cells. *J. Clin. Pharm. Ther.* **2007**, *32*, 41–47. [CrossRef]
78. Dinarvand, R.; Sepehri, N.; Manouchehri, S.; Rouhani, H.; Atyabi, F. Polylactide-co-glycolide nanoparticles for controlled delivery of anticancer agents. *Int. J. Nanomed.* **2011**, *6*, 877–895. [CrossRef] [PubMed]
79. Mirakabad, F.S.T.; Nejati-Koshki, K.; Akbarzadeh, A.; Yamchi, M.R.; Milani, M.; Zarghami, N.; Zeighamian, V.; Rahimzadeh, A.; Alimohammadi, S.; Hanifehpour, Y.; et al. PLGA-Based Nanoparticles as Cancer Drug Delivery Systems. *Asian Pac. J. Cancer Prev.* **2014**, *15*, 517–535. [CrossRef]
80. Fonseca, C.; Simões, S.; Gaspar, R. Paclitaxel-loaded PLGA nanoparticles: Preparation, physicochemical characterization and in vitro anti-tumoral activity. *J. Control. Release* **2002**, *83*, 273–286. [CrossRef]
81. Danhier, F.; Lecouturier, N.; Vroman, B.; Jérôme, C.; Marchand-Brynaert, J.; Feron, O.; Préat, V. Paclitaxel-loaded PEGylated PLGA-based nanoparticles: In vitro and in vivo evaluation. *J. Control. Release* **2009**, *133*, 11–17. [CrossRef]
82. Mo, Y.; Lim, L.-Y. Paclitaxel-loaded PLGA nanoparticles: Potentiation of anticancer activity by surface conjugation with wheat germ agglutinin. *J. Control. Release* **2005**, *108*, 244–262. [CrossRef]
83. Esfandyari-Manesh, M.; Mostafavi, S.H.; Majidi, R.F.; Koopaei, M.N.; Ravari, N.S.; Amini, M.; Darvishi, B.; Ostad, S.N.; Atyabi, F.; Dinarvand, R. Improved anticancer delivery of paclitaxel by albumin surface modification of PLGA nano-particles DARU. *J. Pharm. Sci.* **2015**, *23*, 28.
84. Cerqueira, B.B.S.; Lasham, A.; Shelling, A.N.; Al-Kassas, R. Development of biodegradable PLGA nanoparticles surface engineered with hyaluronic acid for targeted delivery of paclitaxel to triple negative breast cancer cells. *Mater. Sci. Eng. C* **2017**, *76*, 593–600. [CrossRef] [PubMed]
85. Godara, S.; Lather, V.; Kirthanashri, S.V.; Awasthi, R.; Pandita, D. Lipid-PLGA hybrid nanoparticles of paclitaxel: Preparation, characterization, in vitro and in vivo evaluation. *Mater. Sci. Eng. C* **2020**, *109*, 110576. [CrossRef]
86. Kim, C.; Lee, A.S.C.; Kang, S.W.; Kwon, I.C.; Kim, A.Y.-H.; Jeong, S.Y. Synthesis and the Micellar Characteristics of Poly(ethylene oxide)−Deoxycholic Acid Conjugates1. *Langmuir* **2000**, *16*, 4792–4797. [CrossRef]
87. Kim, C.; Lee, S.C.; Kwon, I.C.; Chung, H.; Jeong, S.Y. Complexation of Poly (2-ethyl-2-oxazoline)-b lock-poly (ε-caprolactone) Micelles with Multifunctional Car-boxylic Acids. *Macromolecules* **2002**, *35*, 193–200. [CrossRef]
88. Patel, N.K.; Sinha, V.K. Synthesis, Characterization and Optimization of Water-Soluble Chitosan Derivatives. *Int. J. Polym. Mater. Polym. Biomater.* **2009**, *58*, 548–560. [CrossRef]
89. Chen, X.-G.; Park, H.-J. Chemical characteristics of O-carboxymethyl chitosans related to the preparation conditions. *Carbohydr. Polym.* **2003**, *53*, 355–359. [CrossRef]
90. Cao, J.; Zhou, N. Progress in antitumor studies of chitosan. *Chin. J. Biochem. Pharm.* **2005**, *26*, 127.
91. Gupta, U.; Sharma, S.; Khan, I.; Gothwal, A.; Sharma, A.K.; Singh, Y.; Chourasia, M.K.; Kumar, V. Enhanced apoptotic and anticancer potential of paclitaxel loaded biodegradable nanoparticles based on chitosan. *Int. J. Biol. Macromol.* **2017**, *98*, 810–819. [CrossRef]
92. Lv, P.-P.; Ma, Y.-F.; Yu, R.; Yue, H.; Ni, D.-Z.; Wei, W.; Ma, G.-H. Targeted Delivery of Insoluble Cargo (Paclitaxel) by PEGylated Chitosan Nanoparticles Grafted with Arg-Gly-Asp (RGD). *Mol. Pharm.* **2012**, *9*, 1736–1747. [CrossRef]
93. Ashrafizadeh, M.; Ahmadi, Z.; Mohamadi, N.; Zarrabi, A.; Abasi, S.; Dehghannoudeh, G.; Tamaddondoust, R.N.; Khanbabaei, H.; Mohammadinejad, R.; Thakur, V.K. Chitosan-based advanced materials for docetaxel and paclitaxel delivery: Recent advances and future directions in cancer theranostics. *Int. J. Biol. Macromol.* **2020**, *145*, 282–300. [CrossRef]
94. Zhang, C.; Qu, G.; Sun, Y.; Wu, X.; Yao, Z.; Guo, Q.; Ding, Q.; Yuan, S.; Shen, Z.; Ping, Q.; et al. Pharmacokinetics, biodistribution, efficacy and safety of N-octyl-O-sulfate chitosan micelles loaded with paclitaxel. *Biomaterials* **2008**, *29*, 1233–1241. [CrossRef] [PubMed]
95. Li, F.; Wu, H.; Zhang, H.; Li, F.; Gu, C.H.; Yang, Q. Antitumor drug Paclitaxel-loaded pH-sensitive nanoparticles targeting tumor extracellular pH. *Carbohydr. Polym.* **2009**, *77*, 773–778. [CrossRef]
96. Nag, M.; Gajbhiye, V.; Kesharwani, P.; Jain, N.K. Transferrin functionalized chitosan-PEG nanoparticles for targeted delivery of paclitaxel to cancer cells. *Colloids Surfaces B Biointerfaces* **2016**, *148*, 363–370. [CrossRef] [PubMed]
97. Ursachi, V.C.; Dodi, G.; Rusu, A.G.; Mihai, C.T.; Verestiuc, L.; Balan, V. Paclitaxel-Loaded Magnetic Nanoparticles Based on Biotinylated N-Palmitoyl Chitosan: Synthesis, Characterization and Preliminary In Vitro Studies. *Molecules* **2021**, *26*, 3467. [CrossRef]

98. Rautio, J.; Kumpulainen, H.; Heimbach, T.; Oliyai, R.; Oh, D.; Järvinen, T.; Savolainen, J. Prodrugs: Design and clinical applications. *Nat. Rev. Drug Discov.* **2008**, *7*, 255–270. [CrossRef]
99. Ettmayer, P.; Amidon, G.L.; Clement, A.B.; Testa, B. Lessons Learned from Marketed and Investigational Prodrugs. *J. Med. Chem.* **2004**, *47*, 2393–2404. [CrossRef]
100. Skwarczynski, M.; Hayashi, Y.; Kiso, Y. Paclitaxel Prodrugs: Toward Smarter Delivery of Anticancer Agents. *J. Med. Chem.* **2006**, *49*, 7253–7269. [CrossRef]
101. Li, C.; Yu, D.; Inoue, T.; Yang, D.J.; Milas, L.; Hunter, N.R.; Kim, E.E.; Wallace, S. Synthesis and evaluation of water-soluble polyethylene glycol-paclitaxel conjugate as a paclitaxel prodrug. *Anticancer Drugs* **1996**, *7*, 642–648. [CrossRef]
102. Yu, Y.; Chen, C.-K.; Law, W.-C.; Mok, J.; Zou, J.; Prasad, P.N.; Cheng, C. Well-Defined Degradable Brush Polymer–Drug Conjugates for Sustained Delivery of Paclitaxel. *Mol. Pharm.* **2013**, *10*, 867–874. [CrossRef]
103. Yu, Y.; Zou, J.; Yu, L.; Ji, W.; Li, Y.; Law, W.C.; Cheng, C. Functional polylactide-g-paclitaxel–poly (ethylene glycol) by azide–alkyne click chemistry. *Macromolecules* **2011**, *44*, 4793–4800. [CrossRef]
104. Tong, R.; Cheng, J. Paclitaxel-initiated, controlled polymerization of lactide for the formulation of polymeric nanoparticulate delivery vehicles. *Angew. Chem. Int. Ed.* **2008**, *47*, 4830–4834. [CrossRef]
105. Satsangi, A.; Roy, S.S.; Satsangi, R.K.; Vadlamudi, R.K.; Ong, J.L. Design of a Paclitaxel Prodrug Conjugate for Active Targeting of an Enzyme Upregulated in Breast Cancer Cells. *Mol. Pharm.* **2014**, *11*, 1906–1918. [CrossRef] [PubMed]
106. Erez, R.; Segal, E.; Miller, K.; Satchi-Fainaro, R.; Shabat, D. Enhanced cytotoxicity of a polymer–drug conjugate with triple payload of paclitaxel. *Bioorg. Med. Chem.* **2009**, *17*, 4327–4335. [CrossRef] [PubMed]
107. Singer, J.W. Paclitaxel poliglumex (XYOTAX™, CT-2103): A macromolecular taxane. *J. Control. Release* **2005**, *109*, 120–126. [CrossRef] [PubMed]
108. Karmali, P.P.; Kotramaju, V.R.; Kastantin, M.; Black, M.; Missirlis, D.; Tirrell, M.; Ruoslahti, E. Targeting of albumin-embedded paclitaxel nanoparticles to tumors. *Nanomed. Nanotechnol. Biol. Med.* **2009**, *5*, 73–82. [CrossRef] [PubMed]
109. Shan, L.; Shan, X.; Zhang, T.; Zhai, K.; Gao, G.; Chen, X.; Gu, Y. Transferrin-conjugated paclitaxel prodrugs for targeted cancer therapy. *RSC Adv.* **2016**, *6*, 77987–77998. [CrossRef]
110. Zhang, P.; Cheetham, A.G.; Lin, Y.-A.; Cui, H. Self-Assembled Tat Nanofibers as Effective Drug Carrier and Transporter. *ACS Nano* **2013**, *7*, 5965–5977. [CrossRef]
111. Tian, R.; Wang, H.; Niu, R.; Ding, D. Drug delivery with nanospherical supramolecular cell penetrating peptide–taxol conjugates containing a high drug loading. *J. Colloid Interface Sci.* **2015**, *453*, 15–20. [CrossRef]
112. Bhattacharyya, J.; Bellucci, J.J.; Weitzhandler, I.; McDaniel, J.; Spasojevic, I.; Li, X.; Lin, C.-C.; Chi, J.-T.; Chilkoti, A. A paclitaxel-loaded recombinant polypeptide nanoparticle outperforms Abraxane in multiple murine cancer models. *Nat. Commun.* **2015**, *6*, 7939. [CrossRef]
113. Bradley, M.; Swindell, C.; Anthony, F.; Witman, P.; Devanesan, P.; Webb, N.; Baker, S.; Wolff, A.; Donehower, R. Tumor targeting by conjugation of DHA to paclitaxel. *J. Control. Release* **2001**, *74*, 233–236. [CrossRef]
114. Ke, X.-Y.; Zhao, B.-J.; Zhao, X.; Wang, Y.; Huang, Y.; Chen, X.-M.; Zhao, B.-X.; Zhao, S.-S.; Zhang, X.; Zhang, Q. The therapeutic efficacy of conjugated linoleic acid—Paclitaxel on glioma in the rat. *Biomaterials* **2010**, *31*, 5855–5864. [CrossRef] [PubMed]
115. Tam, T.Y.; Gao, J.; Kwon, G.S. Oligo (lactic acid) n-paclitaxel prodrugs for poly (ethylene glycol)-block-poly (lactic acid) micelles: Loading, release, and backbiting conversion for anticancer activity. *J. Am. Chem. Soc.* **2016**, *138*, 8674–8677. [CrossRef] [PubMed]
116. Su, H.; Koo, J.M.; Cui, H. One-component nanomedicine. *J. Control. Release* **2015**, *219*, 383–395. [CrossRef] [PubMed]
117. Ajaj, K.A.; Biniossek, M.L.; Kratz, F. Development of protein-binding bifunctional linkers for a new generation of dual-acting prodrugs. *Bioconjug. Chem.* **2009**, *20*, 390–396. [CrossRef]
118. Cheetham, A.G.; Zhang, P.; Lin, Y.-A.; Lin, R.; Cui, H. Synthesis and self-assembly of a mikto-arm star dual drug amphiphile containing both paclitaxel and camptothecin. *J. Mater. Chem. B* **2014**, *2*, 7316–7326. [CrossRef]
119. Tan, X.; Lu, X.; Jia, F.; Liu, X.; Sun, Y.; Logan, J.K.; Zhang, K. Blurring the Role of Oligonucleotides: Spherical Nucleic Acids as a Drug Delivery Vehicle. *J. Am. Chem. Soc.* **2016**, *138*, 10834–10837. [CrossRef]
120. Berg, J. *An Introduction to Interfaces of Colloids: The Bridge to Nanoscience*; World Scientific Publishing Co.: Singapore, 2009.
121. Paul, B.K.; Moulik, S.P. Microemulsions: An overview. *J. Dispers. Sci. Technol.* **1997**, *18*, 301–367. [CrossRef]
122. Lawrence, M.J.; Warisnicharoen, W. Recent Advances in Microemulsions as Drug Delivery Vehicles. In *Nanoparticulates Drug Carr*; World Scientific: Singapore, 2006; pp. 125–171.
123. Forgiarini, A.M.; Esquena, J.; Gonzalez, C.; Solans, C. Formation of Nano-emulsions by Low-Energy Emulsification Methods at Constant Temperature. *Langmuir* **2001**, *17*, 2076–2083. [CrossRef]
124. Kunieda, H.; Solans, C. *Nano-Emulsions: Where Macro-and Microemulsions Meet*; Imperial College Press: London, UK, 2001.
125. Ma, P.; Mumper, R.J. Paclitaxel nano-delivery systems: A comprehensive review. *J. Nanomed. Nano Technol.* **2013**, *4*, 1000164. [CrossRef]
126. Shakhwar, S.; Darwish, R.; Kamal, M.M.; Nazzal, S.; Pallerla, S.; Abu Fayyad, A. Development and evaluation of paclitaxel nanoemulsion for cancer therapy. *Pharm. Dev. Technol.* **2020**, *25*, 510–516. [CrossRef]
127. Narang, A.S.; Delmarre, D.; Gao, D. Stable drug encapsulation in micelles and microemulsions. *Int. J. Pharm.* **2007**, *345*, 9–25. [CrossRef] [PubMed]
128. Pouton, C.W. Formulation of self-emulsifying drug delivery systems. *Adv. Drug Deliv. Rev.* **1997**, *25*, 47–58. [CrossRef]

129. Veltkamp, S.A.; Thijssen, B.; Garrigue, J.S.; Lambert, G.; Lallemand, F.; Binlich, F.; Huitema, A.D.; Nuijen, B.; Nol, A.; Beijnen, J.H.; et al. A novel self-microemulsifying formulation of paclitaxel for oral administration to patients with advanced cancer. *Br. J. Cancer* **2006**, *95*, 729–734. [CrossRef]
130. Meher, J.G.; Dixit, S.; Pathan, D.K.; Singh, Y.; Chandasana, H.; Pawar, V.K.; Sharma, M.; Bhatta, R.S.; Konwar, R.; Kesharwani, P.; et al. Paclitaxel-loaded TPGS enriched self-emulsifying carrier causes apoptosis by modulating survivin ex-pression and inhibits tumour growth in syngeneic mammary tumours. *Artif. Cells Nanomed. Biotechnol.* **2018**, *46*, S344–S358. [CrossRef] [PubMed]
131. Park, E.-S.; Maniar, M.; Shah, J.C. Biodegradable polyanhydride devices of cefazolin sodium, bupivacaine, and taxol for local drug delivery: Preparation, and kinetics and mechanism of in vitro release. *J. Control. Release* **1998**, *52*, 179–189. [CrossRef]
132. Bode, C.; Kranz, H.; Siepmann, F.; Siepmann, J. In-situ forming PLGA implants for intraocular dexamethasone delivery. *Int. J. Pharm.* **2018**, *548*, 337–348. [CrossRef]
133. Kamali, H.; Khodaverdi, E.; Hadizadeh, F.; Mohajeri, S.A. In-vitro, ex-vivo, and in-vivo evaluation of buprenorphine HCl release from an in situ forming gel of PLGA-PEG-PLGA using N-methyl-2-pyrrolidone as solvent. *Mater. Sci. Eng. C* **2019**, *96*, 561–575. [CrossRef]
134. Samy, W.M.; I Ghoneim, A.; A Elgindy, N. Novel microstructured sildenafil dosage forms as wound healing promoters. *Expert Opin. Drug Deliv.* **2014**, *11*, 1525–1536. [CrossRef]
135. Kempe, S.; Mäder, K. In situ forming implants—An attractive formulation principle for parenteral depot formulations. *J. Control. Release* **2012**, *161*, 668–679. [CrossRef]
136. Packhaeuser, C.B.; Schnieders, J.; Oster, C.G.; Kissel, T. In situ forming parenteral drug delivery systems: An overview. *Eur. J. Pharm. Biopharm. Drug Dispos.* **2004**, *58*, 445–455. [CrossRef]
137. Amini-Fazl, M.S. Biodegradation study of PLGA as an injectable in situ depot-forming implant for controlled release of paclitaxel. *Polym. Bull.* **2021**, *3*, 1–14. [CrossRef]
138. Hollis, C.P.; Li, T. Nanocrystals Production, Characterization, and Application for Cancer Therapy. In *Pharmaceutical Sciences Encyclopedia: Drug Discovery, Development, Manufacturing*; Wiley & Sons, Inc.: New York, NY, USA, 2013; pp. 181–206.
139. Wong, J.; Brugger, A.; Khare, A.; Chaubal, M.; Papadopoulos, P.; Rabinow, B.; Kipp, J.; Ning, J. Suspensions for intravenous (IV) injection: A review of development, preclinical and clinical aspects. *Adv. Drug Deliv. Rev.* **2008**, *60*, 939–954. [CrossRef] [PubMed]
140. Rabinow, B.E. Nanosuspensions in drug delivery. *Nat. Rev. Drug Discov.* **2004**, *3*, 785–796. [CrossRef]
141. Müller, H.R.; Gohla, S.; Keck, C.M. State of the art of nanocrystals—Special features, production, nanotoxicology aspects and intracellular delivery. *Eur. J. Pharm. Biopharm. Drug Dispos.* **2011**, *78*, 1–9. [CrossRef] [PubMed]
142. Shegokar, R.; Müller, R.H. Nanocrystals: Industrially feasible multifunctional formulation technology for poorly soluble actives. *Int. J. Pharm.* **2010**, *399*, 129–139. [CrossRef] [PubMed]
143. Merisko-Liversidge, E.; Liversidge, G.G.; Cooper, E.R. Nanosizing: A formulation approach for poorly-water-soluble compounds. *Eur. J. Pharm. Sci.* **2003**, *18*, 113–120. [CrossRef]
144. Zhang, H.; Hollis, C.P.; Zhang, Q.; Li, T. Preparation and antitumor study of camptothecin nanocrystals. *Int. J. Pharm.* **2011**, *415*, 293–300. [CrossRef]
145. Wang, J.; Muhammad, N.; Li, T.; Wang, H.; Liu, Y.; Liu, B.; Zhan, H. Hyaluronic Acid-Coated Camptothecin Nanocrystals for Targeted Drug Delivery to Enhance Anticancer Efficacy. *Mol. Pharm.* **2020**, *17*, 2411–2425. [CrossRef]
146. Gao, L.; Liu, G.; Ma, J.; Wang, X.; Zhou, L.; Li, X.; Wang, F. Application of Drug Nanocrystal Technologies on Oral Drug Delivery of Poorly Soluble Drugs. *Pharm. Res.* **2013**, *30*, 307–324. [CrossRef]
147. Sinha, B.; Müller, R.; Möschwitzer, J.P. Bottom-up approaches for preparing drug nanosuspensions: Formulations and factors affecting particle size. *Int. J. Pharm.* **2013**, *453*, 126–141. [CrossRef]
148. Xia, D.; Quan, P.; Piao, H.; Piao, H.; Sun, S.; Yin, Y.; Cui, F. Preparation of stable nitrendipine nanosuspensions using the precipitation–ultrasonication method for enhancement of dissolution and oral bioavailability. *Eur. J. Pharm. Sci.* **2010**, *40*, 325–334. [CrossRef] [PubMed]
149. Hollis, C.P.; Li, T. Nanocrystals production, characterization, and application for cancer therapy. *Pharm. Sci. Encycl. Drug Discov. Dev. Manuf.* **2010**, 1–26.
150. De Castro, M.L.; Priego-Capote, F. Ultrasound-assisted crystallization (sonocrystallization). *Ultrason. Sonochem.* **2007**, *14*, 717–724. [CrossRef] [PubMed]
151. Kakran, M.; Sahoo, N.G.; Tan, I.-L.; Li, L. Preparation of nanoparticles of poorly water-soluble antioxidant curcumin by antisolvent precipitation methods. *J. Nanopart. Res.* **2012**, *14*, 757. [CrossRef]
152. Lonare, A.A.; Patel, S.R. Antisolvent crystallization of poorly water soluble drugs. *Int. J. Chem. Eng. Appl.* **2013**, *4*, 337. [CrossRef]
153. Pawar, N.; Agrawal, S.; Methekar, R. Modeling, Simulation, and Influence of Operational Parameters on Crystal Size and Morphology in Semibatch Antisolvent Crystallization of α-Lactose Monohydrate. *Cryst. Growth Des.* **2018**, *18*, 4511–4521. [CrossRef]
154. Crisp, J.; Dann, S.; Blatchford, C. Antisolvent crystallization of pharmaceutical excipients from aqueous solutions and the use of preferred orientation in phase identification by powder X-ray diffraction. *Eur. J. Pharm. Sci.* **2011**, *42*, 568–577. [CrossRef]
155. Sharma, C.; Desai, M.; Patel, S.R. Effect of surfactants and polymers on morphology and particle size of telmisartan in ultrasound-assisted anti-solvent crystallization. *Chem. Pap.* **2019**, *73*, 1685–1694. [CrossRef]
156. Miao, X.; Yang, W.; Feng, T.; Lin, J.; Huang, P. Drug nanocrystals for cancer therapy. *Nanomed. Nanobiotechnol.* **2018**, *10*, e1499. [CrossRef]

157. Zhou, M.; Zhang, X.; Yu, C.; Nan, X.; Chen, X.; Zhang, X.-H. Shape regulated anticancer activities and systematic toxicities of drug nanocrystals in vivo. *Nanomed. Nanotechnol. Biol. Med.* **2016**, *12*, 181–189. [CrossRef]
158. Lu, Y.; Wang, Z.-H.; Li, T.; McNally, H.; Park, K.; Sturek, M. Development and evaluation of transferrin-stabilized paclitaxel nanocrystal formulation. *J. Control. Release* **2014**, *176*, 76–85. [CrossRef] [PubMed]
159. Liu, J.; Tu, L.; Cheng, M.; Feng, J.; Jin, Y. Mechanisms for oral absorption enhancement of drugs by nanocrystals. *J. Drug Deliv. Sci. Technol.* **2020**, *56*, 101607. [CrossRef]
160. Park, J.; Sun, B.; Yeo, Y. Albumin-coated nanocrystals for carrier-free delivery of paclitaxel. *J. Control. Release* **2017**, *263*, 90–101. [CrossRef] [PubMed]
161. Park, J.; Park, J.E.; Hedrick, V.E.; Wood, K.V.; Bonham, C.; Lee, W.; Yeo, Y. A Comparative In Vivo Study of Albumin-Coated Paclitaxel Nanocrystals and Abraxane. *Small* **2018**, *14*, 1703670. [CrossRef] [PubMed]
162. Sohn, J.S.; Yoon, D.S.; Sohn, J.Y.; Park, J.S.; Choi, J.S. Development and evaluation of targeting ligands surface modified paclitaxel nanocrystals. *Mater. Sci. Eng. C* **2017**, *72*, 228–237. [CrossRef]
163. Sharma, S.; Singh, J.; Verma, A.; Teja, B.V.; Shukla, R.P.; Singh, S.K.; Sharma, V.; Konwar, R.; Mishra, P.R. Hyaluronic acid anchored paclitaxel nanocrystals improves chemotherapeutic efficacy and inhibits lung metastasis in tumor-bearing rat model. *RSC Adv.* **2016**, *6*, 73083–73095. [CrossRef]
164. Zhang, H.; Hu, H.; Zhang, H.; Dai, W.; Wang, X.; Wang, X.; Zhang, Q. Effects of PEGylated paclitaxel nanocrystals on breast cancer and its lung metastasis. *Nanoscale* **2015**, *7*, 10790–10800. [CrossRef]
165. Polomska, A.; Gauthier, M.A.; Leroux, J.C. In Vitro and In Vivo Evaluation of PEGylated Layer-by-Layer Polyelectro-lyte-Coated Paclitaxel Nanocrystals. *Small* **2017**, *13*, 1602066. [CrossRef]
166. Zhao, J.; Du, J.; Wang, J.; An, N.; Zhou, K.; Hu, X.; Dong, Z.; Liu, Y. Folic Acid and Poly(ethylene glycol) Decorated Paclitaxel Nanocrystals Exhibit Enhanced Stability and Breast Cancer-Targeting Capability. *ACS Appl. Mater. Interfaces* **2021**, *13*, 14577–14586. [CrossRef]
167. Wang, D.; Wang, Y.; Zhao, G.; Zhuang, J.; Wu, W. Improving systemic circulation of paclitaxel nanocrystals by surface hybridization of DSPE-PEG2000. *Colloids Surfaces B Biointerfaces* **2019**, *182*, 110337. [CrossRef]
168. Guo, F.; Shang, J.; Zhao, H.; Lai, K.; Li, Y.; Fan, Z.; Hou, Z.; Su, G. Cube-shaped theranostic paclitaxel prodrug nanocrystals with surface functionalization of SPC and MPEG-DSPE for imaging and chemotherapy. *Colloids Surfaces B Biointerfaces* **2017**, *160*, 649–660. [CrossRef] [PubMed]
169. Choi, J.-S.; Park, J.-S. Effects of paclitaxel nanocrystals surface charge on cell internalization. *Eur. J. Pharm. Sci.* **2016**, *93*, 90–96. [CrossRef] [PubMed]
170. Liu, H.; Ma, Y.; Liu, D.; Fallon, J.K.; Liu, F. The Effect of Surfactant on Paclitaxel Nanocrystals: An In Vitro and In Vivo Study. *J. Biomed. Nanotechnol.* **2016**, *12*, 147–153. [CrossRef] [PubMed]
171. Noh, J.-K.; Naeem, M.; Cao, J.; Lee, E.H.; Kim, M.-S.; Jung, Y.; Yoo, J.-W. Herceptin-functionalized pure paclitaxel nanocrystals for enhanced delivery to HER2-postive breast cancer cells. *Int. J. Pharm.* **2016**, *513*, 543–553. [CrossRef] [PubMed]
172. Gao, W.; Chen, Y.; Thompson, D.H.; Park, K.; Li, T. Impact of surfactant treatment of paclitaxel nanocrystals on biodistribution and tumor accumulation in tumor-bearing mice. *J. Control. Release* **2016**, *237*, 168–176. [CrossRef]
173. Han, X.; Su, R.; Huang, X.; Wang, Y.; Kuang, X.; Zhou, S.; Liu, H. Triphenylphosphonium-modified mitochondria-targeted paclitaxel nanocrystals for overcoming multidrug resistance. *Asian J. Pharm. Sci.* **2019**, *14*, 569–580. [CrossRef]
174. Mei, D.; Gong, L.; Zou, Y.; Yang, D.; Liu, H.; Liang, Y.; Sun, N.; Zhao, L.; Zhang, Q.; Lin, Z. Platelet membrane-cloaked paclitaxel-nanocrystals augment postoperative chemotherapeutical efficacy. *J. Control. Release* **2020**, *324*, 341–353. [CrossRef] [PubMed]
175. Huang, Z.-G.; Lv, F.-M.; Wang, J.; Cao, S.-J.; Liu, Z.-P.; Liu, Y.; Lu, W.-Y. RGD-modified PEGylated paclitaxel nanocrystals with enhanced stability and tumor-targeting capability. *Int. J. Pharm.* **2019**, *556*, 217–225. [CrossRef]
176. Wang, J.; Lv, F.M.; Wang, D.L.; Du, J.L.; Guo, H.Y.; Chen, H.N.; Zhao, S.J.; Liu, Z.P.; Liu, Y. Synergistic Antitumor Effects on Drug-Resistant Breast Cancer of Paclitaxel/Lapatinib Composite Nano-crystals. *Molecules* **2020**, *25*, 604. [CrossRef]
177. Sharma, S.; Verma, A.; Pandey, G.; Mittapelly, N.; Mishra, P.R. Investigating the role of Pluronic-g-Cationic polyelectrolyte as functional stabilizer for nanocrystals: Impact on Paclitaxel oral bioavailability and tumor growth. *Acta Biomater.* **2015**, *26*, 169–183. [CrossRef]
178. Sharma, S.; Verma, A.; Teja, B.V.; Shukla, P.; Mishra, P.R. Development of stabilized Paclitaxel nanocrystals: In-vitro and in-vivo efficacy studies. *Eur. J. Pharm. Sci.* **2015**, *69*, 51–60. [CrossRef] [PubMed]
179. Han, S.; Li, X.; Zhou, C.; Hu, X.; Zhou, Y.; Jin, Y.; Liu, Q.; Wang, L.; Li, X.; Liu, Y. Further Enhancement in Intestinal Absorption of Paclitaxel by Using Transferrin-Modified Paclitaxel Nano-crystals. *ACS Appl. Bio Mater.* **2020**, *3*, 4684–4695. [CrossRef] [PubMed]
180. Liu, R.; Chang, Y.-N.; Xing, G.; Li, M.; Zhao, Y. Study on orally delivered paclitaxel nanocrystals: Modification, characterization and activity in the gastrointestinal tract. *R. Soc. Open Sci.* **2017**, *4*, 170753. [CrossRef] [PubMed]
181. Lin, Z.; Xu, S.; Gao, W.; Hu, H.; Chen, M.; Wang, Y.; He, B.; Dai, W.; Zhang, H.; Wang, X.; et al. A comparative investigation between paclitaxel nanoparticle- and nanocrystal-loaded thermosensitive PECT hydrogels for peri-tumoural administration. *Nanoscale* **2016**, *8*, 18782–18791. [CrossRef]
182. Tiet, P.; Li, J.; Abidi, W.; Mooney, R.; Flores, L.; Aramburo, S.; Batalla-Covello, J.; Gonzaga, J.; Tsaturyan, L.; Kang, Y.; et al. Silica Coated Paclitaxel Nanocrystals Enable Neural Stem Cell Loading for Treatment of Ovarian Cancer. *Bioconjug. Chem.* **2019**, *30*, 1415–1424. [CrossRef]

183. Huang, X.; Shi, Q.; Du, S.; Lu, Y.; Han, N. Poly-tannic acid coated paclitaxel nanocrystals for combinational photothermal-chemotherapy. *Colloids Surfaces B Biointerfaces* **2021**, *197*, 111377. [CrossRef]
184. Lin, Z.; Mei, D.; Chen, M.; Wang, Y.; Chen, X.; Wang, Z.; He, B.; Zhang, H.; Wang, X.; Dai, W.; et al. A comparative study of thermo-sensitive hydrogels with water-insoluble paclitaxel in molecule, nanocrystal and microcrystal dispersions. *Nanoscale* **2015**, *7*, 14838–14847. [CrossRef]
185. Sun, B.; Taha, M.S.; Ramsey, B.; Torregrosa-Allen, S.; Elzey, B.D.; Yeo, Y. Intraperitoneal chemotherapy of ovarian cancer by hydrogel depot of paclitaxel nanocrystals. *J. Control. Release* **2016**, *235*, 91–98. [CrossRef]
186. Zhao, D.; Hu, C.; Fu, Q.; Lv, H. Combined chemotherapy for triple negative breast cancer treatment by paclitaxel and niclosamide nano-crystals loaded thermosensitive hydrogel. *Eur. J. Pharm. Sci.* **2021**, *167*, 105992. [CrossRef]

Review

Development of Polymer-Based Nanoformulations for Glioblastoma Brain Cancer Therapy and Diagnosis: An Update

Bijuli Rabha [1,†], Kaushik Kumar Bharadwaj [1,†], Siddhartha Pati [2,3,†], Bhabesh Kumar Choudhury [4], Tanmay Sarkar [5,6], Zulhisyam Abdul Kari [7], Hisham Atan Edinur [8], Debabrat Baishya [1,*] and Leonard Ionut Atanase [9,*]

1. Department of Bioengineering & Technology, GUIST, Gauhati University, Guwahati 781014, India; bijulipep@gmail.com (B.R.); kkbhrdwj01@gmail.com (K.K.B.)
2. Skills Innovation & Academic Network (SIAN) Institute-Association for Biodiversity Conservation and Research (ABC), Balasore 756001, India; patisiddhartha@gmail.com
3. NatNov Bioscience Private Limited, Balasore 756001, India
4. Department of Chemistry, Gauhati University, Guwahati 781014, India; bkcsat@gmail.com
5. Malda Polytechnic, West Bengal State Council of Technical Education, Govt. of West Bengal, Malda 732102, India; tanmays468@gmail.com
6. Department of Food Technology and Biochemical Engineering, Jadavpur University, Kolkata 700032, India
7. Faculty of Agro Based Industry, Universiti Malaysia Kelantan, Jeli 17600, Malaysia; zulhisyam.a@umk.edu.my
8. School of Health Sciences, Health Campus, Universiti Sains Malaysia, Kubang Kerian 16150, Malaysia; edinur@usm.my
9. Faculty of Medical Dentistry, "Apollonia" University of Iasi, 700511 Iasi, Romania
* Correspondence: drdbaishya@gmail.com (D.B.); leonard.atanase@yahoo.com (L.I.A.)
† Marked authors contributed equally.

Abstract: Brain cancers, mainly high-grade gliomas/glioblastoma, are characterized by uncontrolled proliferation and recurrence with an extremely poor prognosis. Despite various conventional treatment strategies, viz., resection, chemotherapy, and radiotherapy, the outcomes are still inefficient against glioblastoma. The blood–brain barrier is one of the major issues that affect the effective delivery of drugs to the brain for glioblastoma therapy. Various studies have been undergone in order to find novel therapeutic strategies for effective glioblastoma treatment. The advent of nanodiagnostics, i.e., imaging combined with therapies termed as nanotheranostics, can improve the therapeutic efficacy by determining the extent of tumour distribution prior to surgery as well as the response to a treatment regimen after surgery. Polymer nanoparticles gain tremendous attention due to their versatile nature for modification that allows precise targeting, diagnosis, and drug delivery to the brain with minimal adverse side effects. This review addresses the advancements of polymer nanoparticles in drug delivery, diagnosis, and therapy against brain cancer. The mechanisms of drug delivery to the brain of these systems and their future directions are also briefly discussed.

Keywords: polymer nanoparticles; glioma/glioblastoma; blood–brain barrier (BBB)/blood brain tumour barrier (BBTB); nanodiagnostics; drug delivery and imaging

1. Introduction

Cancer is one of the serious life-threatening diseases worldwide with a higher risk of mortality, around 10 million new cases are diagnosed every year [1,2]. Among different types of cancer, brain cancer is the most lethal and invasive type of central nervous system (CNS) disorder [3]. Brain cancer is characterised as a heterogeneous group of primary and metastatic cancers in the CNS [4,5]. The average incidence of both malignant and non-malignant brain cancer is reported approximately 28.57 per 100,000 population, mostly affecting 0 to 19 years, with a mean annual morbidity rate of 5.57 per 100,000 population [6,7]. Among these, the malignant primary brain cancers with a 5-year survival rate of less than 33.3–35% and even the rate are still alleviating. The average survival span is still not improved and even lower between 15 to 22 months [8,9]. A recent report from 2020 of the Central Brain Tumor Registry of the United States accounted for primary malignant tumour incidence rate to be 7.08 per 100,000, with 123,484 estimated cases, and 16.71 per 100,000, with 291,927 cases of non-malignant tumour [10]. Malignant primary tumours, i.e., gliomas derived from the glial origin, are newly diagnosed for approximately 70%, mostly in adults [5,11]. The reduced efficacy of brain cancer therapy is mainly attributed to the presence of the blood–brain barrier (BBB) that limits the permeation of systemically applied drugs into the brain [3].

Brain cancers are categorised into two groups, viz., primary brain cancer originated from the brain and resided within the brain, commonly called glioma, and secondary or metastatic brain cancer spreading from primary cancer outside the CNS, originate from systemic neoplasms and further evolved in the interior of brain parenchyma [12,13]. Glial cell originated gliomas include glioblastomas, astrocytomas, schwannomas, oligodendrogliomas, etc. [14]. According to World Health Organization (WHO), glioma tumours of CNS is classified into four grades based on aggressiveness, Grade I pilocytic astrocytoma, Grade II diffuse astrocytoma, Grade III anaplastic astrocytoma, and Grade IV glioblastoma [12]. Glioblastoma (GBM) and its variants were categorised as Grade IV tumours [15]. Grades I and II are considered low-grade glioma, and Grades III and IV are considered high-grade gliomas, i.e., malignant gliomas, and are characterised by poor prognosis [8,16,17]. GBM can either develop from normal brain cells or evolve from pre-existing low-grade astrocytoma [18]. GBM is also termed as glioblastoma multiforme or Grade IV astrocytoma [19]. Excessive penetration and vascular proliferation into brain parenchyma is the indication of aggressive cancer [20].

Conventional glioma therapy includes tumour resection followed by radiotherapy and chemotherapy. Surgical resection is generally considered a standard method for glioblastoma therapy. Yet resection of tumour tissue cannot be entirely removed and hence is limited by the glioblastoma's aggressiveness caused by penetration into surrounding tissue microenvironment and tumour vascularisation [20,21]. Hence, tumour resection is associated with the administration of chemotherapeutic drugs and/or radiation therapy for enhanced efficiency. Radiation therapy can be delivered internally or externally and is regarded as the standard treatment for high-grade gliomas [22]. Chemotherapy drugs such as carmustine (BCNU) can cross the BBB and target glioma cells directly [20]. Further, chemotherapy has undergone some alteration by replacing the use of some alkylating agents, viz., carmustine (BCNU), nimustine (ACNU), and lomustine (CCNU) with temozolomide (TMZ) [23]. Temozolomide is converted to 5-3-(methyl)-1-(triazen-1-yl) imidazole-4-carboxamide, at physiological pH, damages DNA via methylation of the O6-position of guanines, blocks DNA replication and induces tumour cell death. Presently, TMZ, along with surgical resection and radiotherapy, is applied for glioblastoma therapy [17]. Despite that, all the treatment strategies possess some limitations towards survival and thus, the prognosis still remains poor (Table 1).

Although brain cancer resembles to other forms of cancer in the body, the major difference is their intracranial neoplasms, heterogeneity, intricate brain system, and the physiological features of the cranial cavity which restrain the treatment options [10]. Gliomas tend to permeate the surrounding tissue microenvironment, and thereby, it is very difficult to determine the tumour boundaries. This also attributes to several difficulties in conventional therapeutic approaches for a curative outcome. Moreover, the physical and chemical barriers hamper therapeutic drug molecules from reaching tumour locations [11]. The BBB and blood–brain tumour barrier (BBTB) represent the diffusion barrier systems of the brain that regulate the influx of drugs to the brain except owing to certain characteristics [24]. Standard treatments remain ineffective due to poor surgical resection of tumours, mainly the infiltrative ones, poor chemo-therapeutic drug influx to the tumour site, and BBB that restrict them from diffusing toward tumour location [25]. The limitations of radiotherapy also result in incomplete eradication of GBM cells resulting in self-renewal and recurrence [26]. Targeting active anticancer agents to the brain is a challenging task in the area of drug delivery as BBB prevents the transportation of a drug. Hence, higher doses are needed to attain desired therapeutic efficacy which causes undesirable side effects [27].

Table 1. Advantages and limitations of conventional glioblastoma therapy.

Conventional Therapy	Advantage	Limitation
Resection	Local removal of a tumour	• Entire tumour cannot be removed • GBM cannot be fully cured, may relapse within 2 to 3 cm of the original tumour boundary • Invasive in nature
Radiotherapy	Standard treatment protocol for HGGs	• Necrosis of normal brain tissue • Neuronal damage • Resistance to radiation of tumour cells
Chemotherapy	Standard therapy for cancer, cytotoxicity	• High dose • BBB • Low accumulation of the drug • Tumour heterogeneity • Resistance to drug

2. The Blood–Brain Barrier (BBB)

One of the main hurdles for the effective systemic treatment of brain cancer is the presence of the BBB. The BBB is a semipermeable membrane barrier between blood capillaries and cellular components of brain tissues that control the movement of ions, nutrients, and cells. The BBB also serves for the dynamic transport of nutrients, peptides, proteins and immune cells between the brain and blood [28]. The BBB consists of endothelial cells, glial cells (pericytes, astrocytes, and neurons) and basement membrane [29] (Figure 1). The endothelial cells line the interior brain capillaries forming the tight junctions that allow small molecules, gases and curb the influx of harmful toxins or pathogens such as bacteria, lipophilic neurotoxins, xenobiotics and hydrophilic substances from the blood to the brain [30]. Due to the presence of pinocytic vesicles, other carriers, transport proteins, and large numbers of mitochondria, hydrophobic and essential molecules such as O_2, CO_2, glucose, hormones, etc. can infiltrate either by passive diffusion or active transport mechanisms [29]. The presence of several transmembrane proteins characterises the tight junctions between the inter-endothelial cells. These protein complexes are mainly comprised of occludin, claudin, and junctional adhesion molecules. These three specialised proteins interact to develop an intricate, tight barrier that is exclusive to the cerebro-endothelial cells [31]. The apical part of the endothelial cell is exposed to the brain's blood capillaries, and the basolateral part is exposed to the cerebrospinal fluid supported by the basement membrane. The basement membrane with 30–40 nm thickness consists of Type IV collagen, fibronectin, laminin, heparin sulfate proteoglycans and other extracellular matrix proteins that completely covers the endothelial cells and limits the movement of the solutes [29,31,32]. Approximately 98% of smaller molecular weight drugs and 100% of larger molecular weight drugs are reported for their inability to cross the intact BBB [33,34]. Under various brain-related pathological conditions, including brain cancers, glioma cells loose the structural integrity and the function of the BBB [35]. BBB is compromised in human glioma cells because of the leaky inter endothelial tight junction and poorly differentiated astrocytes that are unable to release essential components for BBB function [31,36]. In this case, it is termed as blood–brain tumour barrier (BBTB) or blood–tumour barrier (BTB) [14] (Figure 1).

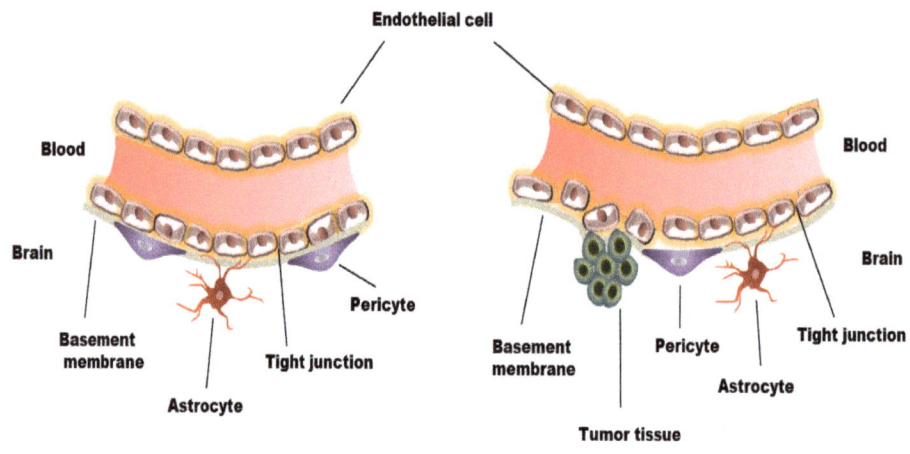

Figure 1. Schematic representation of blood–brain Barrier (BBB) and the blood–brain tumour barrier (BBTB).

In low-grade gliomas, the structure and function of the BBTB resemble normal BBB, while in hi-grade glioma, BBB is significantly altered, disrupted. Although the degree of BBB disruption varies from the tumour malignancy, low-grade glioma is still a hurdle to treat due to intact BBB. Despite high-grade gliomas, the structural disruption of their vascular density and integrity is negligible to drug permeability in tumour cells [37,38]. However, BBTB is more permeable than the BBB and allows heterogeneous permeability to drugs and other components. Thus, it is a more challenging task to combat the difficulties of brain cancer [39]. Therefore, along with the existing therapeutic regimen, new approaches are required to combat the BBB. To combat these difficulties, various techniques were developed, which are mostly invasive and cause serious side effects. Nanotechnology, especially use of polymer nanoparticles, helps address the major hurdle of glioma therapy non-invasively. Polymer nanoparticles aid in the targeting and delivery of potent drug molecules to the brain. In this review paper, we will briefly summarise the up-to-date existing therapies and diagnoses in brain cancer gliomas using polymer nanoparticles.

3. Polymer Nanoparticles for Drug Delivery Strategy to Overcome the BBB

The BBB is the main problem in the treatment of brain cancer glioma. The chemotherapeutic drugs are mostly ineffective due to limiting permeability to BBB as it allows to pass only low molecular weight (<500 Da), electrically neutral hydrophobic drugs with lipophilicity at log P 2–3 [11,36,40]. The majority of chemotherapeutic drugs are larger in size, ionic, hydrophilic molecules and thus cannot cross the BBB that is attributed to the requirement of a higher systemic dose that results in severe side effects [11]. To overcome these drawbacks, nanoparticles can be utilised for the controlled and sustained delivery of drugs. Biodegradable polymer nanoparticles are extensively studied systems in cancer drug delivery and therapy. These nanoparticles are also highly stable and can be tuned in order to obtain the desirable characteristics for a passive or an active targeting [41]. Polymer nanoparticles can induce selective toxicity and can load ample anticancer drugs or other molecules. Various biodegradable polymeric drug delivery systems include nanogels or hydrogels, poly(ε-caprolactone) (PCL), poly (lactic-co-glycolic acid) (PLGA), chitosan [42,43], dendrimers, etc. [44]. Due to versatile tuneable properties, these nanoparticles can open tight junctions of BBB, shield BBB limiting properties of anticancer drugs, release the drug in a sustainable manner, prolong the systemic circulation, and protect against enzymatic degradation [1,45].

Studies showed that Resveratrol loaded PLGA: D-α-tocopheryl polyethylene glycol 1000 succinate blend nanoparticles (RSV-PLGA-BNPs) displayed significant increasing cytotoxicity and enhanced cell penetration in C6 glioma cells. Haemocompatibility evaluation is one of the critical analyses of interaction between nanoparticles and various blood components that determine any adverse effect upon nanoparticle exposure to blood. The nanoparticles should not cause haemolysis during and after infusions. The haemocompatibility analysis of RSV-PLGA-BNPs revealed safe for i.v. administration. The nanoparticles exhibited prolonged systemic circulation up to 36 h. The nanoparticles also showed higher brain accumulation, suggesting a potential system for the betterment of systemic circulation and plasma half-life with a promising anticancer effect against glioma [1]. In another study, L-carnitine-conjugated PLGA NPs were developed to target glioma cells. These NPs were found to significantly cross the BBB and showed a potential anti-glioma effect [46]. Lactoferrin decorated PEG-PLGA NPs was developed for the delivery of shikonin and the treatment of gliomas [47]. Lactoferrin coating promotes internalisation across the BBB. In vitro and in vivo experiments showed the enhanced nanoparticle uptake and distribution of NPs in the brain with effective treatment of glioblastomas.

4. Polymer Nanoparticles for Anticancer Drug Delivery to the Brain: Mechanism

Polymer nanoparticles can cross BBB or BBTB either passively or via active endocytosis mechanisms. The unmodified polymer NPs internalise BBB mainly through passive mechanism, the so-called enhanced permeability and retention (EPR) effect, which depends on nanoparticle size. However, the NPS internalised by a passive mechanism have comparatively lower brain uptake than ligand-functionalised polymer NPs [48]. Various strategies have been undertaken to improve the infiltration of NPs into the brain. These strategies involve modification of NPs with certain moieties or components to take benefit of BBB endocytosis pathways for drug delivery. Polymer nanoparticles are able to cross BBB/BBTB through adsorption-mediated transcytosis (AMT), carrier-mediated transport (CMT), and receptor-mediated transcytosis (RMT) [49–52] (Figure 2). The internalisation of polymer nanoparticles crossing BBB/BBTB is summarized in Table 2. Polymer nanoparticles with positively charged can electrostatically interact with a negatively charged luminal surface that is attributed to cross the BBB/BBTB. The cationic polymer nanoparticles can be achieved by various surface modification strategies, either by coating or conjugation of cationic polymer or surfactant to non-ionic or neutral polymer. These modifications of NPs have been shown to utilise the AMT mechanisms to improve brain uptake. For example, a study of cationic bovine serum albumin (CBSA) conjugated with poly (ethylene glycol)–b-poly(lactide) (PEG–PLA) nanoparticles (CBSA–NPs), loaded with 6-coumarin was reported for brain delivery. Results revealed that CBSA–NPs uptake in rat brain capillary endothelial cells (BCECs) was enhanced as compared to control group BSA conjugated with pegylated nanoparticles (BSA–NP) BSA–NPs. Fluorescent microscopy of coronal brain sections displayed increased accumulation of CBSA–NPs than of BSA–NPs [53].

In the CMT mechanism, polymers NPs are designed to deliver drugs in order to take advantage of carrier molecules present in BBB. Polymer NPs are modified or decorated with membrane-penetrating components such as amino acids, peptides, and nutrients capable of transporting cargo across the BBB endothelial cells by utilising systemic transporters. For example, 2-deoxy-D-glucose modified poly (ethylene glycol)-co-poly (trimethylene carbonate) nanoparticles (DGlu-NPs) were studied for targeting the glioma BBB. The internalisation of DGlu-NP on RG-2 rat glioma cells was significantly higher than that of non-modified nanoparticles. This was attributed to the recognition of NPs by GLUT1 leading to enhanced cellular internalisation in glioma cells than in surrounding normal tissue and thus exhibiting promising in vivo anti-glioma activity [54].

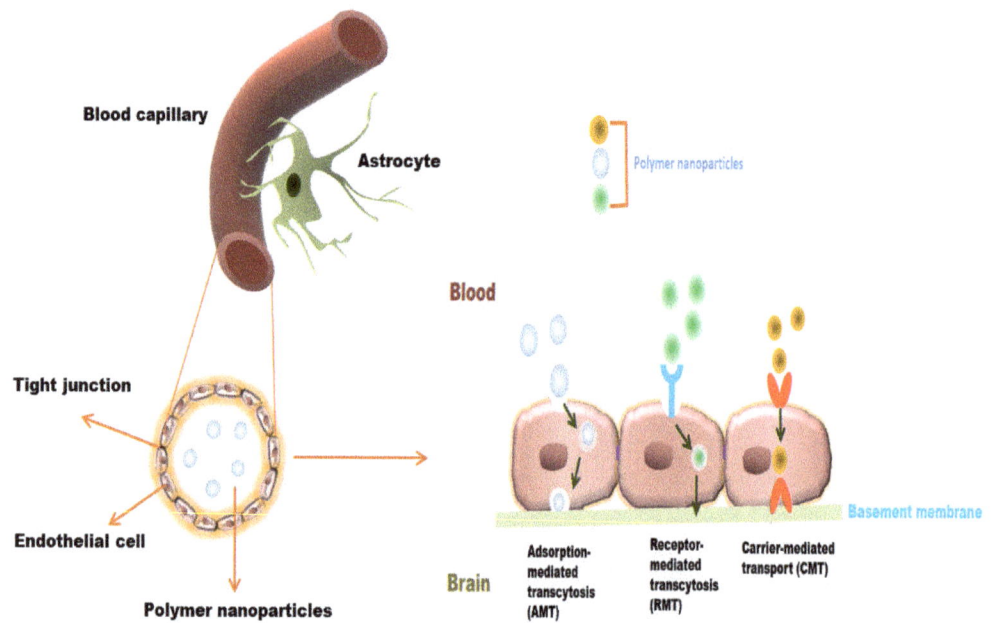

Figure 2. Various transport mechanisms of polymer NPs across blood–brain barrier (BBB).

Similarly, L-carnitine modified PLGA nanoparticles (LC-PLGA NPs) were designed to utilise the advantage of Na-coupled carnitine transporter 2 (OCTN2) expressions on brain capillary endothelial cells as glioma cells for BBB infiltration and targeting. Results showed increased accumulation of NPs in the BBB endothelial cell line (hCMEC/D3) and the glioma cell line (T98G). This revealed the Na dependent cellular uptake that involves OCTN2 in the NPs internalisation process. Moreover, a higher accumulation of LC-PLGA NPs was also observed in the in vivo mouse model study. Furthermore, loading of drugs Taxol and paclitaxel in the LC-PLGA NPs improved anti-glioma activity in both 2D-cell and 3D-spheroid models [46].

With the RMT mechanism, polymer NPs are decorated/designed with targeting ligands that bind to specific cell surface receptors highly expressed in BBB transport pathways. For example, the Transferrin receptor (TfR) is one of the primary targets for investigating RMT across the BBB because of its high expression on BBB/BBTB endothelium [55]. To evaluate in vivo BBB penetration and targeting efficacy, transferrin modified doxorubicin (DOX) and paclitaxel (PTX) loaded magnetic silica PLGA nanoparticles (MNP-MSN-PLGA-Tf NPs) were developed. The nanoparticles were effectively accumulated in the tumour bearing mice suggesting that Tf facilitates NPs delivery across BBB [56].

Table 2. Summary of BBB permeability based on polymer-based nanoparticles.

Polymer Nanoparticles	Cargo	Internalisation Mechanism	Cell Line/Animal Model	Remarks	References
Trimethylated chitosan (TMC)-modified PLGA NPs	Coenzyme Q10 6-coumarin	AMT	SH-SY5Y cells, AD transgenic mouse brains	Increase uptake of PLGA nanoparticles, neuroprotective effects of Q10 observed in TMC-PLGA NPs than PLGA-NP.	[57]
Angiopep-2 modified PLGA NPs	Doxorubicin (DOX), Epidermal growth factor receptor (EGFR) siRNA	RMT	U87MG cells, brain orthotopic U87MG glioma xenograft model	Improved DOX and siRNA cellular uptake, NPs able to cross BBB.	[58]

Table 2. Cont.

Polymer Nanoparticles	Cargo	Internalisation Mechanism	Cell Line/Animal Model	Remarks	References
Lactoferrin, folic acid modified PLGA NPs	Etoposide	RMT	HBMEC/HA monolayer, U87MG cells	PLGA NPs cross BBB and enhanced 2-fold uptake with Lf-and FA.	[59]
RVG29 modified PLGA NPs	Docetaxel	RMT	C6 cells, bEnd3 monolayer BBB model	Better BBB penetration in vitro.	[60]
OX26 Mab modified PLGA NPs	Temozolomide (TMZ)	RMT	U215 and U87, in vitro HBLECs monolayer model	Improved TMZ internalisation in glioblastoma cells.	[61]
T7- modified, magnetic PLGA nanoparticulate system (MNP/T7-PLGA NPs)	paclitaxel (PTX) and curcumin (CUR)	RMT	U87 cells and mouse brain endothelial cell line bEnd.3., mice bearing orthotopic glioma (U87-Luc)	>10-fold increase in cellular uptake studies and a >5-fold enhancement in brain delivery compared to the non-functionalized NPs.	[62]
Angiopep conjugated PEG-PCL nanoparticles (ANG-PEG-NP)	paclitaxel (PTX)	RMT (LRP-mediated transcytosis)	U87 MG, Male BALB/c nude mice and ICR mice	The penetration, distribution, and accumulation into 3D glioma spheroid and in vivo glioma region of ANG-PEG-NP was higher than that of plain PEG-PCL nanoparticles (PEG-NP).	[63]
dCatAlb encrusted DOX-loaded PLGA nanoparticle	Doxorubicin (DOX)	AMT	monolayer bEnd.3 cells	Enhanced BBB permeation	[64]
cRGD/PEG-SS-PCL micelles	Doxorubicin (DOX)	RMT	U87MG glioma xenografts	Efficient accumulation	[65]
DOX-loaded cRGD-SS-NGs	Doxorubicin (DOX)	RMT	U87-MG cells, U87-MG glioblastoma xenograft in nude mice	Facilitated cellular uptake and intracellular DOX release	[66]
T7–PEG–PLGA micelles	Carmustine (BCNU)	RMT	U87-MG cells, BALB/c nude mice	Accumulation in tumour more efficiently than unconjugated one	[67]
PLGA based SSTR2 pep-DIM-NPs	3,3′-diindolylmethane	RMT	C6 glioma cells, rat Glioma model	Accumulation of the NPs into rat brain tumour sites by crossing the BBB	[68]
L-carnitine modified PLGA nanoparticles (LC-PLGA NPs)	Taxol and paclitaxel (PTX)	CMT	hCMEC/D3, U98G cells	Efficient accumulation	[16]

Abbreviation: Adsorption-mediated transcytosis (AMT), carrier-mediated transport (CMT), and receptor-mediated transcytosis (RMT).

5. Polymer Nanoparticles for Brain Cancer Therapy

Polymer NPs are solid colloidal particles that can be utilised as carriers in which the therapeutic drugs or other active components are dissolved, entrapped, encapsulated, or adsorbed on the surface of the polymer matrix [69]. The structure of the polymer NPs can range from nanospheres to nanocapsules depending on the preparation procedure. Various polymers such as chitosan, gelatin, sodium alginate, albumin and polylactides (PLA), polyglycolides (PGA), poly(lactide co-glycolides) (PLGA), polyanhydrides, polyorthoesters, polycyanoacrylates, poly(ε-caprolactone), poly(glutamic acid), poly(malic acid), poly(N-vinyl pyrrolidone), poly(methyl methacrylate), poly(vinyl alcohol), poly(acrylic acid), poly(acrylamide), poly(ethylene glycol), poly(methacrylic acid) are mostly used in nanoparticle formation for both passive and ligand- functionalized actively targeted therapy [70]. Based on the nature of drugs to be loaded and their route of administration, different synthesis methods were implemented for the production of polymer NPs that include solvent evaporation, solvent diffusion, nanoprecipitation, emulsification, reverse salting out, nano-capsules nano-precipitation, layer-by-layer (LbL) method, etc. [71–73]. The molecular weight, crystallinity, and stability of polymers and the drug's physicochemical properties can be analysed to develop polymeric NPs for drug administration to the brain. Polymeric NPs have a unique ability to reach the tumour site through an active targeting route [74]. Researchers have developed docetaxel (DOC)-loaded PCL and its derivative poly (ethylene glycol)-block-poly(ε–caprolactone) methyl ether (mePEG-PCL) nanoparticles that were dispersed in a bioadhesive film and the formulation exhibited sustained release of drugs. Docetaxel-loaded nanoparticles induced more significant cytotoxicity than free docetaxel for glioma treatment [75]. The study reveals that glycopeptide-engineered poly(d,l-lactide-co-glycolide) (PLGA) NPs (g7-NPs) provides in vivo evidence of endocytosis of g7-NPs and transported into the endosomes, which help to cross BBB [76]. Gaudin and co-workers have demonstrated the use of convection-enhanced delivery (CED) of NPs for improved chemotherapeutic drugs to the tumour site. They successfully administered gemcitabine, a nucleoside analogue used for the wide range of solid tumours using squalene-based NPs. The study also revealed that PEGylation of the NPs with PEG dramatically improves the distribution of squalene-gemcitabine NPs in the tumours [77].

Most of the current nanomedicines approved by the FDA for clinical use for solid tumour treatment depend on the EPR effect. The enhanced permeability and retention effect or EPR effect is a feature that allows small sized nanoparticles and other active molecules or drugs to pass due to large pore size through leaky vasculature and accumulate in the tumor location.

The brain endothelial cells and glioblastoma cells generally overexpressed a number of receptors, including the low-density lipoprotein receptor, IL-13 receptor, transferrin receptor (TfR), and nicotine acetylcholine receptor that used as drug delivery targets in the brain [78]. Numerous in vivo studies revealed that polymer NPs could circulate for a longer time and accumulate in the tumour site. It is possible to enhance the retention and accumulation of these useful NPs by decorating NPs with tumour-homing ligands such as peptide, aptamer, polysaccharides, saccharides, antibodies, flic acids, etc. [79]. Recently, pluronic micelles (PEG-PPG-PEG) have evolved as perfect candidates for brain therapy, as they can easily cross the BBB and prove their ability to inhibit drug efflux [17]. For instance, Sun et al. developed TfR-T12 peptide-modified PEG-PLA polymer nanoparticle micelles loaded with paclitaxel (PTX) for glioma therapy. They found that the polymeric micelles (TfR-T12-PMs) could be absorbed by tumour cells, cross across BBB monolayers, and inhibit the proliferation of U87MG cells in vitro. A better antiglioma effect with a prolonged median survival of nude mice-bearing glioma was also observed in comparison with unmodified PMs [80]. This suggests that TfR-T12 peptide-modified micelles can cross the BBB system and target glioma cells. In Tables 3 and 4 are shown various synthetic and natural polymer-based NPs for GBMs therapy or diagnosis.

Recently, much research has been carried out on combined photo-based therapy along with other conventional therapy or imaging for glioblastoma treatment. For example, a novel photoacoustic and photothermal guided semiconducting polymer nanoparticles (SPNs) using poly (ethylene glycol)-block-poly (propylene glycol)-block-poly (ethylene glycol) (PEG-b-PPG-b-PEG) and SP were reported. The SPNs displayed efficient cellular internalisation for PAI and PTT toward U87 cells and accumulated in subcutaneous as well as brain tumours upon intravenous injection and induced efficient cell death upon NIR-II light irradiation [81].

The recent updates reveal that conjugated polymer nanoparticles (CPNs) are performing well as photosensitiser (PS) in photodynamic therapy (PDT). This efficiency is achieved by CPNs due to their uniform size, biocompatibility, and outstanding ROS production due to extraordinary photo-physical properties as well as fluorescence emission. It is found that porphyrin doped CPNs can eliminate GBMs through ROS-induced apoptotic damage [82].

Table 3. Synthetic polymer-based nanoparticles for brain cancer glioma therapy.

Polymers	Method of Preparation	Therapeutic Drug/Other	Targeting Receptor/Molecule	Diagnostic Component	Cell Line/Animal Model	Remark	References
Synthetic protein nanoparticle (SPNP)	Electrohydrodynamic (EHD) jetting	siRNA	STAT3i	Alexa Fluor 647-labeled albumin	GL26 syngeneic mouse glioma model	Five-fold increase in iRGD loaded SPNP in glioma cell observed in comparison to NPs without iRGD. A total of 87.5% of mice developed anti-GBM immunological memory.	[83]
Porphyrin doped conjugated polymer nanoparticles (CPNs)	Controlled nanoaggregation	-	m-RNA	DCF-DA	U-87 MG, T98G and MO59K	NPs enhance the efficacy of PDT to eliminate tumor via ROS generation. MO59K and U-87 MG cells are died with CPN having IC50 values 8 mg/L and 9 mg/L, however, T98G cells are found resistant to CNP-PDT.	[84]
PLGA	Single-emulsion, solvent evaporation technique	Paclitaxel	-	-	U87MG with rats and pigs' model	Enhanced in vivo efficacy	[85]
PBAEs	Step-wise synthesis	DNA	-	Cy3 dye	BTICs from patient	More than 60% transfection efficacy is observed.	[86]
cRGD-conjugated PGNRs	Ligand exchange method	-	$\alpha_v \beta_v$-integrin	-	U87MG	cRGD-PGNRs is proved having excellent tumor targeting ability, no cytotoxicity, and sufficient cellular uptake.	[87]
Aptamer/gold nanorod conjugate	Step-wise synthesis	Sgc8 aptamer	Cell protein	Fluorescein	Rat or mouse model	A total of 99.09% binding affinity due to the aptamers. Complete destruction of GMB on exposure to LAER is observed.	[88]
Poly(N-isopropylacrylamide)-based nanogels and magnetic NPs composite	Co-polymerisation and co-evaporation	Ferrrofluid	-	Sodium fluorescein	Rat model	The drug dose delivered to tumor site is directly proportional to the duration of the "on" pulse.	[89]
PEG–PBAE/ ePBAE nanoparticles (NPs)	Step wise synthesis, Michael addition	Plasmid DNA, pHSV-tk, ganciclovir	-	Hoechst 33342 dye	GBM1A and BTIC375 cells/Mice model	PEG–PBAE/ ePBAE NP shows 54 and 82% transfection efficacies in GBM1A and BTIC375 cells while it is 37 and 66% for optimised PBAE NPs without PEG. Death of cancer cell with enhancement of mice life time was observed.	[90]

Table 3. Cont.

Polymers	Method of Preparation	Therapeutic Drug/Other	Targeting Receptor/Molecule	Diagnostic Component	Cell Line/Animal Model	Remark	References
TEB	Co-precipitation	-	Transferrin (TfR), lactoferrin (LfR) and lipoprotein (LRP)	-	bEnd.3/Mouse model	Ligand-coated TEB nanoparticles are transported across BBB with high efficacy.	[91]
PEG-PLA	Emulsion/solvent evaporation technique		Neuropilin (NRP), tLyp-1 peptide		Human umbilical vein endothelial cells and Rat C6 glioma cells	tLyp-1 peptide functionalised NPs show better performance in paclitaxel glioma therapy. Observed inhibition of avascular C6 glioma spheroids. Interestingly tLyp-1-NP-PTX formulations shows higher antiproliferation ability with IC50 0.087 mg/mL in comparison to NP-PTX and Taxol.	[92]
Transferrin modified PEG-PLA	Double emulsion and solvent evaporation method.	Resveratrol (RSV)	-	-	C6 and U87 glioma cells	RSV-conjugates decreased brain tumor volume and accumulated well in comparison to free RSV.	[93]
Polysorbate-coated NPs	Surfactant mediated ultrasonication	Doxorubicin (DOX)	-	Evans Blue solution	Glioblastoma 101/8-bearing rats	Enhanced permeability and retention effect	[94]
PCL	Solvent evaporation technique	Irinotecan hydrochloride trihydrate (IRH)	-	-	HGG cells	IRH-loaded PCL NPs has excellent anti-brain tumor activity. PCL shows better drug encapsulation than PLGA.	[95]
cRGD-directed AuNR/PEG–PCL hybrid NPs	Nanoprecipitation	Doxorubicin (DOX)		Cy7	Human U87MG glioma	Controlled release of doxorubicin into human glioblastoma using mice model is achieved that leads to inhibition of 100 % tumour growth.	[96]
PCL-Diol-b-PU/gold nanofiber composite		Temozolomide (TMZ)			U-87 MG human glioblastoma cells	Slower release of TMZ showing its high potential as implantable device for drug release. Enhanced activity against the U-87 cell.	[97]
PEG-PCL NPs conjugated with ALMWP	Emulsion/solvent evaporation method	Paclitaxel (PTX), Taxol	-	coumarin-6	C6 cells	Animals treated for C6 gliomas with ALMWP-NP-PTX survive longer than those treated with Taxol-NP-PTX.	[98]

Abbreviation: PLG: poly(lactide-coglycolide), DCF-DA: 2′,7′-dichlorofluorescin-diacetate PBAEs: poly (β-amino ester) s, cRGD: cyclic RGD peptides, PGNRs: PEGylated gold nanorods, PEG: polyethylene glycol, PSMA: prostate-specific membrane antigen, NR: nanorods, PCL-diol: poly (ε-caprolactone diol), PU: polyurethane, ALMWP: activatable low molecular weight protamine.

Table 4. Natural polymer-based nanoparticles for brain cancer glioma therapy.

Natural Polymer-Based Nanoparticles	Method of Preparation	Therapeutic Drug/Other	Targeting Receptor/Molecule	Diagnostic Component	Cell Line/Animal Model	Remark	References
Den-angio nanoprobe	Step-wise synthesis	-	LRP receptor-mediated endocytosis		U87MG	Den-Angio shows localisation in the brain tumours and makes image-guided tumour resection possible.	[99]
CDP-NP	Single-step synthesis at room temperature, self-assembly method	-	Proteins	e-GFP, luciferin	BV2, N9 microglia (MG) cells and GL261 glioma cells/mice model	CDP-NPs were efficiently taken up by BV2 and N9 microglia (MG) cells compared to GL261 glioma cells.	[100]
Silver NPs impregnated alginate–chitosan-blended nanocarrier	Polyelectrolyte complex formation reaction		DNA	Acridine Orange/Ethidium Bromide dual stain	U87MG	Extensive DNA damage was observed on cell cycle analysis.	[101]
Hyaluronan (HA)-grafted lipid-based NPs (LNPs)	Amine coupling strategy	rRNA interference (RNAi), doxorubicin and BCNU	CD44 receptor	DAPI (blue)	T98G, U87MG, and U251	Prolonged survival of treated mice in the orthotopic model was observed.	[102]
Cardamom extract-loaded gelatine NPs (CE-loaded GNPs)	Two-step de-solvation method	Cardamom extract	-	-	U87MG	Extract to polymer ratio as 1:20 was found to be the best with entrapment. efficiency close to 70%	[103]
NK@AIEdots (natural-killer-cell-mimic nanorobots with aggregation-induced emission)	Step-wise synthesis, assembly process	-	-	-	U-87 MG, bEnd.3	The tumour growth was also successfully inhibited by NK@AIEdots on exposure to NIR light.	[104]
Heparin-based polymer)–SWL–(cRGD) NPs (S = serine, W = tryptophan, L = leucine)	Coupling reaction		$\alpha_v \beta_v$ and EphA2 in glioma	f Oregon-green488	U87 and U251	NPs easily pass-through BBB to the tumour site. In addition, inhibition of glioma cell proliferation is noticed.	[105]
poly-L-arginine-chitosan-triphosphate matrix (ACSD)	Green co-precipitation method	Doxorubicin, SPIONs	-	Prussian blue staining and inductively coupled plasma	Rat glioma C6 cells	ACSD NPs are proved as promising theranostic formulation MRI analysis shows uptake of NPs in C6 glioma cells. There observed 38.6% drug release in neutral pH while 58% in acidic pH. A 44-fold increase in IC_{50} value of doxorubicin was found when the drug was loaded in NPs.	[106]

Table 4. Cont.

Natural Polymer-Based Nanoparticles	Method of Preparation	Therapeutic Drug/Other	Targeting Receptor/Molecule	Diagnostic Component	Cell Line/Animal Model	Remark	References
Albumin nanoparticles (NPs)	Two-step synthesis, grafting	Paclitaxel (PTX)	Substance P (SP) peptide	Cou-6 dye	Glioma U87 cells	Albumin nanoparticles are found satisfactory for drug delivery vehicles for the treatment of GBM. The targeting effect of SP, and efficient cellular uptake of SP-HSA-PTX NPs into brain capillary endothelial cells (BCECs) and U87 cells is improved.	[107]
Human serum albumin (HSA) NPs	High-pressure homogeniser technique	Doxorubicin	-	LysoTracker	bEnd.3 cells as well as U87MG	Anti-glioma efficacy is improved due to the dual-enhanced system of dual cationic absorptive transcytosis and glucose-transport by using c- and m-HSA together.	[108]
Albumin NPs	Green synthesis	Paclitaxel and fenretinide	-	CY5 dye	Human glioma U87, U251 cells, mouse glioma C6, GL261 cells,	The albumin-binding proteins are found to be overexpressed in the tumour/glioma cells, where epithelium cells are responsible for delivering NPs to brain tumours.	[109]
Menthol-modified casein NPs(M-CA-NP)	Self-assembled micelle formation	10-Hydroxycamptothecin,-methanol		Cou-6	C6 cells	Resulted in enhanced drug accumulation in the tumour site.	[110]
Transferrin-functionalised NPs (Tf-NP)	Functionalisation	Temozolomide and the bromodomain		Cy5.5	U87MG and GL261 cells	Therapy showed 1.5- to 2-fold decrease in tumor burden and corresponding increase in survival in tumor bearing mice	[111]

Abbreviation: Den: dendrimer, Angio: angiopep-2, PDT: photodynamic therapy, CDP-NP: cyclodextrin-based nanoparticle, TEB: triphenylamine-4-vinyl- (P-methoxy-benzene), DAPI: 4′,6-diamidino-2-phenylindole, SPIONs: superparamagnetic iron oxide nanoparticles.

6. Polymer Nanoparticles in the Diagnosis of Brain Cancer

Before surgery, a high-resolution image using imaging modalities is required for glioma detection. Owing to the invasiveness of glioma cells, determining the exact tumour boundary by eye is challenging. Proper imaging of a tumour is essential for assessing the extent of tumour distribution before surgery and the response to a treatment regimen after surgery [5]. Several available techniques for visualisation and diagnosis of brain cancer glioma include optical and ultrasound (US) imaging, photoacoustic (PA) imaging, computed tomography (CT), positron emission tomography (PET), single-photon emission computed tomography (SPECT) and fluorescence (FL) imaging techniques (Figure 3) [112]. Currently, magnetic resonance imaging (MRI), a non-invasive technique that can detect the size, shape, and tumour location, is initially employed diagnostic method for patients with suspected GBM [113]. MRI can determine the boundaries of the tumour tissues and/or intraoperative to elucidate tumour outline during surgical resection by applying gadolinium (Gd). Due to a shorter half-life, Gd must be administered often to maintain blood levels for efficient scanning. The use of intraoperative ultrasonography to obtain integrated brain tissue imaging is another non-optical method. However, this approach does not provide enough information for detecting smaller or superficial brain tumours. Other invasive techniques for analysing brain tumour tissues include Raman spectroscopy, optical coherence tomography, fluorescence spectroscopy, and thermal imaging [114]. Computed tomography (CT) can also be used to determine the presence of the tumour. Still, its use is relatively lesser in clinics for diagnosing GBM due to poor resolution compared to MRI [115]. Likewise, positron emission tomography (PET) imaging with 11C-methionine could be an effective diagnostic tool for GBM patients' prognosis [116,117]. To understand cancer tumours, precise preoperative imaging and painless sensitive post-imaging techniques to provide real-time data are demanded. Current imaging modalities, however, lack accuracy, sensitivity, and specificity. Nanotechnology has sparked interest in bioimaging and biosensing in recent years.

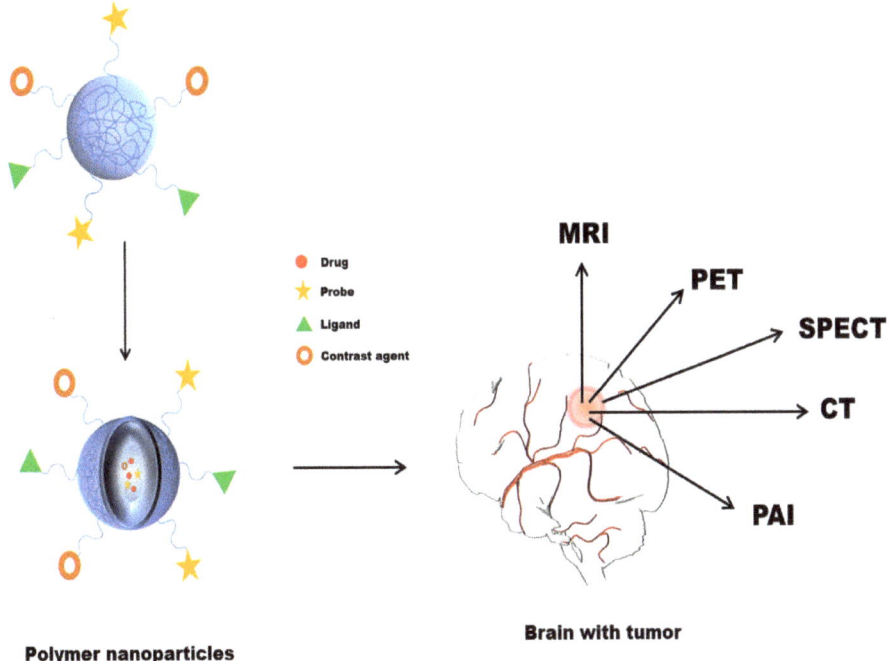

Figure 3. Polymer NPs in imaging for improved diagnosis of brain cancer.

'Nanodiagnostics' combined with nanotechnology could provide a drug delivery system with traditional diagnostic and imaging procedures [118,119]. Nanotechnology has made it easier to acquire data with great precision and accuracy while avoiding invasive procedures. NPs with tunable optical, magnetic, and electrical properties are able to provide diagnostic tools for detection and imaging brain cancer/tumours [120]. Biocompatible NPs owing ideal physical characteristics, such as surface chemistry, morphology, solubility, stability, etc., facilitate drug delivery and imaging as it acts as image contrast agents [121]. Polymer NPs could be a good reservoir system for drugs and a platform for additional modification for efficient tumour targeting or imaging [122]. Polymer NPs possess various advantages in drug delivery to the brain that can entrap or carry drugs that prevent them from metabolism and excretion. Moreover, NPs can easily transport drugs across the BBB without changing the barrier properties [31,123,124]. In this section, polymer NPs utilised in the diagnosis and detection of brain cancer glioma until now are primarily focused. The imaging and diagnosis techniques currently being investigated with reference to polymer NPs are listed in Table 5.

Polymer-based superparamagnetic NPs have mainly been employed as drug delivery systems and contrast agents in MRI imaging. These NPs are highly stable and biocompatible, can prolong systemic circulation time, have drug loading ability and control of drug release, and combine with their magnetic performance for MRI [125]. Ganipineni et al. synthesised paclitaxel (PTX) and superparamagnetic iron oxide (SPIO)-loaded PEGylated PLGA-based NPs (PTX/SPIONPs) and analysed for therapeutic efficacy in an orthotopic U87MG model. The cellular internalisation of these NPs was found to be concentration dependent. The MRI scanning displayed the blood–brain barrier disruption in the glioma affected location. Moreover, enhanced accumulation was also observed in ex vivo biodistribution analysis of GBM-bearing mice with magnetic targeting [126]. Researchers have evaluated SPIO-loaded brain penetrating PLGA NPs by CED administration on rat models and visualised using positron emission tomography (PET) and MRI [127]. SPIO-loaded NPs showed excellent transverse (T2) relaxivity. After CED of NPs, the biodistribution in the brain was analysed using MRI, which revealed a period of one month longer signal attenuation of SPIO-loaded brain-penetrating PLGA NPs. The co-administration of SPIO-loaded PLGA NPs allows intraoperative monitoring of biodistribution in the brain in order to ensure the delivery to tumour location and therapeutic effect over time [127]. Researchers have developed Polysorbate 80 coated temozolomide-loaded PLGA-based superparamagnetic nanoparticles (P80- TMZ/SPIO-NPs), evaluated for anti-glioma activity and analysed as a diagnostic agent for MRI [128]. The superparamagnetic P80-TMZ/SPIO-NPs showed a significant antiproliferative effect and remarkable cellular internalisation on C6 glioma cells. Moreover, the in vitro MRI scanning revealed that P80-TMZ/SPIO-NPs could also serve as a good contrast agent [128].

Table 5. Polymer nanoparticles in imaging and diagnosis of brain cancer therapy.

Nanoparticles	Detection Method	Cell Line	Animal Model	Therapy/Drug	References
SPIONs and DOX loaded poly-l-arginine-chitosan-triphosphate matrix (ACSD) NPs	MRI	C6 glioma cells	-	DOX	[106]
P80- TMZ/SPIO-NPs (PLGA coating)	MRI	C6 glioma cells	-	TMZ	[128]
Micelles SPION and Au NPs (PEG-PCL coating)	MRI, CT	-	U251 xenograft and orthotopic brain tumour models.	Radiotherapy	[129]
Chitosan-dextran superparamagnetic NPs (CS-DX-SPIONs)	MRI	C6 glioma, U87	orthotopic C6 gliomas in rats	-	[130]
DOX-Ps@80-SPIONs	MRI	glioblastoma C6 cells	Glioma-bearing rats	DOX	[131]
Paclitaxel (PTX) and superparamagnetic iron oxide (SPIO)-loaded PEGylated poly (lactic-co-glycolic acid) (PLGA)-based NPs(PTX/SPIONPs)	MRI	-	orthotopic U87MG model	PTX	[126]
SPIO-loaded brain penetrating PLGA NPs	PET, MRI	-	rat model	-	[127]
[18F] NPB4-labeled and C6-loaded PLGA NPs	PET	-	rats bearing BCSC-derived xenografts	-	[85]
TMZ and iron oxide-containing polymer NPs(PMNPs)	MRI	U87 glioma cells	rodent model	TMZ	[132]

Abbreviation: N-(4-[18F] fluorobenzyl) propanamido-PEG$_4$-Biotin, brain cancer stem cells (BCSCs).

7. Limitations and Challenges

From the past times, tremendous developments have been evidenced in brain cancer therapy. Yet, there have not been emerged significant changes in mortality rate and improving patients' quality of life. Although nanoparticle-based drug delivery systems have brought a new horizon, many challenges remain and need to be solved in the future. The development of effective polymeric NPs for drug delivery and targeting is a challenging task for clinical translations. The advantage and limitations are summarised in Figure 4. The toxicity of these systems is one of the main challenges. The slow degradation rate of polymer NPs induce a longer circulation time in the body and could cause unknown complications.

Further, extensive investigations are required for optimization of the NPs. One of the major obstacles in clinical translation is the interaction of NPs and biological systems. Upon entering the complicated biological system, the designed polymeric NPs will instantly interact with neighboring biomolecules, leading to the formation of protein corona that alters their properties. This affects NPs size, stability, surface properties and determines the pharmacokinetics, biodistribution, cellular internalisation, intracellular trafficking, immune system, and toxicity [133–136]. In addition, more in vitro and in vivo studies are required to better understand the mechanisms in targeted nanoparticle-based therapy. Several essential factors related to the in vivo behaviour of NPs and their effect on other healthy brain cells are hence required to be extensively examined. Currently, there is still insufficient pre-clinical data of polymer-based NPs on brain delivery, data to correlate in vitro-in vivo observation, which makes it difficult to conclude about their therapeutic efficacy.

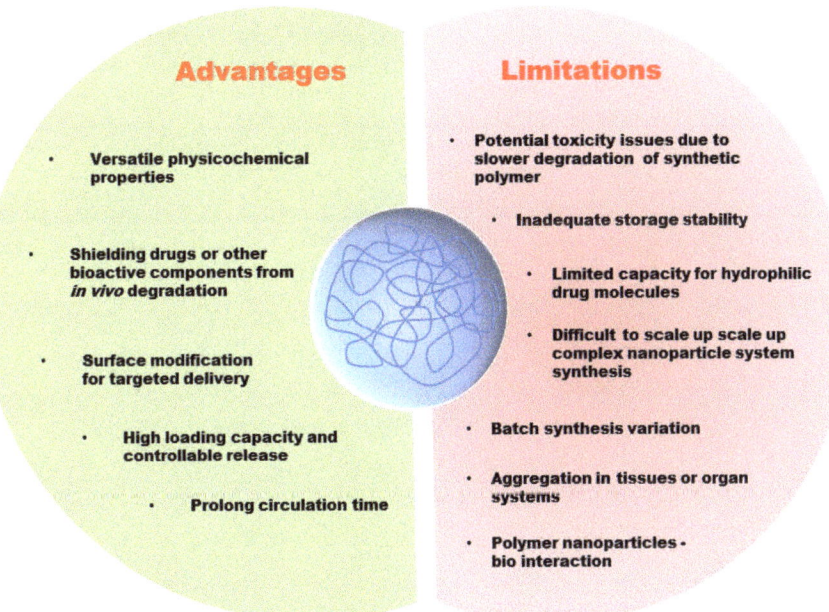

Figure 4. The advantages and limitations of polymer NPs in drug delivery and therapy.

8. Future Perspective and Conclusions

Glial originated brain cancers are the most aggressive gliomas that depict a threat to humans. The conventional therapies are still inefficient to overcome due to tumor heterogeneity and, specifically, the blood–brain barrier (BBB) of malignant gliomas. The polymeric nanoparticles-based brain cancer therapy approaches are currently gaining interest due to the drug safety, controllable drug release, and efficient targeting in tumors. Most importantly, reports revealed that polymer NPs could even transport across BBB. In this review article, we summarize the newest breakthroughs in the use of polymer nanocarriers for drug delivery, therapy and diagnosis of brain cancer are explored, emphasizing how they are a critical aspect of modern anticancer drug delivery strategies. Various polymer NPs have been generated to reduce anticancer drug losses, premature degradation, enhance drug availability, and reduce drug toxicity by improving drug accumulation in specific organs and tissues. Although the potential impact of polymer NPs in cancer therapy is exceedingly promising, numerous obstacles that currently limit their widespread clinical usage must be solved. For polymer NPs to be used in clinical trials, long-term safety investigations must be conducted in various animal models to eliminate the possibility of non-endogenous components accumulating in the body causing any harm. As a result, huge costs must be provided when conducting in vivo pharmacokinetic studies to evaluate the applicability in the human body. Another factor to consider is the challenges that may arise when transitioning from laboratory to large-scale production. The scaling up of the preparatory process is a major obstacle that must be surmounted. A significant number of polymer NPs are currently in the pre-clinical stage of development, but only one system has entered a clinical study. This is primarily because several challenges impede further development, such as a lack of potency in animal models and toxicity concerns. To overcome the aforementioned concerns, researchers need to focus more on new therapeutic innovations such as revising fabrication processes to modify and improve polymeric NPs in order to accommodate the demand for various anticancer drugs for effective clinical feasibility. New therapeutic innovations also include novel therapeutic strategies for combination therapy and stimuli-activated drug delivery. For example, delivering two or more anticancer drugs simultaneously might enhance the treatment of various cancer developments by targeting different tumour related signalling pathways, resulting in a synergistic therapeutic impact. In addition, the researcher needs to improve the targeting of cancer stem cells (CSCs) for effective cancer therapeutic effect as CSCs is a critical factor for tumour recurrence. In conclusion, pre-clinical experimentation and clinical trials are mandatory for an efficient polymer nanoparticle-based anticancer therapy. Hopefully, all of these developments will lead to more patient-specific and targeted anticancer therapies.

Author Contributions: Conceptualization, D.B., K.K.B. and B.R.; methodology, K.K.B., B.R., T.S., and S.P.; formal analysis, K.K.B., B.R., T.S., B.K.C. and D.B.; investigation S.P., T.S., K.K.B. and L.I.A.; writing—original draft preparation, K.K.B., B.R., T.S., S.P., B.K.C., D.B., Z.A.K., H.A.E. writing—review and editing, K.K.B., B.R., T.S., S.P., B.K.C., D.B. and L.I.A. All authors have read and agreed to the published version of the manuscript.

Funding: This work was supported in part by a research grant from the University Grants Commission (NFST) vide Grant No. F1-17.1/2015-16/NFST-2015-17-ST-ASS-3863.

Institutional Review Board Statement: Not applicable.

Informed Consent Statement: Not applicable.

Data Availability Statement: The study did not report any data.

Conflicts of Interest: The authors declare no conflict of interest.

References

1. Vijayakumar, M.R.; Kosuru, R.; Singh, S.K.; Prasad, C.B.; Narayan, G.; Muthu, M.S.; Singh, S. Resveratrol loaded PLGA:d-α-tocopheryl polyethylene glycol 1000 succinate blend nanoparticles for brain cancer therapy. *RSC Adv.* **2016**, *6*, 74254–74268. [CrossRef]
2. Cadinoiu, A.N.; Rata, D.M.; Atanase, L.I.; Mihai, C.T.; Bacaita, S.E.; Popa, M. Formulations Based on Drug Loaded Aptamer-Conjugated Liposomes as a Viable Strategy for the Topical Treatment of Basal Cell Carcinoma—In Vitro Tests. *Pharmaceutics* **2021**, *13*, 886. [CrossRef]
3. Saenz del Burgo, L.; Hernández, R.M.; Orive, G.; Pedraz, J.L. Nanotherapeutic approaches for brain cancer management. *Nanomedicine* **2014**, *10*, 905–919. [CrossRef]
4. Dolecek, T.A.; Propp, J.M.; Stroup, N.E.; Kruchko, C. CBTRUS statistical report: Primary brain and central nervous system tumors diagnosed in the United States in 2005–2009. *Neuro-Oncology* **2012**, *14* (Suppl. 5), v1–v49. [CrossRef]
5. Cheng, Y.; Morshed, R.A.; Auffinger, B.; Tobias, A.L.; Lesniak, M.S. Multifunctional nanoparticles for brain tumor imaging and therapy. *Adv. Drug Deliv. Rev.* **2014**, *66*, 42–57. [CrossRef]

6. Ostrom, Q.T.; Gittleman, H.; Farah, P.; Ondracek, A.; Chen, Y.; Wolinsky, Y.; Stroup, N.E.; Kruchko, C.; Barnholtz-Sloan, J.S. CBTRUS statistical report: Primary brain and central nervous system tumors diagnosed in the United States in 2006–2010. *Neuro-Oncol.* **2013**, *15* (Suppl. 2), ii1–ii56. [CrossRef] [PubMed]
7. Atanase, L.I. Micellar Drug Delivery Systems Based on Natural Biopolymers. *Polymers* **2021**, *13*, 447. [CrossRef]
8. Sona, M.M.; Viswanadh, M.K.; Singh, R.P.; Agrawal, P.; Mehata, A.K.; Pawde, D.M.; Narendra; Sonkar, R.; Muthu, M.S. Nanotheranostics: Emerging Strategies for Early Diagnosis and Therapy of Brain Cancer. *Nanotheranostics* **2018**, *2*, 70–86. [CrossRef]
9. Lapointe, S.; Perry, A.; Butowski, N.A. Primary brain tumours in adults. *Lancet* **2018**, *392*, 432–446. [CrossRef]
10. Stawicki, B.; Schacher, T.; Cho, H. Nanogels as a Versatile Drug Delivery System for Brain Cancer. *Gels* **2021**, *7*, 63. [CrossRef] [PubMed]
11. Mahmoud, B.S.; AlAmri, A.H.; McConville, C. Polymeric Nanoparticles for the Treatment of Malignant Gliomas. *Cancers* **2020**, *12*, 175. [CrossRef] [PubMed]
12. Cerna, T.; Stiborova, M.; Adam, V.; Kizek, R.; Eckschlager, T. Nanocarrier drugs in the treatment of brain tumors. *J. Cancer Metastasis Treat.* **2016**, *2*, 407–416. [CrossRef]
13. Kabitha, K.; Rajan, M.S.; Hegde, K.; Koshy, S.; Shenoy, A. A comprehensive review on brain tumor. *Int. J. Pharm. Chem. Biol. Sci.* **2013**, *3*, 1165–1171.
14. ELAmrawy, F.; Othman, A.A.; Adkins, C.; Helmy, A.; Nounou, M.I. Tailored nanocarriers and bioconjugates for combating glioblastoma and other brain tumors. *J. Cancer Metastasis Treat.* **2016**, *2*, 112–122. [CrossRef]
15. Louis, D.N.; Perry, A.; Wesseling, P.; Brat, D.J.; Cree, I.A.; Figarella-Branger, D.; Hawkins, C.; Ng, H.K.; Pfister, S.M.; Reifenberger, G.; et al. The 2021 WHO Classification of Tumors of the Central Nervous System: A summary. *Neuro-Oncol.* **2021**, *23*, 1231–1251. [CrossRef]
16. Paw, I.; Carpenter, R.C.; Watabe, K.; Debinski, W.; Lo, H.-W. Mechanisms regulating glioma invasion. *Cancer Lett.* **2015**, *362*, 1–7. [CrossRef] [PubMed]
17. Glaser, T.; Han, I.; Wu, L.; Zeng, X. Targeted Nanotechnology in Glioblastoma Multiforme. *Front. Pharmacol.* **2017**, *8*, 166. [CrossRef]
18. Taiarol, L.; Formicola, B.; Magro, R.D.; Sesana, S.; Re, F. An update of nanoparticle-based approaches for glioblastoma multiforme immunotherapy. *Nanomedicine* **2020**, *15*, 1861–1871. [CrossRef]
19. Scheithauer, B.W. Development of the WHO classification of tumors of the central nervous system: A historical perspective. *Brain Pathol.* **2009**, *19*, 551–564. [CrossRef] [PubMed]
20. Ozdemir-Kaynak, E.; Qutub, A.A.; Yesil-Celiktas, O. Advances in Glioblastoma Multiforme Treatment: New Models for Nanoparticle Therapy. *Front. Physiol.* **2018**, *9*, 170. [CrossRef]
21. Alifieris, C.; Trafalis, D.T. Glioblastoma multiforme: Pathogenesis and treatment. *Pharmacol. Ther.* **2015**, *152*, 63–82. [CrossRef]
22. van den Bent, M.J.; Afra, D.; de Witte, O.; Ben Hassel, M.; Schraub, S.; Hoang-Xuan, K.; Malmström, P.-O.; Collette, L.; Piérart, M.; Mirimanoff, R.; et al. Long-term efficacy of early versus delayed radiotherapy for low-grade astrocytoma and oligodendroglioma in adults: The EORTC 22845 randomised trial. *Lancet* **2005**, *366*, 985–990. [CrossRef]
23. Beier, D.; Schulz, J.B.; Beier, C.P. Chemoresistance of glioblastoma cancer stem cells—Much more complex than expected. *Mol. Cancer* **2011**, *10*, 128. [CrossRef] [PubMed]
24. Cook, L.; Freedman, J. *Brain Tumors*; The Rosen Publishing Group: New York, NY, USA, 2012.
25. Kim, S.-S.; Harford, J.B.; Pirollo, K.F.; Chang, E.H. Effective treatment of glioblastoma requires crossing the blood-brain barrier and targeting tumors including cancer stem cells: The promise of nanomedicine. *Biochem. Biophys. Res. Commun.* **2015**, *468*, 485–489. [CrossRef]
26. Binello, E.; Germano, I.M. Targeting glioma stem cells: A novel framework for brain tumors. *Cancer Sci.* **2011**, *102*, 1958–1966. [CrossRef]
27. Muthu, M.S.; Mei, L.; Feng, S.-S. Nanotheranostics: Advanced nanomedicine for the integration of diagnosis and therapy. *Nanomedicine* **2014**, *9*, 1277–1280. [CrossRef] [PubMed]
28. Jain, K.K. Nanobiotechnology-based strategies for crossing the blood-brain barrier. *Nanomedicine* **2012**, *7*, 1225–1233. [CrossRef] [PubMed]
29. Mohanta, B.C.; Palei, N.N.; Surendran, V.; Dinda, S.C.; Rajangam, J.; Deb, J.; Sahoo, B.M. Lipid Based Nanoparticles: Current Strategies for Brain Tumor Targeting. *Curr. Nanomater.* **2019**, *4*, 84–100. [CrossRef]
30. Dinda, S.C.; Pattnaik, G. Nanobiotechnology-based drug delivery in brain targeting. *Curr. Pharm. Biotechnol.* **2013**, *14*, 1264–1274. [CrossRef] [PubMed]
31. Zhang, W.; Mehta, A.; Tong, Z.; Esser, L.; Voelcker, N.H. Development of Polymeric Nanoparticles for Blood–Brain Barrier Transfer—Strategies and Challenges. *Adv. Sci.* **2021**, *8*, 2003937. [CrossRef]
32. Serlin, Y.; Shelef, I.; Knyazer, B.; Friedman, A. Anatomy and physiology of the blood-brain barrier. *Semin. Cell Dev. Biol.* **2015**, *38*, 2–6. [CrossRef]
33. Pardridge, W.M. The blood-brain barrier: Bottleneck in brain drug development. *NeuroRx* **2005**, *2*, 3–14. [CrossRef]
34. Jena, L.; McErlean, E.; McCarthy, H. Delivery across the blood-brain barrier: Nanomedicine for glioblastoma multiforme. *Drug Deliv. Transl. Res.* **2020**, *10*, 304–318. [CrossRef] [PubMed]

35. Ising, C.; Heneka, M.T. Functional and structural damage of neurons by innate immune mechanisms during neurodegeneration. *Cell Death Dis.* **2018**, *9*, 120. [CrossRef]
36. Bhowmik, A.; Khan, R.; Ghosh, M.K. Blood brain barrier: A challenge for effectual therapy of brain tumors. *BioMed Res. Int.* **2015**, *2015*, 320941. [CrossRef]
37. Stephenson, J.; Nutma, E.; van der Valk, P.; Amor, S. Inflammation in CNS neurodegenerative diseases. *Immunology* **2018**, *154*, 204–219. [CrossRef]
38. Belykh, E.; Shaffer, K.V.; Lin, C.; Byvaltsev, V.A.; Preul, M.C.; Chen, L. Blood-Brain Barrier, Blood-Brain Tumor Barrier, and Fluorescence-Guided Neurosurgical Oncology: Delivering Optical Labels to Brain Tumors. *Front. Oncol.* **2020**, *10*, 739. [CrossRef]
39. Reddy, S.; Tatiparti, K.; Sau, S.; Iyer, A.K. Recent advances in nano delivery systems for blood-brain barrier (BBB) penetration and targeting of brain tumors. *Drug Discov. Today* **2021**, *26*, 1944–1952. [CrossRef] [PubMed]
40. Wei, X.; Chen, X.; Ying, M.; Lu, W. Brain tumor-targeted drug delivery strategies. *Acta Pharm. Sin. B* **2014**, *4*, 193–201. [CrossRef] [PubMed]
41. Latasha, L.P. Nanomedicines and the future of glioma. *Neuro. Oncol.* **2015**, *10*, 16–22.
42. Pati, S.; Chatterji, A.; Dash, B.P.; Nelson, B.R.; Sarkar, T.; Shahimi, S.; Edinur, H.A.; Abd Manan, T.S.B.; Jena, P.; Mohanta, Y.K.; et al. Structural characterization and antioxidant potential of chitosan by γ-irradiation from the carapace of horseshoe crab. *Polymers* **2020**, *12*, 2361. [CrossRef] [PubMed]
43. Pati, S.; Sarkar, T.; Sheikh, H.I.; Bharadwaj, K.K.; Mohapatra, P.K.; Chatterji, A.; Dash, B.P.; Edinur, H.A.; Nelson, B.R. γ-Irradiated Chitosan from Carcinoscorpius rotundicauda (Latreille, 1802) Improves the Shelf Life of Refrigerated Aquatic Products. *Front. Mar. Sci.* **2021**, *8*, 498. [CrossRef]
44. Rabha, B.; Bharadwaj, K.K.; Baishya, D.; Sarkar, T.; Edinur, H.A.; Pati, S. Synthesis and Characterization of Diosgenin Encapsulated Poly-ε-Caprolactone-Pluronic Nanoparticles and Its Effect on Brain Cancer Cells. *Polymers* **2021**, *13*, 1322. [CrossRef]
45. Lo, Y.-L.; Lin, H.-C.; Hong, S.-T.; Chang, C.-H.; Wang, C.-S.; Lin, A.M.-Y. Lipid polymeric nanoparticles modified with tight junction-modulating peptides promote afatinib delivery across a blood–brain barrier model. *Cancer Nanotechnol.* **2021**, *12*, 13. [CrossRef]
46. Kou, L.; Hou, Y.; Yao, Q.; Guo, W.; Wang, G.; Wang, M.; Fu, Q.; He, Z.; Ganapathy, V.; Sun, J. L-Carnitine-conjugated nanoparticles to promote permeation across blood-brain barrier and to target glioma cells for drug delivery via the novel organic cation/carnitine transporter OCTN2. *Artif. Cells Nanomed. Biotechnol.* **2018**, *46*, 1605–1616. [CrossRef] [PubMed]
47. Li, H.; Tong, Y.; Bai, L.; Ye, L.; Zhong, L.; Duan, X.; Zhu, Y. Lactoferrin functionalized PEG-PLGA nanoparticles of shikonin for brain targeting therapy of glioma. *Int. J. Biol. Macromol.* **2018**, *107*, 204–211. [CrossRef]
48. Golombek, S.K.; May, J.-N.; Theek, B.; Appold, L.; Drude, N.; Kiessling, F.; Lammers, T. Tumor targeting via EPR: Strategies to enhance patient responses. *Adv. Drug Deliv. Rev.* **2018**, *130*, 17–38. [CrossRef]
49. Patel, T.; Zhou, J.; Piepmeier, J.M.; Saltzman, W.M. Polymeric nanoparticles for drug delivery to the central nervous system. *Adv. Drug Deliv. Rev.* **2012**, *64*, 701–705. [CrossRef]
50. Pulgar, V.M. Transcytosis to Cross the Blood Brain Barrier, New Advancements and Challenges. *Front. Neurosci.* **2018**, *12*, 1019. [CrossRef] [PubMed]
51. Kou, L.; Bhutia, Y.D.; Yao, Q.; He, Z.; Sun, J.; Ganapathy, V. Transporter-Guided Delivery of Nanoparticles to Improve Drug Permeation across Cellular Barriers and Drug Exposure to Selective Cell Types. *Front. Pharmacol.* **2018**, *9*, 27. [CrossRef]
52. Zhi, K.; Raji, B.; Nookala, A.R.; Khan, M.M.; Nguyen, X.H.; Sakshi, S.; Pourmotabbed, T.; Yallapu, M.M.; Kochat, H.; Tadrous, E.; et al. PLGA Nanoparticle-Based Formulations to Cross the Blood-Brain Barrier for Drug Delivery: From R&D to cGMP. *Pharmaceutics* **2021**, *13*, 500. [CrossRef]
53. Lu, W.; Zhang, Y.; Tan, Y.-Z.; Hu, K.-L.; Jiang, X.-G.; Fu, S.-K. Cationic albumin-conjugated pegylated nanoparticles as novel drug carrier for brain delivery. *J. Control. Release* **2005**, *107*, 428–448. [CrossRef]
54. Jiang, X.; Xin, H.; Ren, Q.; Gu, J.; Zhu, L.; Du, F.; Feng, C.; Xie, Y.; Sha, X.; Fang, X. Nanoparticles of 2-deoxy-d-glucose functionalized poly(ethylene glycol)-co-poly(trimethylene carbonate) for dual-targeted drug delivery in glioma treatment. *Biomaterials* **2014**, *35*, 518–529. [CrossRef]
55. Bray, N. Biologics: Transferrin' bispecific antibodies across the blood-brain barrier. *Nat. Rev. Drug Discov.* **2015**, *14*, 14–15. [CrossRef]
56. Cui, Y.; Xu, Q.; Chow, P.K.-H.; Wang, D.; Wang, C.-H. Transferrin-conjugated magnetic silica PLGA nanoparticles loaded with doxorubicin and paclitaxel for brain glioma treatment. *Biomaterials* **2013**, *34*, 8511–8520. [CrossRef]
57. Wang, Z.H.; Wang, Z.Y.; Sun, C.S.; Wang, C.Y.; Jiang, T.Y.; Wang, S.L. Trimethylated chitosan-conjugated PLGA nanoparticles for the delivery of drugs t the brain. *Biomaterials* **2010**, *31*, 908–915. [CrossRef]
58. Wang, L.; Hao, Y.; Li, H.; Zhao, Y.; Meng, D.; Li, D.; Shi, J.; Zhang, H.; Zhang, Z.; Zhang, Y. Co-delivery of doxorubicin and siRNA for glioma therapy by a brain targeting system: Angiopep-2-modified poly(lactic-co-glycolic acid) nanoparticles. *J. Drug Target.* **2015**, *23*, 832–846. [CrossRef] [PubMed]
59. Kuo, Y.-C.; Chen, Y.-C. Targeting delivery of etoposide to inhibit the growth of human glioblastoma multiforme using lactoferrin- and folic acid-grafted poly(lactide-co-glycolide) nanoparticles. *Int. J. Pharm.* **2015**, *479*, 138–149. [CrossRef] [PubMed]
60. Hua, H.; Zhang, X.; Mu, H.; Meng, Q.; Jiang, Y.; Wang, Y.; Lu, X.; Wang, A.; Liu, S.; Zhang, Y.; et al. RVG29-modified docetaxel-loaded nanoparticles for brain-targeted glioma therapy. *Int. J. Pharm.* **2018**, *543*, 179–189. [CrossRef]

61. Ramalho, M.J.; Sevin, E.; Gosselet, F.; Lima, J.; Coelho, M.A.N.; Loureiro, J.A.; Pereira, M.C. Receptor-mediated PLGA nanoparticles for glioblastoma multiforme treatment. *Int. J. Pharm.* **2018**, *545*, 84–92. [CrossRef] [PubMed]
62. Cui, Y.; Zhang, M.; Zeng, F.; Jin, H.; Xu, Q.; Huang, Y. Dual-Targeting Magnetic PLGA Nanoparticles for Codelivery of Paclitaxel and Curcumin for Brain Tumor Therapy. *ACS Appl. Mater. Interfaces* **2016**, *8*, 32159–32169. [CrossRef]
63. Xin, H.; Sha, X.; Jiang, X.; Zhang, W.; Chen, L.; Fang, X. Anti-glioblastoma efficacy and safety of paclitaxel-loading Angiopep-conjugated dual targeting PEG-PCL nanoparticles. *Biomaterials* **2012**, *33*, 8167–8176. [CrossRef]
64. Muniswamy, V.J.; Raval, N.; Gondaliya, P.; Tambe, V.; Kalia, K.; Tekade, R.K. 'Dendrimer-Cationized-Albumin' encrusted polymeric nanoparticle improves BBB penetration and anticancer activity of doxorubicin. *Int. J. Pharm.* **2019**, *555*, 77–99. [CrossRef]
65. Zhu, Y.; Zhang, J.; Meng, F.; Deng, C.; Cheng, R.; Feijen, J.; Zhong, Z. cRGD-functionalized reduction-sensitive shell-sheddable biodegradable micelles mediate enhanced doxorubicin delivery to human glioma xenografts in vivo. *J. Control. Release* **2016**, *233*, 29–38. [CrossRef] [PubMed]
66. Chen, W.; Zou, Y.; Zhong, Z.; Haag, R. Cyclo(RGD)-Decorated Reduction-Responsive Nanogels Mediate Targeted Chemotherapy of Integrin Overexpressing Human Glioblastoma In Vivo. *Small* **2017**, *13*, 1601997. [CrossRef] [PubMed]
67. Bi, Y.; Liu, L.; Lu, Y.; Sun, T.; Shen, C.; Chen, X.; Chen, Q.; An, S.; He, X.; Ruan, C.; et al. T7 Peptide-Functionalized PEG-PLGA Micelles Loaded with Carmustine for Targeting Therapy of Glioma. *ACS Appl. Mater. Interfaces* **2016**, *8*, 27465–27473. [CrossRef]
68. Bhowmik, A.; Chakravarti, S.; Ghosh, A.; Shaw, R.; Bhandary, S.; Bhattacharyya, S.; Sen, P.C.; Ghosh, M.K. Anti-SSTR2 peptide based targeted delivery of potent PLGA encapsulated 3,3′-diindolylmethane nanoparticles through blood brain barrier prevents glioma progression. *Oncotarget* **2017**, *8*, 65339–65358. [CrossRef]
69. Prabhu, R.H.; Patravale, V.B.; Joshi, M.D. Polymeric nanoparticles for targeted treatment in oncology: Current insights. *Int. J. Nanomed.* **2015**, *10*, 1001–1018. [CrossRef]
70. von Roemeling, C.; Jiang, W.; Chan, C.K.; Weissman, I.L.; Kim, B.Y.S. Breaking Down the Barriers to Precision Cancer Nanomedicine. *Trends Biotechnol.* **2017**, *35*, 159–171. [CrossRef]
71. Zielińska, A.; Carreiró, F.; Oliveira, A.M.; Neves, A.; Pires, B.; Venkatesh, D.N.; Durazzo, A.; Lucarini, M.; Eder, P.; Silva, A.M.; et al. Polymeric Nanoparticles: Production, Characterization, Toxicology and Ecotoxicology. *Molecules* **2020**, *25*, 3731. [CrossRef] [PubMed]
72. Guzmán, E.; Mateos-Maroto, A.; Ruano, M.; Ortega, F.; Rubio, R.G. Layer-by-Layer polyelectrolyte assemblies for encapsulation and release of active compounds. *Adv. Colloid Interface Sci.* **2017**, *249*, 290–307. [CrossRef] [PubMed]
73. Mateos-Maroto, A.; Abelenda-Núñez, I.; Ortega, F.; Rubio, R.G.; Guzmán, E. Polyelectrolyte Multilayers on Soft Colloidal Nanosurfaces: A New Life for the Layer-By-Layer Method. *Polymer* **2021**, *13*, 1221. [CrossRef]
74. Peer, D.; Karp, J.M.; Hong, S.; Farokhzad, O.C.; Margalit, R.; Langer, R. Nanocarriers as an emerging platform for cancer therapy. *Nat. Nanotechnol.* **2007**, *2*, 751–760. [CrossRef]
75. Varan, C.; Bilensoy, E. Cationic PEGylated polycaprolactone nanoparticles carrying post-operation docetaxel for glioma treatment. *Beilstein J. Nanotechnol.* **2017**, *8*, 1446–1456. [CrossRef] [PubMed]
76. Pinto, M.P.; Arce, M.; Yameen, B.; Vilos, C. Targeted brain delivery nanoparticles for malignant gliomas. *Nanomedicine* **2017**, *12*, 59–72. [CrossRef]
77. Gaudin, A.; Song, E.; King, A.R.; Saucier-Sawyer, J.K.; Bindra, R.; Desmaële, D.; Couvreur, P.; Saltzman, W.M. PEGylated squalenoyl-gemcitabine nanoparticles for the treatment of glioblastoma. *Biomaterials* **2016**, *105*, 136–144. [CrossRef] [PubMed]
78. Yang, J.; Li, Y.; Zhang, T.; Zhang, X. Development of bioactive materials for glioblastoma therapy. *Bioact. Mater.* **2016**, *1*, 29–38. [CrossRef]
79. Zhong, Y.; Meng, F.; Deng, C.; Zhong, Z. Ligand-Directed Active Tumor-Targeting Polymeric Nanoparticles for Cancer Chemotherapy. *Biomacromolecules* **2014**, *15*, 1955–1969. [CrossRef]
80. Sun, P.; Xiao, Y.; Di, Q.; Ma, W.; Ma, X.; Wang, Q.; Chen, W. Transferrin Receptor-Targeted PEG-PLA Polymeric Micelles for Chemotherapy Against Glioblastoma Multiforme. *Int. J. Nanomed.* **2020**, *15*, 6673–6688. [CrossRef]
81. Wen, G.; Li, X.; Zhang, Y.; Han, X.; Xu, X.; Liu, C.; Chan, K.W.Y.; Lee, C.-S.; Yin, C.; Bian, L.; et al. Effective Phototheranostics of Brain Tumor Assisted by Near-Infrared-II Light-Responsive Semiconducting Polymer Nanoparticles. *ACS Appl. Mater. Interfaces* **2020**, *12*, 33492–33499. [CrossRef]
82. Ibarra, L.E.; Beaugé, L.; Arias-Ramos, N.; Rivarola, V.A.; Chesta, C.A.; López-Larrubia, P.; Palacios, R.E. Trojan horse monocyte-mediated delivery of conjugated polymer nanoparticles for improved photodynamic therapy of glioblastoma. *Nanomedicine* **2020**, *15*, 1687–1707. [CrossRef] [PubMed]
83. Gregory, J.V.; Kadiyala, P.; Doherty, R.; Cadena, M.; Habeel, S.; Ruoslahti, E.; Lowenstein, P.R.; Castro, M.G.; Lahann, J. Systemic brain tumor delivery of synthetic protein nanoparticles for glioblastoma therapy. *Nat. Commun.* **2020**, *11*, 5687. [CrossRef]
84. Caverzán, M.D.; Beaugé, L.; Chesta, C.A.; Palacios, R.E.; Ibarra, L.E. Photodynamic therapy of Glioblastoma cells using doped conjugated polymer nanoparticles: An in vitro comparative study based on redox status. *J. Photochem. Photobiol. B Biol.* **2020**, *212*, 112045. [CrossRef] [PubMed]
85. Zhou, J.; Patel, T.R.; Sirianni, R.W.; Strohbehn, G.; Zheng, M.-Q.; Duong, N.; Schafbauer, T.; Huttner, A.J.; Huang, Y.; Carson, R.E.; et al. Highly penetrative, drug-loaded nanocarriers improve treatment of glioblastoma. *Proc. Natl. Acad. Sci. USA* **2013**, *110*, 11751–11756. [CrossRef]

86. Guerrero-Cázares, H.; Tzeng, S.Y.; Young, N.P.; Abutaleb, A.O.; Quiñones-Hinojosa, A.; Green, J.J. Biodegradable Polymeric Nanoparticles Show High Efficacy and Specificity at DNA Delivery to Human Glioblastoma in Vitro and in Vivo. *ACS Nano* **2014**, *8*, 5141–5153. [CrossRef]
87. Choi, J.; Yang, J.; Park, J.; Kim, E.; Suh, J.-S.; Huh, Y.-M.; Haam, S. Specific Near-IR Absorption Imaging of Glioblastomas Using Integrin-Targeting Gold Nanorods. *Adv. Funct. Mater.* **2011**, *21*, 1082–1088. [CrossRef]
88. Oli, M. Aptamer conjugated gold nanorods for targeted nanothermal radiation of Glioblastoma cancer cells (A novel selective targeted approach to cancer treatment). *Young Sci. J.* **2009**, *2*, 18. [CrossRef]
89. Hoare, T.; Santamaria, J.; Goya, G.F.; Irusta, S.; Lin, D.; Lau, S.; Padera, R.; Langer, R.; Kohane, D.S. A Magnetically Triggered Composite Membrane for On-Demand Drug Delivery. *Nano Lett.* **2009**, *9*, 3651–3657. [CrossRef]
90. Kim, J.; Mondal, S.K.; Tzeng, S.Y.; Rui, Y.; Al-kharboosh, R.; Kozielski, K.K.; Bhargav, A.G.; Garcia, C.A.; Quiñones-Hinojosa, A.; Green, J.J. Poly(ethylene glycol)–Poly(beta-amino ester)-Based Nanoparticles for Suicide Gene Therapy Enhance Brain Penetration and Extend Survival in a Preclinical Human Glioblastoma Orthotopic Xenograft Model. *ACS Biomater. Sci. Eng.* **2020**, *6*, 2943–2955. [CrossRef]
91. Lu, Q.; Cai, X.; Zhang, X.; Li, S.; Song, Y.; Du, D.; Dutta, P.; Lin, Y. Synthetic Polymer Nanoparticles Functionalized with Different Ligands for Receptor-mediated Transcytosis across Blood-Brain Barrier. *ACS Appl. Bio Mater.* **2018**, *1*, 1687–1694. [CrossRef]
92. Hu, Q.; Gao, X.; Gu, G.; Kang, T.; Tu, Y.; Liu, Z.; Song, Q.; Yao, L.; Pang, Z.; Jiang, X.; et al. Glioma therapy using tumor homing and penetrating peptide-functionalized PEG–PLA nanoparticles loaded with paclitaxel. *Biomaterials* **2013**, *34*, 5640–5650. [CrossRef]
93. Guo, W.; Li, A.; Jia, Z.; Yuan, Y.; Dai, H.; Li, H. Transferrin modified PEG-PLA-resveratrol conjugates: In vitro and in vivo studies for glioma. *Eur. J. Pharmacol.* **2013**, *718*, 41–47. [CrossRef]
94. Ambruosi, A.; Khalansky, A.S.; Yamamoto, H.; Gelperina, S.E.; Begley, D.J.; Kreuter, J. Biodistribution of polysorbate 80-coated doxorubicin-loaded [14C]-poly(butyl cyanoacrylate) nanoparticles after intravenous administration to glioblastoma-bearing rats. *J. Drug Target.* **2006**, *14*, 97–105. [CrossRef]
95. Mahmoud, B.S.; McConville, C. Development and Optimization of Irinotecan-Loaded PCL Nanoparticles and Their Cytotoxicity against Primary High-Grade Glioma Cells. *Pharmaceutics* **2021**, *13*, 541. [CrossRef]
96. Zhong, Y.; Wang, C.; Cheng, R.; Cheng, L.; Meng, F.; Liu, Z.; Zhong, Z. cRGD-directed, NIR-responsive and robust AuNR/PEG-PCL hybrid nanoparticles for targeted chemotherapy of glioblastoma in vivo. *J. Control. Release* **2014**, *195*, 63–71. [CrossRef] [PubMed]
97. Irani, M.; Sadeghi, G.M.M.; Haririan, I. The sustained delivery of temozolomide from electrospun PCL-Diol-b-PU/gold nanocomposite nanofibers to treat glioblastoma tumors. *Mater. Sci. Eng. C Mater. Biol. Appl.* **2017**, *75*, 165–174. [CrossRef]
98. Gu, G.; Xia, H.; Hu, Q.; Liu, Z.; Jiang, M.; Kang, T.; Miao, D.; Tu, Y.; Pang, Z.; Song, Q.; et al. PEG-co-PCL nanoparticles modified with MMP-2/9 activatable low molecular weight protamine for enhanced targeted glioblastoma therapy. *Biomaterials* **2013**, *34*, 196–208. [CrossRef]
99. Yan, H.; Wang, J.; Yi, P.; Lei, H.; Zhan, C.; Xie, C.; Feng, L.; Qian, J.; Zhu, J.; Lu, W.; et al. Imaging brain tumor by dendrimer-based optical/paramagnetic nanoprobe across the blood-brain barrier. *Chem. Commun.* **2011**, *47*, 8130–8132. [CrossRef] [PubMed]
100. Alizadeh, D.; Zhang, L.; Hwang, J.; Schluep, T.; Badie, B. Tumor-associated macrophages are predominant carriers of cyclodextrin-based nanoparticles into gliomas. *Nanomedicine* **2010**, *6*, 382–390. [CrossRef] [PubMed]
101. Sharma, S.; Chockalingam, S.; Sanpui, P.; Chattopadhyay, A.; Ghosh, S.S. Silver nanoparticles impregnated alginate-chitosan-blended nanocarrier induces apoptosis in human glioblastoma cells. *Adv. Healthc. Mater.* **2014**, *3*, 106–114. [CrossRef] [PubMed]
102. Cohen, Z.R.; Ramishetti, S.; Peshes-Yaloz, N.; Goldsmith, M.; Wohl, A.; Zibly, Z.; Peer, D. Localized RNAi Therapeutics of Chemoresistant Grade IV Glioma Using Hyaluronan-Grafted Lipid-Based Nanoparticles. *ACS Nano* **2015**, *9*, 1581–1591. [CrossRef]
103. Nejat, H.; Rabiee, M.; Varshochian, R.; Tahriri, M.; Jazayeri, H.E.; Rajadas, J.; Ye, H.; Cui, Z.; Tayebi, L. Preparation and characterization of cardamom extract-loaded gelatin nanoparticles as effective targeted drug delivery system to treat glioblastoma. *React. Funct. Polym.* **2017**, *120*, 46–56. [CrossRef]
104. Deng, G.; Peng, X.; Sun, Z.; Zheng, W.; Yu, J.; Du, L.; Chen, H.; Gong, P.; Zhang, P.; Cai, L.; et al. Natural-Killer-Cell-Inspired Nanorobots with Aggregation-Induced Emission Characteristics for Near-Infrared-II Fluorescence-Guided Glioma Theranostics. *ACS Nano* **2020**, *14*, 11452–11462. [CrossRef]
105. Wang, J.; Yang, Y.; Zhang, Y.; Huang, M.; Zhou, Z.; Luo, W.; Tang, J.; Wang, Y.; Xiao, Q.; Chen, H.; et al. Dual-Targeting Heparin-Based Nanoparticles that Re-Assemble in Blood for Glioma Therapy through Both Anti-Proliferation and Anti-Angiogenesis. *Adv. Funct. Mater.* **2016**, *26*, 7873–7885. [CrossRef]
106. Gholami, L.; Tafaghodi, M.; Abbasi, B.; Daroudi, M.; Kazemi Oskuee, R. Preparation of superparamagnetic iron oxide/doxorubicin loaded chitosan nanoparticles as a promising glioblastoma theranostic tool. *J. Cell. Physiol.* **2019**, *234*, 1547–1559. [CrossRef] [PubMed]
107. Ruan, C.; Liu, L.; Lu, Y.; Zhang, Y.; He, X.; Chen, X.; Zhang, Y.; Chen, Q.; Guo, Q.; Sun, T.; et al. Substance P-modified human serum albumin nanoparticles loaded with paclitaxel for targeted therapy of glioma. *Acta Pharm. Sin. B* **2018**, *8*, 85–96. [CrossRef] [PubMed]
108. Byeon, H.J.; Thao, L.Q.; Lee, S.; Min, S.Y.; Lee, E.S.; Shin, B.S.; Choi, H.-G.; Youn, Y.S. Doxorubicin-loaded nanoparticles consisted of cationic- and mannose-modified-albumins for dual-targeting in brain tumors. *J. Control. Release* **2016**, *225*, 301–313. [CrossRef]

109. Lin, T.; Zhao, P.; Jiang, Y.; Tang, Y.; Jin, H.; Pan, Z.; He, H.; Yang, V.C.; Huang, Y. Blood-Brain-Barrier-Penetrating Albumin Nanoparticles for Biomimetic Drug Delivery via Albumin-Binding Protein Pathways for Antiglioma Therapy. *ACS Nano* **2016**, *10*, 9999–10012. [CrossRef]
110. Gao, C.; Liang, J.; Zhu, Y.; Ling, C.; Cheng, Z.; Li, R.; Qin, J.; Lu, W.; Wang, J. Menthol-modified casein nanoparticles loading 10-hydroxycamptothecin for glioma targeting therapy. *Acta Pharm. Sin. B* **2019**, *9*, 843–857. [CrossRef]
111. Lam, F.C.; Morton, S.W.; Wyckoff, J.; Vu Han, T.-L.; Hwang, M.K.; Maffa, A.; Balkanska-Sinclair, E.; Yaffe, M.B.; Floyd, S.R.; Hammond, P.T. Enhanced efficacy of combined temozolomide and bromodomain inhibitor therapy for gliomas using targeted nanoparticles. *Nat. Commun.* **2018**, *9*, 1991. [CrossRef]
112. Kuhnline Sloan, C.D.; Nandi, P.; Linz, T.H.; Aldrich, J.V.; Audus, K.L.; Lunte, S.M. Analytical and biological methods for probing the blood-brain barrier. *Annu. Rev. Anal. Chem.* **2012**, *5*, 505–531. [CrossRef] [PubMed]
113. Carlsson, S.K.; Brothers, S.P.; Wahlestedt, C. Emerging treatment strategies for glioblastoma multiforme. *EMBO Mol. Med.* **2014**, *6*, 1359–1370. [CrossRef] [PubMed]
114. Vasefi, F.; MacKinnon, N.; Farkas, D.L.; Kateb, B. Review of the potential of optical technologies for cancer diagnosis in neurosurgery: A step toward intraoperative neurophotonics. *Neurophotonics* **2017**, *4*, 11010. [CrossRef] [PubMed]
115. Fakhoury, M. Drug delivery approaches for the treatment of glioblastoma multiforme. *Artif. Cells Nanomed. Biotechnol.* **2016**, *44*, 1365–1373. [CrossRef]
116. Dhermain, F.G.; Hau, P.; Lanfermann, H.; Jacobs, A.H.; van den Bent, M.J. Advanced MRI and PET imaging for assessment of treatment response in patients with gliomas. *Lancet. Neurol.* **2010**, *9*, 906–920. [CrossRef]
117. Galldiks, N.; Dunkl, V.; Kracht, L.W.; Vollmar, S.; Jacobs, A.H.; Fink, G.R.; Schroeter, M. Volumetry of [^{11}C]-methionine positron emission tomographic uptake as a prognostic marker before treatment of patients with malignant glioma. *Mol. Imaging* **2012**, *11*, 516–527. [CrossRef]
118. Zottel, A.; Videtič Paska, A.; Jovčevska, I. Nanotechnology Meets Oncology: Nanomaterials in Brain Cancer Research, Diagnosis and Therapy. *Materials* **2019**, *12*, 1588. [CrossRef] [PubMed]
119. d'Angelo, M.; Castelli, V.; Benedetti, E.; Antonosante, A.; Catanesi, M.; Dominguez-Benot, R.; Pitari, G.; Ippoliti, R.; Cimini, A. Theranostic Nanomedicine for Malignant Gliomas. *Front. Bioeng. Biotechnol.* **2019**, *7*, 325. [CrossRef]
120. Tang, W.; Fan, W.; Lau, J.; Deng, L.; Shen, Z.; Chen, X. Emerging blood-brain-barrier-crossing nanotechnology for brain cancer theranostics. *Chem. Soc. Rev.* **2019**, *48*, 2967–3014. [CrossRef]
121. Reddy, L.H.; Couvreur, P. Nanotechnology for therapy and imaging of liver diseases. *J. Hepatol.* **2011**, *55*, 1461–1466. [CrossRef]
122. Gauger, A.J.; Hershberger, K.K.; Bronstein, L.M. Theranostics Based on Magnetic Nanoparticles and Polymers: Intelligent Design for Efficient Diagnostics and Therapy. *Front. Chem.* **2020**, *8*, 561. [CrossRef]
123. van Vlerken, L.E.; Amiji, M.M. Multi-functional polymeric nanoparticles for tumour-targeted drug delivery. *Expert Opin. Drug Deliv.* **2006**, *3*, 205–216. [CrossRef]
124. Qu, J.; Zhang, L.; Chen, Z.; Mao, G.; Gao, Z.; Lai, X.; Zhu, X.; Zhu, J. Nanostructured lipid carriers, solid lipid nanoparticles, and polymeric nanoparticles: Which kind of drug delivery system is better for glioblastoma chemotherapy? *Drug Deliv.* **2016**, *23*, 3408–3416. [CrossRef] [PubMed]
125. Luk, B.T.; Zhang, L. Current Advances in Polymer-Based Nanotheranostics for Cancer Treatment and Diagnosis. *ACS Appl. Mater. Interfaces* **2014**, *6*, 21859–21873. [CrossRef] [PubMed]
126. Ganipineni, L.P.; Ucakar, B.; Joudiou, N.; Bianco, J.; Danhier, P.; Zhao, M.; Bastiancich, C.; Gallez, B.; Danhier, F.; Préat, V. Magnetic targeting of paclitaxel-loaded poly(lactic-co-glycolic acid)-based nanoparticles for the treatment of glioblastoma. *Int. J. Nanomed.* **2018**, *13*, 4509–4521. [CrossRef] [PubMed]
127. Strohbehn, G.; Coman, D.; Han, L.; Ragheb, R.R.T.; Fahmy, T.M.; Huttner, A.J.; Hyder, F.; Piepmeier, J.M.; Saltzman, W.M.; Zhou, J. Imaging the delivery of brain-penetrating PLGA nanoparticles in the brain using magnetic resonance. *J. Neurooncol.* **2015**, *121*, 441–449. [CrossRef]
128. Ling, Y.; Wei, K.; Zou, F.; Zhong, S. Temozolomide loaded PLGA-based superparamagnetic nanoparticles for magnetic resonance imaging and treatment of malignant glioma. *Int. J. Pharm.* **2012**, *430*, 266–275. [CrossRef]
129. Sun, L.; Joh, D.Y.; Al-Zaki, A.; Stangl, M.; Murty, S.; Davis, J.J.; Baumann, B.C.; Alonso-Basanta, M.; Kaol, G.D.; Tsourkas, A.; et al. Theranostic Application of Mixed Gold and Superparamagnetic Iron Oxide Nanoparticle Micelles in Glioblastoma Multiforme. *J. Biomed. Nanotechnol.* **2016**, *12*, 347–356. [CrossRef] [PubMed]
130. Shevtsov, M.; Nikolaev, B.; Marchenko, Y.; Yakovleva, L.; Skvortsov, N.; Mazur, A.; Tolstoy, P.; Ryzhov, V.; Multhoff, G. Targeting experimental orthotopic glioblastoma with chitosan-based superparamagnetic iron oxide nanoparticles (CS-DX-SPIONs). *Int. J. Nanomed.* **2018**, *13*, 1471–1482. [CrossRef]
131. Xu, H.-L.; Mao, K.-L.; Huang, Y.-P.; Yang, J.-J.; Xu, J.; Chen, P.-P.; Fan, Z.-L.; Zou, S.; Gao, Z.-Z.; Yin, J.-Y.; et al. Glioma-targeted superparamagnetic iron oxide nanoparticles as drug-carrying vehicles for theranostic effects. *Nanoscale* **2016**, *8*, 14222–14236. [CrossRef] [PubMed]
132. Bernal, G.M.; LaRiviere, M.J.; Mansour, N.; Pytel, P.; Cahill, K.E.; Voce, D.J.; Kang, S.; Spretz, R.; Welp, U.; Noriega, S.E.; et al. Convection-enhanced delivery and in vivo imaging of polymeric nanoparticles for the treatment of malignant glioma. *Nanomedicine* **2014**, *10*, 149–157. [CrossRef]
133. Mahmoudi, M.; Bertrand, N.; Zope, H.; Farokhzad, O.C. Emerging understanding of the protein corona at the nano-bio interfaces. *Nano Today* **2016**, *11*, 817–832. [CrossRef]

134. Rosenblum, D.; Joshi, N.; Tao, W.; Karp, J.M.; Peer, D. Progress and challenges towards targeted delivery of cancer therapeutics. *Nat. Commun.* **2018**, *9*, 1410. [CrossRef]
135. Neerooa, B.N.; Ooi, L.-T.; Shameli, K.; Dahlan, N.A.; Islam, J.M.M.; Pushpamalar, J.; Teow, S.-Y. Development of Polymer-Assisted Nanoparticles and Nanogels for Cancer Therapy: An Update. *Gels* **2021**, *7*, 60. [CrossRef] [PubMed]
136. Bharadwaj, K.K.; Rabha, B.; Pati, S.; Sarkar, T.; Choudhury, B.K.; Barman, A.; Bhattacharjya, D.; Srivastava, A.; Baishya, D.; Edinur, H.A.; et al. Green Synthesis of Gold Nanoparticles Using Plant Extracts as Beneficial Prospect for Cancer Theranostics. *Molecules* **2021**, *26*, 6389. [CrossRef] [PubMed]

Article

Preclinical Evaluation of Polymeric Nanocomposite Containing Pregabalin for Sustained Release as Potential Therapy for Neuropathic Pain

Rafaela Figueiredo Rodrigues [1,*], Juliana Barbosa Nunes [2], Sandra Barbosa Neder Agostini [1], Paloma Freitas dos Santos [1], Juliana Cancino-Bernardi [3], Rodrigo Vicentino Placido [1], Thamyris Reis Moraes [4], Jennifer Tavares Jacon Freitas [1], Gislaine Ribeiro Pereira [1], Flávia Chiva Carvalho [1], Giovane Galdino [4] and Vanessa Bergamin Boralli [1,*]

[1] Faculdade de Ciências Farmacêuticas, Universidade Federal de Alfenas (UNIFAL-MG), Alfenas 371300-001, Brazil; sbneder@gmail.com (S.B.N.A.); paloma.nut13@gmail.com (P.F.d.S.); rodrigov_placido@hotmail.com (R.V.P.); jenniferjaconfreitas@gmail.com (J.T.J.F.); gislaine.pereira@unifal-mg.edu.br (G.R.P.); flaviachiva@gmail.com (F.C.C.)
[2] Instituto de Ciências Médicas, Universidade de São Paulo (USP), São Paulo 01246-903, Brazil; juliana_bnunes@yahoo.com.br
[3] Instituto de Física de São Carlos, Universidade de São Paulo (USP), São Carlos 13566-590, Brazil; jcancinobernardi@gmail.com
[4] Instituto de Ciências da Motricidade, Universidade Federal de Alfenas (UNIFAL-MG), Alfenas 37133-840, Brazil; thamoraes@yahoo.com.br (T.R.M.); giovane.souza@yahoo.com.br (G.G.)
* Correspondence: rafaelafigueiredor@gmail.com (R.F.R.); vboralli@gmail.com (V.B.B.); Tel.: +55-35-3701-9508 (V.B.B.)

Citation: Rodrigues, R.F.; Nunes, J.B.; Agostini, S.B.N.; dos Santos, P.F.; Cancino-Bernardi, J.; Placido, R.V.; Moraes, T.R.; Freitas, J.T.J.; Pereira, G.R.; Carvalho, F.C.; et al. Preclinical Evaluation of Polymeric Nanocomposite Containing Pregabalin for Sustained Release as Potential Therapy for Neuropathic Pain. *Polymers* **2021**, *13*, 3837. https://doi.org/10.3390/polym13213837

Academic Editor: Leonard Atanase

Received: 14 October 2021
Accepted: 3 November 2021
Published: 6 November 2021

Publisher's Note: MDPI stays neutral with regard to jurisdictional claims in published maps and institutional affiliations.

Copyright: © 2021 by the authors. Licensee MDPI, Basel, Switzerland. This article is an open access article distributed under the terms and conditions of the Creative Commons Attribution (CC BY) license (https://creativecommons.org/licenses/by/4.0/).

Abstract: This study offers a novel oral pregabalin (PG)-loaded drug delivery system based on chitosan and hypromellose phthalate-based polymeric nanocomposite in order to treat neuropathic pain (PG-PN). PG-PN has a particle size of 432 ± 20 nm, a polydispersity index of 0.238 ± 0.001, a zeta potential of $+19.0 \pm 0.9$ mV, a pH of 5.7 ± 0.06, and a spherical shape. Thermal and infrared spectroscopy confirmed nanocomposite generation. PG-PN pharmacokinetics was studied after a single oral dose in male Wistar rats. PG-PN showed greater distribution and clearance than free PG. The antinociceptive effect of PG-PN in neuropathic pain rats was tested by using the chronic constriction injury model. The parameter investigated was the mechanical nociceptive threshold measured by the von Frey filaments test; PG-PN showed a longer antinociceptive effect than free PG. The rota-rod and barbiturate sleep induction procedures were used to determine adverse effects; the criteria included motor deficit and sedative effects. PG-PN and free PG had plenty of motors. PG-PN exhibited a less sedative effect than free PG. By prolonging the antinociceptive effect and decreasing the unfavorable effects, polymeric nanocomposites with pregabalin have shown promise in treating neuropathic pain.

Keywords: neuropathy; polymeric nanoparticles; preclinical investigation; pharmacokinetics of pregabalin; antinociceptive effect; induced sleep

1. Introduction

Pregabalin (PG) (S-[+]-3-isobutyl GABA or (S)-3-aminomethyl-5-methylhexanoic acid) is an anticonvulsant, antihyperalgesic, and anxiolytic drug that acts by binding to the alpha-2-delta-1 proteins of voltage-dependent calcium channels in the Central Nervous System (CNS), reducing the release of excitatory neurotransmitters [1]. PG is one of the first-choice medicines for the treatment of neuropathic pain that has been authorized by the FDA [2]. Due to its short half-life, PG is sold as an instant release (IR) tablet, with a daily dosage of 150 to 600 mg split into two or three administrations [1,3]. In addition, sleepiness, dizziness, and loss of consciousness are common adverse effects of PG [1]. According to studies, around 15% of patients using pregabalin for neuropathic pain discontinue their therapy due to adverse effects, even when the dosages are tolerable [4,5]. PG is a class I molecule with good solubility and permeability, according to the Biopharmaceutical Classification System (BSC) [6], indicating that it has no physicochemical issues that would require a change in pharmaceutical form. As neuropathic pain is persistent and needs lengthy therapy, these two variables (short half-life and adverse effects) might be a barrier for appropriate treatment compliance [3].

The FDA has approved the commercialization of PG extended-release coated tablets (Lyrica CR®) as a means to avoid the discomfort of numerous doses [7]. Even if formulation development was successful, three flaws may be addressed in terms of once-daily dosing: (a) controlled-release (CR) tablets must be taken after an evening meal; (b) this evening meal must be hypercaloric (800 to 1000 kcal) in order to achieve the same level of absorption as PG IR; (c) the CR formulation has essentially the same side effects as IR tablets, with similar user incidences [8].

Some studies have proposed modified-release pharmaceutical formulations containing PG for once-daily administration, such as transdermal delivery of PG [9]; PG microspheres [10–12]; PG formulations with longer stomach duration [6,13–20]; and PG suppository [21]. The treatment of neuropathic pain has been the focus of several of these formulations.

These formulations, which were created for once-daily PG delivery, demonstrated that controlled release methods may be used to improve treatment adherence. However, none of the formulations that were previously provided assessed the reduction in adverse effects. This is an essential aspect to address because side effects are linked to treatment adherence [4,5]. Furthermore, the majority of studies have not evaluated the formulation's effectiveness in terms of pain/nociception reduction. As a result, there is a significant research gap as well as an opportunity to enhance therapy in experimental studies with the potential to be used in clinical practice.

Many investigations have been conducted recently on the use of natural polysaccharides (e.g., alginate, chitosan, hyaluronic acid, cellulose, and starch) for various biological, biomedical, functional food, and tissue engineering applications due to their biocompatibility and biodegradability [22–25]. These natural polysaccharides have been widely used as carriers for the delivery of various therapeutic molecules (e.g., proteins, peptides, and drugs), mainly for anti-cancer therapies, diabetes, and other chronic diseases [24,26,27]. Polymeric nanoparticles are one form of drug delivery system that can be based on natural polymers such as chitosan (CS) and hydroxypropylmethylcellulose phthalate (HPMCP) or hypromellose phthalate, and they are advantageous due to desirable properties such as stability, safety, non-toxicity, hydrophilicity, and biodegradability in addition to being abundant in nature and having low processing costs [28,29]. Drug carrier nanoparticles can be used as an alternative pharmaceutical form because they allow a significant increase in the drug's bioavailability in CNS, increased specificity of the drug at its site of action, increased distribution in the body, dose reduction, and reduced adverse effects [30,31].

In a model of chronic sciatic nerve constriction, we aimed to develop a PG-loaded polymeric nanocomposite formulation utilizing CS and HPMCP that would extend antinociceptive effects, improve nociception perception, and enable administration once daily without generating adverse effects.

2. Materials and Methods

2.1. Materials

Pregabalin was purchased from Pfizer® (Karlsrufe, Germany). Low viscosity shrimp chitosan (150 kDa) was from Sigma-Aldrich® (Saint Louis, MS, USA). Hypromellose phthalate (HPMCP, type HP-55) was kindly donated by Shin-Etsu Chemical Co.® (Tokyo, Japan). Ketamine hydrochloride 10% injectable was purchased from Linavet® (Rio de Janeiro, Brazil). Injectable xylazine was purchased from Hertape Calier Saúde Animal S.A. (Juataba, Brazil). Isoflurane and sodium thiopental were purchased from Cristália

(Itapira, Brazil). Injectable sodium heparin was purchased from Blau Farmacêutica S.A. (São Paulo, Brazil). Metformin was purchased from USP Reference Standard (Betheseda, Rockville, MD, USA). Sodium phosphate dibasic heptahydrate PA and sodium phosphate tribasic dodecahydrate PA were purchased from Vetec® (Duque de Caxias, Brazil). Sodium phosphate monobasic monohydrate PA was purchased from Proquimios® (Rio de Janeiro, Brazil). Sodium hydroxide PA was purchased from Sigma-Aldrich® (Saint Louis, MS, USA). Glacial acetic acid PA was purchased from Dinâmica Química Contemporânea Ltd.a. (Indaiatuba, Brazil). Hydrochloric acid 37% PA and formic acid were purchased from Alphatec® (Macaé, Brazil). Acetonitrile and Methanol HPLC grade were purchased from J.T. Baker® (Phillipsburg, NJ, USA). Ammonium formate HPLC was purchased from Sigma-Aldrich® (Saint Louis, MS, USA).

2.2. Obtaining Pregabalin-Loaded Polymeric Nanocomposite (PG-PN)

Polymeric nanocomposites were prepared by ionic crosslinking of CS dispersion with HPMCP aqueous solution, according to the ionotropic gelation method [32].

The solution of CS was prepared by the dispersion of 4 mg/mL of chitosan in acetic acid solution (0.1 M, pH = 5.5) at room temperature (24 °C) and mechanical stirring at 800 rpm (Velp Scientifica®, Usmate Velate, Italy). PG solution, made of PG (8 mg/mL), was dissolved in a sodium phosphate buffer solution (0.2 M, pH = 6.0) at room temperature. The solution of HPMCP was prepared by dissolving HPMCP (2 mg/mL) in sodium hydroxide solution (0.1 M, pH = 5.5) at room temperature, and mechanical stirring was conducted at 800 rpm.

The PG solution was added dropwise to CS dispersion (3.18 PG: 3 QS % m/m) under mechanical stirring at 800 rpm at room temperature. Then, the HPMCP solution (3 QS: 1 HPMCP, % m/m) was added dropwise into the mixture under mechanical stirring at 800 rpm at room temperature. The nanocomposite dispersion was kept under mechanical stirring at 800 rpm for 30 min at room temperature (24 °C). For dripping, an insulin syringe (100 U.I.) (Descarpack, São Paulo, Brazil) with a 26.5 g $\frac{1}{2}$" needle attached (São Paulo, Descarpack, Brazil) was used for each solution at the speed of 100 drops per minute. The ratio of polymers (3 QS: 1 HPMCP, % m/m) was chosen based on a previous study by our research group [33]. The same method was used to make empty polymeric nanocomposites (EMP-PN) or PG-free by using CS solution, HPMCP solution, and 0.2 M sodium phosphate buffer pH = 6.0.

2.3. Experimental Animals

This study utilized conventional heterogeneous male Wistar rats (*Rattus norvegicus albinus*) weighing 220 to 250 g, 7 weeks old, and housed in groups of five per cage with shavings as bedding material under a 12 h light/dark cycle at a constant room temperature (24 ± 2 °C) and humidity (60%). Water and standard chow were available ad libitum. All experiments were carried out within the animals' circadian cycle, respecting the light cycle, between 7 and 19 h. If animal manipulation was necessary after the light cycle, a 15-watt red lamp was used to illuminate the room. All procedures followed the Committee for Research and Ethical issues of the International Association for the Study of Pain [34], and the study was approved by the Ethics Committee on the Use of Animals of Universidade Federal de Alfenas, Brazil (protocol number 57/2016).

All animals were randomly assigned to an experimental group based on the treatment they received orally: water (animals treated with ultra-pure water–control), pregabalin (PG) (animals treated with free pregabalin), PG-PN (animals treated with polymeric nanocomposite containing pregabalin), and EMP-PN (animals treated with polymeric nanocomposite without pregabalin). For this, free PG was prepared by dissolving PG in distilled water. PG-PN and EMP-PN were prepared according to 2.2 item. All formulations were prepared shortly before use and administrated orally in a single dose by gavage (2.5 mL/kg). The dose was chosen based on the research group's earlier investigations [35], which used a dose of 10 mg pregabalin/kg animal weight.

2.4. Characterization of PNs

Dynamic light scattering (DLS) and electrophoretic light scattering on a Nano Zs (Malvern Instruments®, Worcestershire, England) were used to assess particle size, polydispersity index (PDI), and zeta potential. After being kept in an ultrasonic bath for 2 min, all samples were examined without previous dilution. The tests were carried out in triplicate at 25 °C, with an attenuation value of 8 and a detection angle of 173°.

Nanoparticle tracking analysis (NTA) on the Nanosight NS 300 (Malvern Instruments®, Worcestershire, England) was also used to measure particle sizes. The samples have previously been diluted in distilled water (1:5000). Three measurements were made in sequence at 25 °C.

An Mpa-210 pH meter (Tecnopon®, Piracicaba, Brazil) was used to determine the pH without previous dilution.

The samples were diluted (1:100) in ultrapure water before being dripped onto previously cleaned silicon support, dried in a vacuum drying apparatus, and metalized with carbon for morphological evaluation. After that, the samples were examined by using a high-resolution field emission scanning electron microscope SEM-FEG JEOL® JSM-7500F (Tokyo, Japan) with an energy dispersive spectroscopy (EDS) Ultra Dry model detector (Thermo Scientific®, Waltham, MA, USA).

The possible interactions between PG and the polymers in the nanostructured system were analyzed by Fourier transform infrared spectroscopy (FT-IR) (FT-IR Affinity^{-1}, Shimadzu®, Tokyo, Japan) scanning at 4000 to 600 cm^{-1}, with 64 scans at 4 cm^{-1} resolution, using the dropping technique on potassium bromide (KBr) pellets. The samples were concentrated 4 times; 3 drops of PG-PN or EMP-PN (15 µL each drop) were dropped onto a KBr pellet, with a 30 min interval between drops. After adding 3 drops, the KBr pellets were placed to dry in a desiccator containing silica under vacuum for 19 h. After drying, the KBr pellet was analyzed. The KBr pellets containing the solutions of the polymers and the drug were prepared and analyzed in the same manner.

A differential exploratory calorimeter DSC 3500 Sirius (Netzsch®, Selb, Germany) was used to analyze freeze-dried PN samples, calibrated with Indium, Tin, Bismuth, and Zinc standards, by placing the sample (5–7 mg) in a closed aluminum sample holder with a perforated lid under a dynamic nitrogen atmosphere (50 mL/min) with a heat flow of 10 °C/min.

Thermogravimetric measurements (TG) and Differential Thermal Analysis (DTA) were taken by using freeze-dried samples with masses ranging from 3.4 to 8.2 mg (depending on sample particle size) in aluminum sample holders and a heating rate of 10 °C/min in a simultaneous thermogravimetric TG/DTA 7300 module (Star®, Kyoto, Japan) calibrated with Indium standard. The tests were conducted at temperatures ranging from 30 to 550 °C under an inert atmosphere of nitrogen at a steady flow rate of 50 mL/min.

2.5. Pharmacokinetic Study

For the pharmacokinetic study, the rats were split into two groups (n = 12): PG and PG-PN. Before the treatments, each rat was cannulated in the jugular vein to collect blood and was housed in its cage [36]. Drugs were administered 12 h after cannulation, and 500 µL of blood samples was taken after 0.16, 0.5, 1, 1.5, 2.5, 4, 6, 8, 12, 24, 36, and 48 h, and the volume was reposed with sterile saline. After the last blood collection, the animals were euthanized by an excess of anesthetic (isoflurane 8%, inhalation route). The lack of vital signs and mucosal stains indicated death. In order to avoid interference with absorption, the animals received water ad libitum during the experiment and were fasted for at least six hours before and up to two hours following drug administration.

The samples were collected in heparinized tubes and centrifuged ($2500 \times g$ for 10 min). The plasma was separated and stored at −70 °C for the pregabalin assay.

Pregabalin was measured by using high-performance liquid chromatography-mass spectrometry (LCMS-8030, Shimadzu®, Tokyo, Japan) in positive electrospray ionization (ESI) mode, with the following mass transitions monitored: Pregabalin was measured at

159.85 > 142.10; 159.85 > 97.20; and 159.85 > 83.20. Metformin was used as an internal standard (IS) and measured at 130.10 > 60.05; 130.10 > 71.05; and 130.10 > 83.20. The method was validated by using a pool of blank plasma (free of any chemical). The sample preparation for analysis included precipitation with acetonitrile (1:10 plasma: acetonitrile) and subsequent high-speed centrifugation (17,800× g for 10 min). After centrifugation, 900 microliters of the supernatant was collected and vacuum evaporated at 80 °C. The residue was resuspended in 200 microliters of mobile phase (the gradient's initial concentration), and 50 microliters was chromatographically analyzed.

The mobile phase was acetonitrile: 2 mM ammonium formate solution pH 3.0 in a gradient flow (initial condition 70:30% v/v, maintained until 7 min, followed by a linear reduction of 1 min of the organic phase ratio to 50:50% v/v, which was maintained for 5 min to clean the column and followed by a linear return to the initial conditions for 1 min and stabilization of the column at the initial condition for 5 min) with a constant flow rate of 0.2 mL/min. Formic acid was used to adjust the pH of the ammonium formate solution. As a stationary phase, a BEH HILIC ACQUITY UPLC column (1.7 µm, 2.1 mm × 50 mm) was used with an oven set to 50 °C, a UV detector set at 190 nm, and a chromatographic analysis time of 19 min. The nebulizer gas flow was 1.5 L/min, the DL temperature was 250 °C, the heating block temperature was 400 °C, and the drying gas flow was 15 L/min in the mass spectrometer.

The method was validated according to the Food and Drug Administration's (FDA) validation guidelines [37], with a detection threshold of 1.17 µg/mL of plasma. Aside from that, the technique was precise and accurate, with a linear range of 0.1–12.50 µg/mL of plasma.

Pharmacokinetic parameters were calculated based on plasma concentrations. In order to evaluate differences between the groups' PG and PG-PN, bioavailability (measured by the area under the curve—AUC), distribution (represented by the volume of distribution—Vd), and elimination (expressed by the half-life—$t_{1/2}$ and clearance) parameters were used.

The curve was constructed by using plasma concentration versus time ($AUC^{0-\infty}$) and estimated using the trapezoid technique [38]. The PKSolver add-in in Microsoft Excel® was used to perform pharmacokinetic analysis [39].

2.6. Antinociceptive Effect Study

The sciatic nerve's chronic constriction injury (CCI) was employed as a model for the induction of neuropathic pain [40]. In summary, animals were anesthetized with ketamine (90 mg/kg, intraperitoneal route—IP) and xylazine (10 mg/kg, IP), and the sciatic nerve was exposed and loosely ligated with 4-0 chronic gut thread at four sites with a 1 mm interval.

The rats were randomly assigned to one of four groups (n = 12 each group): water; PG; PG-PN; or EMP-PN. Each group consisted of six CCI rats (n = 6) and six sham rats (n = 6) with the sciatic nerve exposed but not ligated.

In order to assess mechanical allodynia, rats were placed in cages with an elevated metal mesh floor and given at least 30 min to acclimate. Mechanical allodynia was assessed by recording paw withdrawal in response to increasing stimulation with a series of calibrated von Frey filaments (Aesthesio, San Jose, CA, USA) ranging between 0.6 and 60 g in the ipsilateral paw's medial plantar area. The 50% paw withdrawal threshold was determined using the method previously described by Dixon [41].

The nociceptive threshold was determined before the development of neuropathic pain (baseline latency–BL, on day 0). Thus, based on a previous investigation [35], mechanical allodynia was evaluated on the fourteenth day of CCI. On the fourteenth day, the drugs were delivered in a single dosage. In order to prevent interfering with absorption, animals were fasted for at least six hours before and up to two hours following drug administration. In order to determine the effect in hours, the nociceptive threshold was determined before and after drug administration at 1, 2:15, 4, 8, 24, 48, and 72 h (Figure 1). The measurements were always taken throughout the 12 h light cycle.

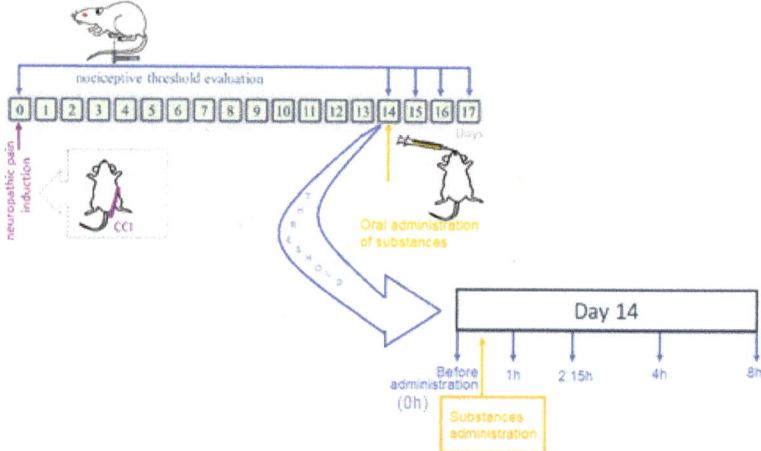

Figure 1. The experimental design was adopted to evaluate pregabalin delivery systems based on polyelectrolyte nanocomposites in comparison to the free drug for oral treatment of neuropathic pain. CCI—chronic constriction injury.

2.7. Motor Coordination and Balance Evaluation

The rats were split into four groups (n = 6): water (control), PG, PG-PN, and EMP-PN. A rotarod device (Insight, Ribeirão Preto, Brazil) was used to test motor coordination and balance. All rats were subjected to a two-day training session, during which time they achieved a consistent baseline level of performance [42]. Rats were trained to walk against the rotation of a revolving drum at a speed of 5 to 37 revolutions per minute (R.P.M.) for a maximum of 4 min (min) at that time. Following the training days, a one-day test was conducted by utilizing the apparatus's accelerating speed level (5 to 37 R.P.M.) mode for four minutes. For the beginning, the baseline latency (BL) was determined, then new tests were conducted 1, 2:15, 4, 8, 24, and 48 h after the drugs were administered. The average time it took for the rotarod to fall off was then recorded.

2.8. Assessment of Barbiturate-Induced Sleep

The animals (n = 12 per group) were split into four groups: water (control); PG; PG-PN; and EMP-PN. They were administered thiopental (40 mg/kg IP) 1 h after receiving the drugs [35]. Anesthesia time was calculated as the time between the loss of straightening reflex and the time it would take for the reflex to return [43,44].

2.9. Statistical Analysis

In pharmacokinetic analysis, data are given as a median, and in behavior experiments, data are reported as a mean ± standard error of the mean (S.E.M.). The Mann–Whitney two-tailed test was used to analyze pharmacokinetic data. Two-way analysis of variance (ANOVA) with repeated measurements followed by a Bonferroni test was used to analyze mechanical allodynia. Motor test and barbiturate-induced sleep were compared using a two-way ANOVA with repeated measurements followed by a Newman–Keuls test. Statistical tests were conducted with a 95 percent confidence interval. Statistical tests were performed with Statistica 7.0 (StatSoft Power Solutions, Inc., Hamburg, Germany); the threshold of significance for all statistical tests was set at 5%.

3. Results

3.1. Characterization of PNs

The PNs suspensions were homogenous and opalescent to a slight degree. The average particle size of PG-PN was determined using the DLS method, PdI, and the zeta potential of

432 ± 20 nm; 0.238 ± 0.001; and +19.0 ± 0.9 mV. For EMP-PN, these values are 425 ± 24 nm; 0.234 ± 0.016; and +18.6 ± 1.0 mV, respectively (Figures 2 and 3).

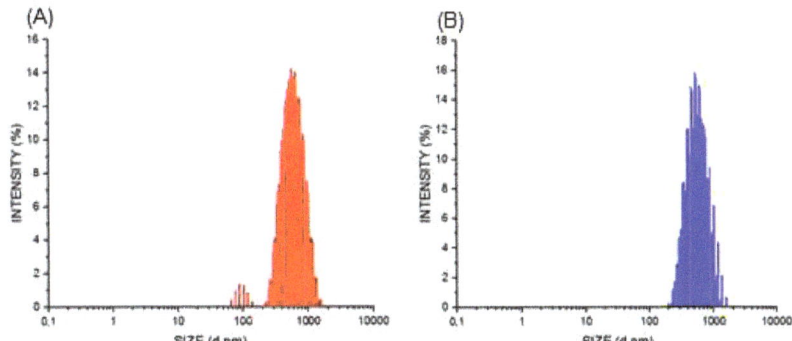

Figure 2. Distribution of PN's size by intensity analyzed by DLS: EMP-PN (**A**) and PG-PN (**B**).

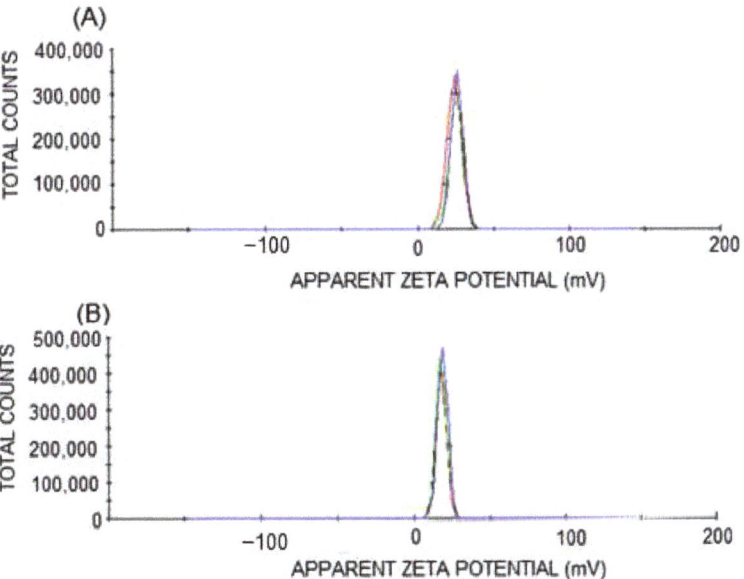

Figure 3. Distribution of PN's zeta potential: EMP-PN (**A**) and PG-PN (**B**).

PG-PN had a pH of 5.7 ± 0.06 while EMP-PN had a pH of 5.6 ± 0.08. Due to the fact that this drug is easily absorbed by oral route of administration in the pH range of 5.5–6.3 [45] and that it is required for the formulation's equilibrium and stability, this pH of the PG-PN formulation can be considered adequate for oral PG absorption.

The size distribution of PNs using the NTA method (Figure 4) is as follows: around 70% (EMP-PN or PG-PN) between 100 and 200 nm (average size: EMP-PN = 174 nm and PG-PN = 193 nm) and also between 200 and 300 nm and 300–500 nm (about 30 percent). These bigger size populations are compatible with the DLS results because the laser incidence angle is fixed in the latter approach; thus, the larger particles scatter lighter and are identified as a monodisperse population [46].

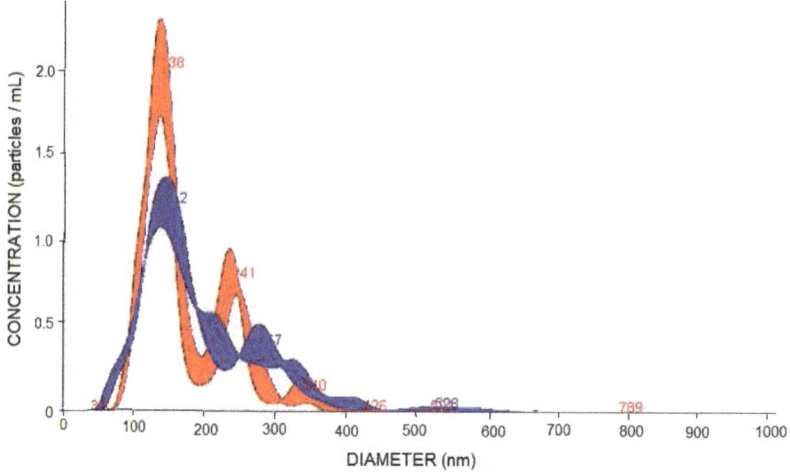

Figure 4. Distribution of PN's size by concentration analyzed by NTA: EMP-PN (red) and PG-PN (blue).

Figure 5 shows SEM pictures of PNs with rounded shapes and nanometric sizes, which correspond well with NTA and DLS results. Furthermore, due to the intumescence of the polymers in an aqueous medium, the EMP-PN particles have a lighter halo around them (Figure 5B), indicating that the particles are breaking apart after 30 days in the aqueous medium. This halo surrounding the EMP-PN was not observed in PG-PN (Figure 5A), suggesting that the particles become more stable when PG is added.

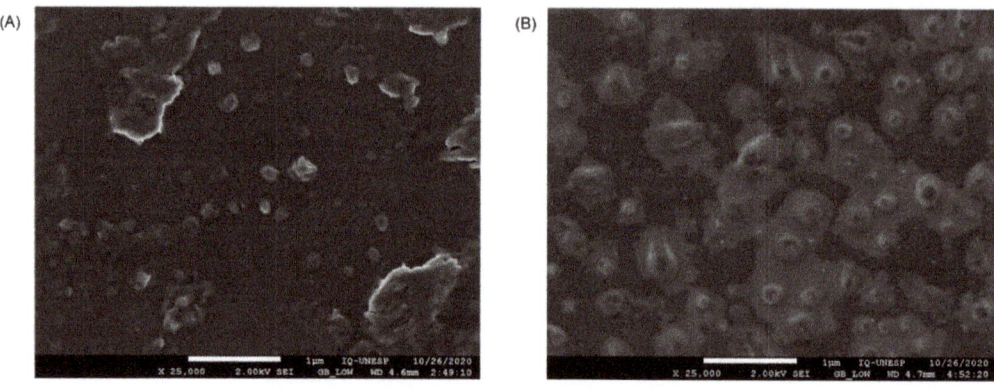

Figure 5. High-resolution field emission scanning electron microscopy (SEM) of PG-PN (**A**) and EMP-PN (**B**).

The development of a deformation band at 1411 cm^{-1} (EMP-PN) and 1413 cm^{-1} (PG-PN) in the FT-IR spectra (Figure 6) suggested polyelectrolyte complexation since ammonium ion and carboxylate ion enhances absorption in this spectral region. Furthermore, there were more strong bands in the PG-PN and EMP-PN spectrums in the range of 1700–1500 cm^{-1} regarding the interacting carboxylate ion and ammonium ion, producing carboxylic acid and amine, as predicted. Furthermore, there was displacement and higher absorption intensity in the N-H stretching band overlapping the O-H stretching in the PG-PN and EMP-PN spectrums (3697–3032 cm^{-1} for PG-PN and 3658–3035 cm^{-1} for EMP-PN), which is consistent with polymer complexation. Furthermore, the presence of the most intense band in the PG-PN spectrum between 2970 and 2912 cm^{-1}, which refers to the stretching of the CH_3 group, shows that pregabalin interacts with the polymers and alters the conformation of the PN. Furthermore, a band between 2295 and 2102 cm^{-1} emerged in the PG-PN spectra, corresponding to NH_3^+ overtones, which indicates that PG or QS amine groups are more exposed as a result of the drug's change in particle shape. Figure 7 depicts the structural formula of the isolated compounds as well as the potential interactions between the polymers and the polymers and PG.

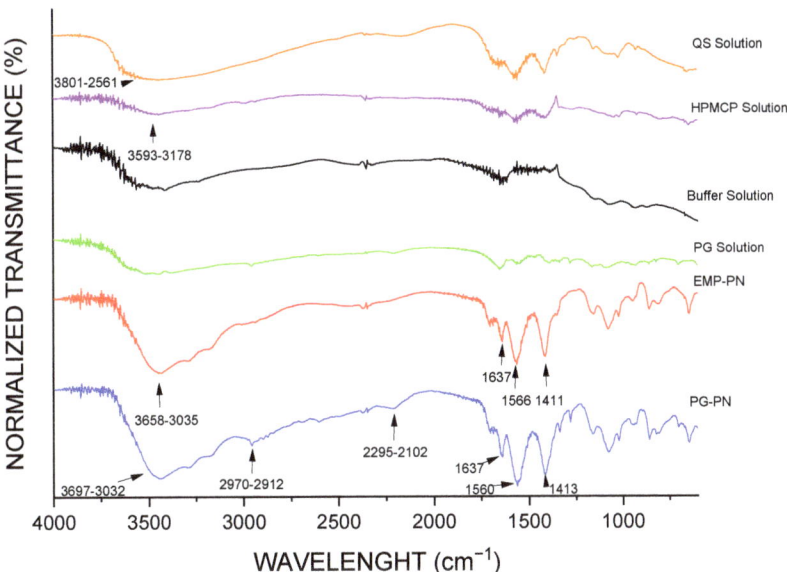

Figure 6. FT-IR by KBr dropping of nanocomplexes and formulation components.

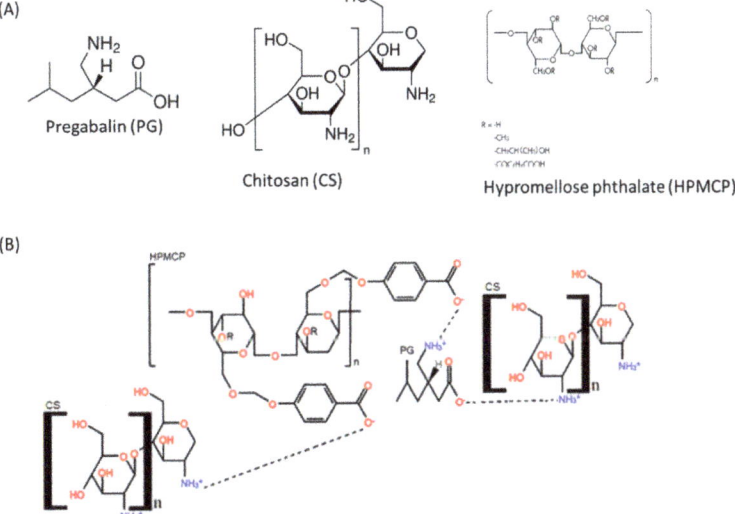

Figure 7. (A) Structural formula of pregabalin, chitosan, and hypromellose phthalate. (B) Possible interaction between the polymers and pregabalin.

DSC (Figure 8) and TG/DTA (Figure 9) were also used to demonstrate nanocomposite formation. The presence of high-temperature events (297.9 °C and 324.5 °C, respectively), which are characteristics of thermal decomposition, suggest polymeric nanocomposite formation. While the endothermic events at 94.8 °C in EMP-PN and 83.7 °C in PG-PN correspond to the polymers, these peaks shifted relative to the solution of isolated polymers (116.4 °C for QS and 116.7 °C for HPMCP), which suggests nanocomposite formation. Furthermore, PG-PN differed from EMP-PN by exhibiting a peak corresponding to PG melting (176.6 °C), which was also displaced when compared to PG in the solution (164.2 °C), implying that the PG was not entirely integrated into the polymer matrix or did not fully interact with the polymers. Additionally, the peaks at 59 °C and 94.8 °C (EMP-PN) and 59.9 °C and 83.7 °C (PG-PN) indicated a glass transition with hysteresis followed by the release of water molecules from polymeric interstice.

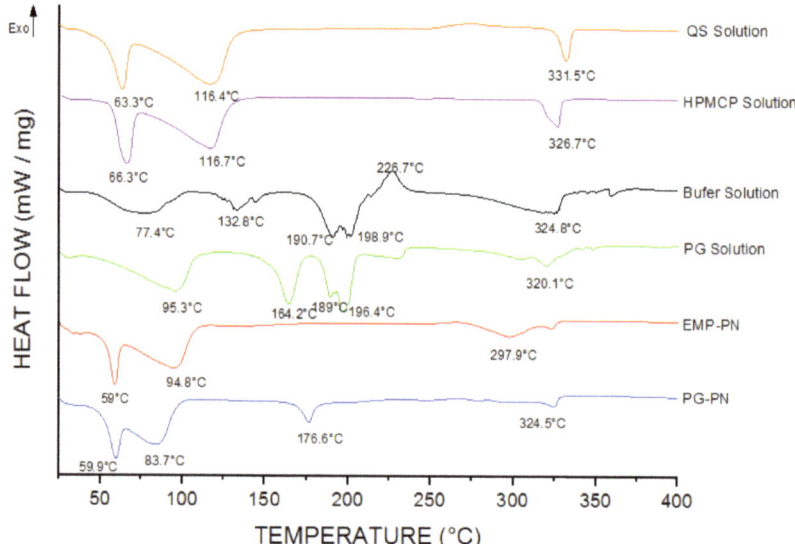

Figure 8. DSC curves of the polyelectrolytic nanocomposites and formulation components obtained in the heating range of 25–400 °C at a heat flow of 10 °C/min under dynamic nitrogen atmosphere flow (50 mL/min).

According to the TG thermograms (Figure 9), the EMP-PN showed an initial mass loss of 20%, indicating dehydration, and a sudden mass loss from 195 °C, indicating breakdown of the polymeric matrix produced (Figure 9A). PG-PN showed an initial mass loss of 25%, implying dehydration, as well as a mass loss of 15% between 150 °C and 175 °C, indicating the decomposition of PG that did not interact with the polymeric matrix, as well as significant mass loss from 200 °C, denoting degradation of the polymeric matrix formed (Figure 9B). These findings are congruent with the DSC findings.

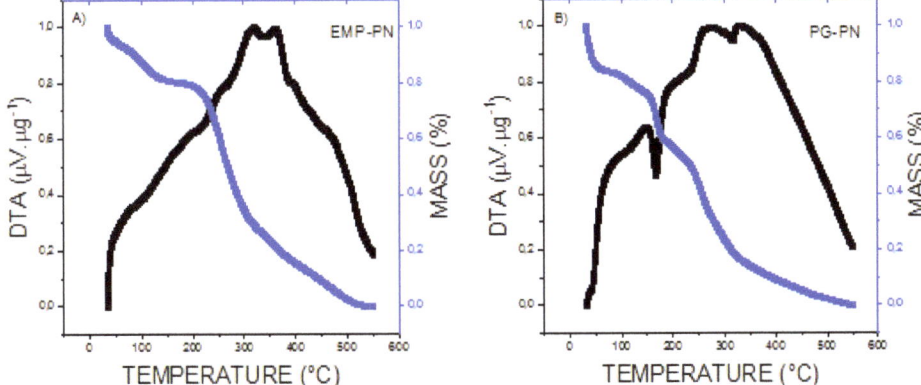

Figure 9. TG (in blue line) and DTA (in black line) curves of EMP-PN (**A**) and PG-PN (**B**) obtained due to heating at 10 °C/min in the range from 25 to 550 °C, under dynamic nitrogen atmosphere.

3.2. Pharmacokinetic Study

Table 1 shows that when the median of each pharmacokinetic parameter (AUC, Cmax, Tmax, Vd, $t_{1/2}$, and clearance) was assessed, PG-PN had a significant decrease in AUC and Cmax ($p < 0.05$ and $p < 0.01$, respectively) compared to PG. Tmax and $t_{1/2}$ of PG-PN were not different ($p > 0.05$) from PG. Furthermore, there was a substantial increase in Vd and the clearance of PG-PN ($p < 0.05$ for both parameters) than compared to PG. When compared to PG, PG-PN raised Vd by 3.3 times, showing that PG-PN allowed for higher PG distribution throughout the organism.

Table 1. Estimated pharmacokinetic parameters after oral administration of PG and PG-PN (10 mg/kg) in rats (n = 6, for each profile). Data expressed as median, CI = 95%.

Parameter	PG	PG-PN
AUC $(^{0-\infty})$ (µg/mL h)	39.60 (82.55–24.25)	22.08 * (32.16–13.53)
Cmax (µg/mL)	6.56 (5.06–13.88)	3.48 ** (5.70–1.40)
Tmax (h)	0.75 (0.5–4)	1.5 (0.5–2.5)
Vd (L/kg)	2.51 (1.82–6.44)	8.21 * (98.02–3.33)
$t_{1/2}$ (h)	7.03 (6.68–10.83)	10.01 (178.35–7.42)
Cl (L/kg/h)	0.25 (0.12–0.41)	0.46 * (0.31–0.74)

Legend: Mann–Whitney test two-tailed. * $p < 0.05$, ** $p < 0.01$.

Furthermore, a representative figure of mean values obtained in each non-compartmental profile demonstrates that PG achieved greater plasma concentrations than PG-PN (Figure 10). Figure 10 further showed that the release of PG from the PN happened in two stages. The first has a maximum peak time of one hour, and the second has a maximum peak time of eight hours. The first step is associated with the instantaneous release of PG that was not previously present in the polymer matrix, whereas the second stage is associated with the release of PG from the polymer matrix.

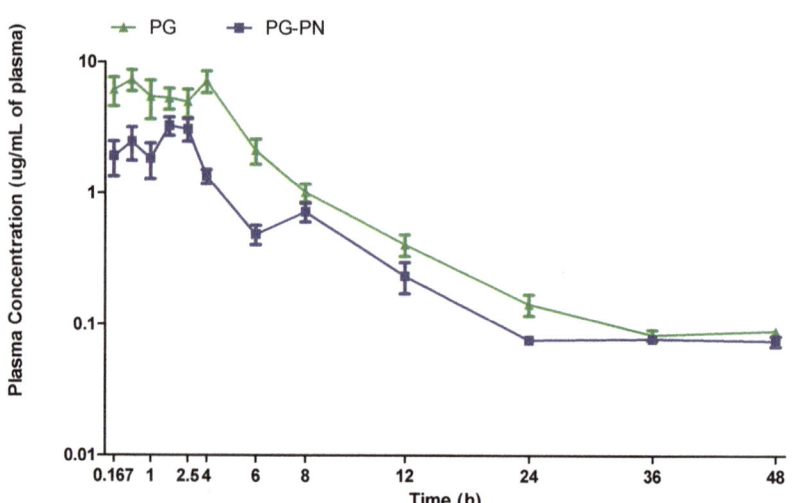

Figure 10. Mean pharmacokinetic profile after oral administration of PG and PG-PN (10 mg/kg) in rats (n = 6, for each profile) administered as a single dose.

3.3. Antinociceptive Effect Study

When compared to sham rats, CCI caused mechanical allodynia in all groups on the fourteenth day before drug administration ($p < 0.01$, $F_{4,25} = 4.05$), (Figure 11).

Figure 11 also shows that the animals with neuropathic pain pretreated with PG or PG-PN (CCI-PG or CCI-PG-PN) showed lower mechanical allodynia than the CCI-WATER group after 1, 2:15, and 4 h ($p < 0.001$, $p < 0.001$ and $p < 0.01$, $F_{4,25} = 4.05$, respectively) and 1, 4, 8, and 48 h ($p < 0.05$. $p < 0.001$, $p < 0.01$ and $p < 0.01$, $F_{4,25} = 4.05$, respectively) of its administration, respectively. In addition, this antinociceptive effect was longer (up to 48 h) in the PG-PN group compared to the PG group ($p < 0.01$).

Furthermore, the CCI-EMP-PN group showed no significant differences from the CCI-WATER group at any time, showing that EMP-PN had no antinociceptive impact.

However, the CCI-PG-PN group had a significantly higher antinociceptive impact than the CCI-PG, CCI-EMP-PN, and CCI-WATER groups at 48 h ($p < 0.01$, $F_{4,25} = 4.05$), indicating a sustained antinociceptive effect caused by pregabalin release from the polymeric matrix. Furthermore, when the CCI-PG and CCI-PG-PN groups were compared, the antinociceptive profile was similar as there were no significant differences, except 48 h after administration of the substances $p < 0.01$, $F_{4,25} = 4.05$), as previously mentioned, indicating that PG was released for a longer time.

The CCI-PG group only had a significantly greater antinociceptive effect than the CCI-EMP-PN group 1 h ($p < 0.05$, $F_{4,25} = 4.05$) after the substances were administered; the rest of the evaluations had a similar antinociceptive profile because there were no significant differences, indicating that EMP-PN has a similar antinociceptive effect potential to PG. Furthermore, when comparing the CCI-PG-PN groups with the CCI-EMP-PN groups, there was a significant difference only at times 4 h ($p < 0.01$, $F_{4,25} = 4.05$) and 48 h ($p < 0.01$, $F_{4,25} = 4.05$), which showed that EMP-PN has antinociceptive potential owing to the presence of chitosan.

Figure 12 shows that the nociceptive threshold did not change in the sham groups that received substances ($p > 0.05$, $F_{3,20} = 0.91$), suggesting that the sham procedure did not generate induced neuropathic pain and that the substances had no hypoalgesic impact.

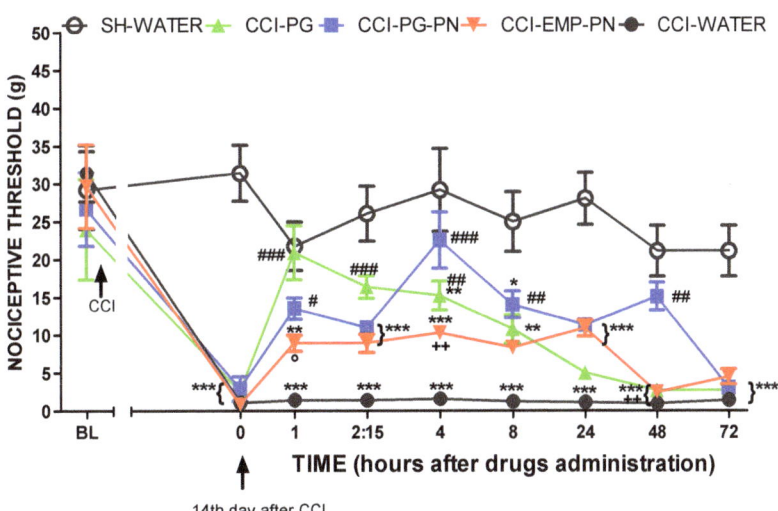

Figure 11. The antinociceptive impact of the drugs—operated animals and sham—compared 72 h after treatment. Legend: Pregabalin: PG (10 mg/kg, Orally); WATER: vehicle of PG; PN: polymeric nanocomposite; PG-PN: PG-loaded polymeric nanocomposite (10 mg PG/kg, Orally); EMP-PN: polymeric nanocomposite without PG (vehicle of PG-PN); CCI: chronic sciatic nerve constriction injury; SH: sham; BL: baseline latency; 0: on 14th day after CCI, immediately before administration of the substances; experimental time: 72 h after administration of the substances. Two-way ANOVA test with repeated measures with Bonferroni post-test was applied for the comparison between the different groups, * $p < 0.05$; ** $p < 0.01$; *** $p < 0.001$ when compared with the SH-WATER group, ## $p < 0.01$; ### $p < 0.001$ when compared with the CCI-WATER group; ++ $p < 0.01$ when compared with the CCI-PG-PN group; ° $p < 0.05$ when compared with the CCI-PG group. Each point represents the mean ± standard error of the mean (±S.E.M.) of six animals.

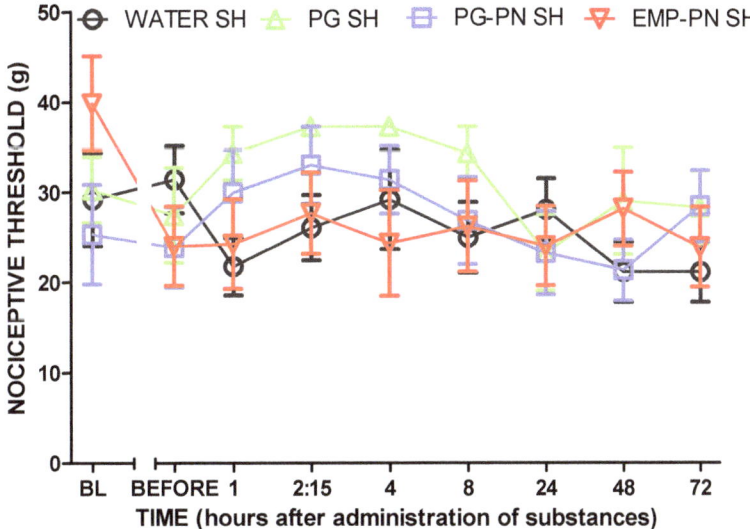

Figure 12. Antinociceptive effect of the comparison between the substances—sham animals—during 72 h after administration. Legend: Pregabalin: PG (10 mg/kg, Orally); WATER: vehicle of PG; PN:

polymeric nanocomposite; PG-PN: PG-loaded polymeric nanocomposite (10 mg PG/kg, Orally); EMP-PN: polymeric nanocomposite without PG (vehicle of PG-PN); SH: sham; BL: baseline latency; BEFORE: on day 14 after CCI, immediately before administration of the substances; experimental time: 72 h after administration of the substances. A two-way ANOVA test with repeated measures with Bonferroni post-test was applied for the comparison between the different groups; there was no significant difference between the groups. Each point represents the mean ± standard error of the mean (±S.E.M.) of six animals.

3.4. Motor Coordination and Balance Evaluation

According to Figure 13, all animals were able to maintain balance on the rotating bar for the established period (4 min) and at all tested times (BL, 1, 2:15, 4, 8, 24, and 48 h) ($p > 0.05$), demonstrating that the administration of the substances did not cause motor coordination loss in the animals and implying that the administered substances do not cause motor neurological deficit. Furthermore, this finding allows us to conclude that the antinociceptive impact of PG, PG-PN, and EMP-PN (item 3.2) was sensory rather than motor.

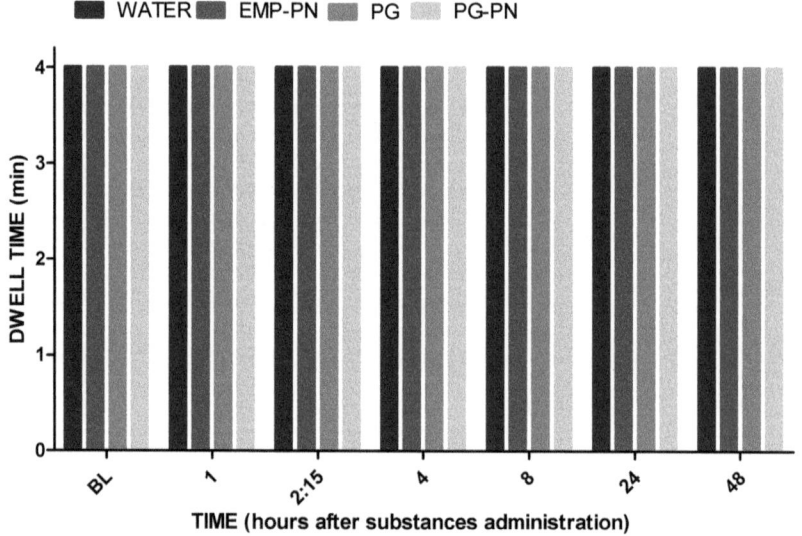

Figure 13. Motor coordination and balance evaluation. Legend: Pregabalin: PG (10 mg/kg, Orally); WATER: vehicle of PG; PN: polymeric nanocomposite; PG-PN: PG-loaded polymeric nanocomposite (10 mg PG/kg, Orally); EMP-PN: polymeric nanocomposite without PG (vehicle of PG-PN); BL: baseline latency; experimental time: 48 h. A two-way ANOVA test with repeated measures with Newman–Keuls post-test was applied for comparison between the different groups. No significant difference was observed. Each bar represents the mean ± standard error of the mean (±S.E.M.) of six animals.

3.5. Evaluation of Barbiturate-Induced Sleep

Figure 14 shows that the WATER and EMP-PN groups had the same sleep time since there was no significant difference ($p > 0.05$), indicating that the EMP-PN components (polymers) do not have nonspecific CNS depressive activity. Furthermore, when compared to the WATER group, the PG and PG-PN groups showed a significant increase in sleep time ($p < 0.001$ and $p < 0.01$, respectively), demonstrating that both substances have CNS depressant action because the barbiturate-induced sleep was potentiated 1.81 and 1.50 times, respectively, as expected and because the test principle states that substances with CNS depressant action, in general, reduce latency anesthesia [1].

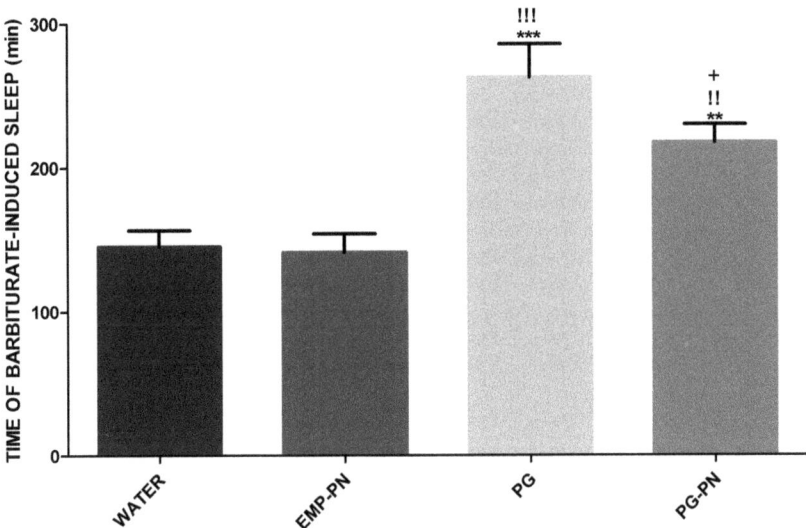

Figure 14. Evaluation of barbiturate-induced sleep time. Legend: Pregabalin: PG (10 mg/kg^{-1}, Orally); WATER: vehicle of PG; PN: polymeric nanocomposite; PG-PN: PG-loaded polymeric nanocomposite (10 mg PG/kg, Orally); EMP-PN: polymeric nanocomposite without PG (vehicle of PG-PN). Administration of thiopental (40 mg/kg, I.P.) 1 h after the administration of substances for sleep induction. One-way ANOVA test with Newman–Keuls post-test was applied for comparison of the different groups, ** $p < 0.01$ and *** $p < 0.001$ when compared with the WATER group; !! $p < 0.01$ and !!! $p < 0.001$ when compared with the EMP-PN group; + $p < 0.05$ when compared with the PG group. Each bar represents the mean ± standard error of the mean (±S.E.M.) of 12 animals.

Furthermore, when compared to the EMP-PN group, there was a significant increase in sleep time in the PG and PG-PN groups ($p < 0.001$ and $p < 0.01$, respectively), demonstrating that, in addition to PG and PG-PN being CNS depressants, the depressant action occurs only due to PG, as EMP-PN has the same sleep time as WATER.

The most significant outcome of this study was a 0.83-fold reduction in sleep time in the PG-PN group compared to the PG group ($p < 0.05$), demonstrating that when PG is delivered in a polymeric nanocomposite, its major adverse effect, which is drowsiness, is reduced.

4. Discussion

The lack of effectiveness and limitations of existing therapies, as well as side effects and tolerance to medicines, are known to render treating neuropathic pain challenging [47–49]. Pregabalin (PG) is the first-line therapy for neuropathic pain [50]. However, this drug's adverse effects cause treatment abandonment [4,5]. With this in mind, we created a controlled release pregabalin system (PG-PN) and tested it in vivo.

For this, we considered the use of the natural polymers HPMCP and chitosan. Chitosan is depolarized by the action of lysozyme and N-acetylglucosaminase, generating the sugars N-acetylglucosamine and glucosamine, which explains its biodegradable characteristic. Due to the fact that it has low toxicity, chitosan can be considered biocompatible [51]. HPMCP is a cellulose derivative that is not absorbed by the body and has low toxicity, and it is also considered biocompatible and biodegradable [52]. In a previous study of the group [53], polymeric nanocomposites obtained by ionic cross-linking of chitosan and HPMCP presented low toxicity against fibroblast cells of the CCD-1059sk strain. For these reasons, polymeric nanocomposites of chitosan and HPMCP are considered safe for oral administration.

PG-PN and EMP-PN exhibited similar mean diameters, indicating that the medication did not appreciably alter PN size. Nanoparticles smaller than 500 nm can penetrate the blood–brain barrier and operate on the central nervous system [54,55]. As a result, these particles are smaller than other studies' suggested microspheres containing PG [10–12]; therefore, they are suited for the intended goal.

The average PdI values obtained for PG-PN and EMP-PN were less than 0.250, which is below the maximum permissible limit for polymeric nanoparticles of 0.300. However, mini-emulsion eye drops containing PG [56] with a PdI smaller than 0.38 and the size distribution were also found to be homogeneous because we know that uncoiled polymer chains scatter light, affecting the overall size measurement [57,58].

Moreover, PG-PN and EMP-PN had similar zeta potentials (+20 mV). The polymers in the formulation determined the particle charge [59]. Due to the fact that sample pH (5.7) is lower than the chitosan pKa (6.5), the positive zeta potential is related to the chitosan contained in the polymer matrix [60]. The positive zeta potential in chitosan-covered mini-emulsion with PG [56] was observed. The zeta potential values of PG-PN were close to EMP-PN in our investigation, indicating that PG is entangled in the polymer matrix.

The pH of PG-PN was 5.7, and the pH EMP-PN was 5.6. The pH readings were not substantially different, showing that the presence of PG does not affect the formulation's final pH. Moreover, the pH of the PG-PN formulations is suitable for oral PG absorption, since this medication is efficiently absorbed in a pH range of 5.5–6.3 [45].

Furthermore, spherical particles have more surface area than shapeless particles, which improves interaction with the intestinal epithelium and, hence, oral absorption [61]. The fact that the PNs in this investigation were spherical and nanosized implies that they are suitable for oral absorption and CNS activity.

FT-IR analysis revealed that PG crosslinks with QS and HPMCP polymers in an ammonium-carboxylate reaction. In thermal analysis, the absence of peaks corresponding to the drug's melting point indicates that the drug is in amorphous form and, hence, molecularly dispersed in the polymer matrix [62,63]. Endothermic peaks around the melting of PG were observed in microspheres [12], showing that PG does not interact with mucilage. This is also observed in the present study. However, other studies noticed the absence of the endothermic peak associated with PG melting in the produced microspheres [10,11].

Pregabalin contains the same functional groups as chitosan (Figure 7A), making it difficult to separate these components and to determine encapsulation efficiency. However, the current study's drug to polymer ratio is known to be the majority (3.18 parts PG: 3 parts QS: 1 part HPMCP, % $m/m/m$), suggesting that PG was not entirely absorbed into the polymer matrix. The percentage of PG that is not linked with the polymeric matrix, on the other hand, plays an essential role in reaching the initial therapeutic plasma concentration of PG.

The bioavailability and maximum plasma levels of PG-PN were found to be lower than PG following a single oral dosage. A strong ionic connection between PG and polymers may explain why PG is not completely released from the polymer matrix. Other investigations employing pregabalin-containing pharmaceuticals [6,20,64] found similar bioavailability and maximum plasma concentration reductions. Other investigations employing pregabalin-containing pharmaceuticals found improved bioavailability and maximum plasma levels [12,19].

The results suggest that PG-PN allowed for more extensive distributions of the substance and deposition in tissues and organs, modifying the Vd [65]. As the volume of distribution grows, so does clearance, as these two factors are proportional. Moreover, higher clearance reduces bioavailability [66]. In contrast, none of the authors that proposed modified-release pharmacological forms for PG published the estimated Vd and Cl values [6,10–12,20,64].

Oral PG-PN and PG were used in this work to alleviate CCI-induced neuropathic pain in rats. As predicted, PG and PG-PN had antinociceptive effects, as PG has been demonstrated to decrease mechanical allodynia in rats [34,67–70]. PG-PN had a longer profile of action than PG, lasting up to 48 h after administration, whereas PG only lasted up to 4 h. The current study's 4 h impact is in line with another study that showed an antinociceptive effect of pregabalin for 8 h [70]. Thus, customized release systems such as microspheres and polymeric nanocomposites are efficient in delaying the release of the active [10]. An intranasal PG microsphere demonstrated better anticonvulsant efficacy than PG given peripherally [11], indicating that modified-release pharmacological formulations of PG are superior to traditional therapy.

The analgesic action of chitosan is due to the absorption of proton ions by ionization of the amino group to NH_3^+ [71]. Furthermore, because pregabalin and chitosan have the same functional groups at Figure 7A, we hypothesized that chitosan acts on the same therapeutic targets as pregabalin (voltage-dependent calcium channels) with respect to exerting the antinociceptive effect, but there is no evidence in the literature to support this hypothesis. However, EMP-PN's antinociceptive effect is weaker than the sham control animals' (which received water). The fact that EMP-PN is antinociceptive is beneficial since it permits the PG-PN vehicle to reduce nociception.

However, the results show that EMP-PN has an antinociceptive impact, whereas PG-PN has a synergistic antinociceptive effect of PG and chitosan. More research is needed to confirm synergistic impacts (additive or potentiating).

Furthermore, as no animals died or showed signs of intoxication after receiving EMP-PN or PG-PN, it appears that the formulation is safe at this dose and for the time observed, although more research is required [72].

In our study, rats were given one oral dosage of PG-PN or PG: Neither drug affected motor behavior nor caused motor impairments. Other researchers found the same PG motor coordination profile [70,73], proving that PG does not impair animal motor coordination. Thus, the efficiency of PG-PN in decreasing one of PG's adverse effects in people, which is the loss of motor coordination, could not be evaluated [1].

PG-PN and PG were also examined in rats following a single oral dosage in order to measure sleepiness reduction [1]. Drowsiness was reduced in our study by 0.83 times in rats receiving PG-PN compared to animals receiving PG, which supports another study [31].

The open-field test was also performed to examine the drugs' sedative effects (data not displayed due to lack of impact). Upon re-exposure to the test, the animals displayed a profile of indifference in exploration that was independent of the drug provided, preventing the detection of CNS depressive impact of PG or PG-PN.

5. Conclusions

The present study used natural polymeric nanocomposites (PN) containing pregabalin (PG) for oral administration in order to minimize administration frequency and adverse effects. The polymeric nanocomposites showed acceptable physicochemical properties for prolonged PG release, making them suitable for in vivo testing. Due to the fact that the in vivo formulation has extended circulation, it has a longer profile of action. In animal trials, PG-PN increased the antinociceptive efficacy of PG in solution while reducing sleepiness, the drug's major adverse effect. The oral administration of PG via natural polymeric nanocomposites may improve treatment compliance and patient well-being when used in clinical practice since it allows less frequent administration with fewer side effects.

Author Contributions: R.F.R.: methodology, investigation, writing—original draft and review; J.B.N.: methodology and investigation; S.B.N.A.: methodology and writing—review; P.F.d.S.: investigation; J.C.-B.: methodology, investigation, and writing—review; R.V.P.: investigation; T.R.M.: investigation; J.T.J.F.: investigation and writing—review; G.R.P.: supervision; F.C.C.: supervision. G.G.: supervision and writing—review and editing; V.B.B.: project administration, supervision, and writing—review and editing. All authors have read and agreed to the published version of the manuscript.

Funding: CAPES (Brazil: Finance code 001) and FAPEMIG for scholarships.

Institutional Review Board Statement: Not applicable.

Informed Consent Statement: Not applicable.

Data Availability Statement: The data presented in this study are available on request from the corresponding author. The data are not publicly available once we can keep these data confidential, as per university policy, until the publication of articles.

Acknowledgments: The authors thank UNIFAL-MG for support (materials and laboratories used for experiments). The authors thank the company, Pensabio (São Paulo, Brazil), for kindly performing NTA analyses.

Conflicts of Interest: The authors declare no conflict of interest. The funders had no role in the design of the study; in the collection, analyses, or interpretation of data; in the writing of the manuscript; or in the decision to publish the results.

References

1. Dworkin, R.H.; O'Connor, A.B.; Backonja, M.; Farrar, J.T.; Finnerup, N.B.; Jensen, T.S.; Kalso, E.A.; Loeser, J.D.; Miaskowski, C.; Nurmikko, T.J.; et al. Pharmacologic management of neuropathic pain: Evidence-based recommendations. *Pain* **2007**, *132*, 237–251. [CrossRef] [PubMed]
2. Food and Drug Administration (FDA). Approval of Lyrica ®(Pregabalin). 2004. Available online: https://www.accessdata.fda.gov/drugsatfda_docs/nda/2004/021446_Lyrica%20Capsules_approv.PDF (accessed on 1 February 2021).
3. Buoli, M.; Caldiroli, A.; Serati, M. Pharmacokinetic evaluation of pregabalin for the treatment of generalized anxiety disorder. *Expert Opin. Drug Metab. Toxicol.* **2017**, *13*, 351–359. [CrossRef]
4. Echterhoff, R.R. A eficácia da pregabalina no tratamento do transtorno de ansiedade generalizada. In *Course Conclusion Paper (Psychiatry Specialization)*; Universidade Federal do Paraná: Curitiba, Brazil, 2012.
5. Moore, R.A.; Straube, S.; Wiffen, P.J.; Derry, S.; McQuay, H.J. Pregabalin for acute and chronic pain in adults. *Cochrane Database Syst. Rev.* **2009**, *8*, CD007076. [CrossRef]
6. Qin, C.; Wu, M.; Xu, S.; Wang, X.; Shi, W.; Dong, Y.; Yang, L.; He, W.; Han, X.; Yin, L. Design and optimization of gastro-floating sustained-release tablet of pregabalin: In vitro and in vivo evaluation. *Int. J. Pharm.* **2018**, *545*, 37–44. [CrossRef] [PubMed]
7. Food and Drug Administration (FDA). Approval of Lyrica CR ®(Pregabalin). 2017. Available online: https://www.accessdata.fda.gov/drugsatfda_docs/nda/2017/209501Orig1s000Approv.pdf (accessed on 1 February 2021).
8. Pfizer. Lyrica Package Insert CR®. 2018. Available online: Labeling.pfizer.com/ShowLabeling.aspx?id=9678 (accessed on 20 January 2021).
9. Fukasawa, H.; Muratake, H.; Nagae, M.; Sugiyama, K.; Shudo, K. Transdermal administration of aqueous pregabalin solution as a potential treatment option for patients with neuropathic pain to avoid central nervous system-mediated side effects. *Biol. Pharm. Bull.* **2014**, *37*, 1816–1819. [CrossRef]
10. Aydogan, E.; Comoglu, T.; Pehlivanoglu, B.; Dogan, M.; Comoglu, S.; Dogan, A.; Basci, N. Process and formulation variables of pregabalin microspheres prepared by w/o/o double emulsion solvent diffusion method and their clinical application by animal modeling studies. *Drug Dev. Ind. Pharm.* **2015**, *41*, 1311–1320. [CrossRef]
11. Taksande, J.B.; Umekar, M.J. Preparation of intranasal pregabalin microspheres: In vitro, ex vivo and in vivo pharmacodynamic evaluation. *J. Pharm. Res.* **2018**, *12*, 112–121.
12. Ghumman, S.A.; Bashir, S.; Noreen, S.; Khan, A.M.; Malik, M.Z. Taro-corms mucilage-alginate microspheres for the sustained release of pregabalin: In vitro &in vivo evaluation. *Int. J. Biol. Macromol.* **2019**, *139*, 1191–1202. [CrossRef]
13. Jeong, K.H.; Woo, H.S.; Kim, C.J.; Lee, K.H.; Jeon, J.Y.; Lee, S.Y.; Kang, J.H.; Lee, S.; Choi, Y.W. Formulation of a modified-release pregabalin tablet using hot-melt coating with glyceryl behenate. *Int. J. Pharm.* **2015**, *495*, 1–8. [CrossRef]
14. Cevik, O.; Gidon, D.; Kizilel, S. Visible-light-induced synthesis of pH-responsive composite hydrogels for controlled delivery of the anticonvulsant drug pregabalin. *Acta Biomater.* **2015**, *11*, 151–161. [CrossRef] [PubMed]
15. Madan, J.R.; Adokar, B.R.; Dua, K. Development and evaluation of in situ gel of pregabalin. *Int. J. Pharm. Investig.* **2015**, *5*, 226–233. [CrossRef] [PubMed]
16. Kanwar, N.; Kumar, R.; Sarwal, A.; Sinha, V.R. Preparation and evaluation of floating tablets of pregabalin. *Drug Dev. Ind. Pharm.* **2016**, *42*, 654–660. [CrossRef]
17. Cinay, G.E.; Erkoc, P.; Alipour, M.; Hashimoto, Y.; Sasaki, Y.; Akiyoshi, K.; Kizilel, S. Nanogel-integrated pH-responsive composite hydrogels for controlled drug delivery. *ACS Biomater. Sci. Eng.* **2017**, *3*, 370–380. [CrossRef]
18. Arafa, M.G.; Ayoub, B.M. DOE optimization of nano-based carrier of pregabalin as hydrogel: New therapeutic &chemometric approaches for controlled drug delivery systems. *Sci. Rep.* **2017**, *7*, 41503. [CrossRef]
19. Kim, S.; Hwang, K.M.; Park, Y.S.; Nguyen, T.T.; Park, E.S. Preparation and evaluation of non-effervescent gastroretentive tablets containing pregabalin for once-daily administration and dose proportional pharmacokinetics. *Int. J. Pharm.* **2018**, *550*, 160–169. [CrossRef] [PubMed]
20. Kim, K.H.; Lim, S.H.; Shim, C.R.; Park, J.; Song, W.H.; Kwon, M.C.; Lee, J.H.; Park, J.S.; Choi, H.G. Development of a novel controlled-release tablet of pregabalin: Formulation variation and pharmacokinetics in dogs and humans. *Drug Des. Devel. Ther.* **2020**, *14*, 445–456. [CrossRef]

21. Yajima, R.; Matsumoto, K.; Yokono, K.; Watabe, Y.; Enoki, Y.; Taguchi, K.; Ise, Y.; Katayama, S.; Kizu, J. Pharmacokinetic and pharmacodynamic studies of pregabalin suppositories based on pharmacological research. *J. Pharm. Pharmacol.* **2019**, *71*, 746–752. [CrossRef] [PubMed]
22. Venkatesan, J.; Anil, S.; Kim, S.K.; Shim, M.S. Seaweed polysaccharide-based nanoparticles: Preparation and applications for drug delivery. *Polymers* **2016**, *26*, 30. [CrossRef]
23. Thambi, T.; Li, Y.; Lee, D.S. Injectable hydrogels for sustained release of therapeutic agents. *J. Control. Release* **2017**, *267*, 57–66. [CrossRef]
24. Thambi, T.; Phan, V.H.G.; Kim, S.H.; Le, T.M.D.; Duong, H.T.T.; Lee, D.S. Smart injectable biogels based on hyaluronic acid bioconjugates finely substituted with poly(β-amino ester urethane) for cancer therapy. *Biomater. Sci.* **2019**, *7*, 5424–5437. [CrossRef]
25. Phan, V.H.G.; Duong, H.T.T.; Tran, P.T.; Thambi, T.; Ho, D.K.; Murgia, X. Self-assembled amphiphilic starch based drug delivery platform: Synthesis, preparation, and interactions with biological barriers. *Biomacromolecules* **2021**, *22*, 572–585. [CrossRef] [PubMed]
26. Rehman, A.; Jafari, S.M.; Tong, Q.; Riaz, T.; Assadpour, E.; Aadil, R.M.; Niazi, S.; Khan, I.M.; Shehzad, Q.; Ali, A.; et al. Drug nanodelivery systems based on natural polysaccharides against different diseases. *Adv. Colloid Interface Sci.* **2020**, *284*, 102251. [CrossRef]
27. Shah, B.M.; Palakurthi, S.S.; Khare, T.; Khare, S.; Palakurthi, S. Natural proteins and polysaccharides in the development of micro/nano delivery systems for the treatment of inflammatory bowel disease. *Int. J. Biol. Macromol.* **2020**, *165*, 722–737. [CrossRef]
28. Liu, Z.; Jiao, Y.; Wang, Y.; Zhou, C.; Zhang, Z. Polysaccharides-based nanoparticles as drug delivery systems. *Adv. Drug Deliv. Rev.* **2008**, *60*, 1650–1662. [CrossRef] [PubMed]
29. Wang, Y.; Chen, B.Z.; Liu, Y.J.; Wu, Z.M.; Guo, X.D. Application of mesoscale simulation to explore the aggregate morphology of pH-sensitive nanoparticles used as the oral drug delivery carriers under different conditions. *Colloids Surf. B.* **2017**, *151*, 280–286. [CrossRef]
30. Kreuter, J. Drug delivery to the central nervous system by polymeric nanoparticles: What do we know? *Adv. Drug Deliv. Rev.* **2014**, *71*, 2–14. [CrossRef] [PubMed]
31. Arachchige, M.C.M.; Reshetnyak, Y.K.; Andreev, O.A. Advanced targeted nanomedicine. *J. Biotechnol.* **2015**, *202*, 88–97. [CrossRef]
32. Calvo, P.; Remuñán-López, C.; Vila-Jato, J.L.; Alonso, M.J. Novel hydrophilic chitosan-polyethylene oxide nanoparticles as protein carriers. *J. Appl. Polym. Sci.* **1997**, *63*, 125–132. [CrossRef]
33. Barbi, M.D.S.; Carvalho, F.C.; Kiill, C.P.; Barud, H.D.S.; Santagneli, S.H.; Ribeiro, S.J.; Gremião, M.P. Preparation and Characterization of Chitosan Nanoparticles for Zidovudine Nasal Delivery. *J. Nanosci. Nanotechnol.* **2015**, *15*, 865–874. [CrossRef]
34. Zimmermann, M. Ethical guidelines for investigations of experimental pain in conscious animals. *Pain* **1983**, *16*, 109–110. [CrossRef]
35. Rodrigues, R.F.; Kawano, T.; Placido, R.V.; Costa, L.H.; Podestá, M.H.M.C.; Santos, R.S.; Galdino, G.; Barros, C.M.; Boralli, V.B. Investigation of the combination of pregabalin with duloxetine or amitriptyline on the pharmacokinetics and antiallodynic effect during neuropathic pain in rats. *Pain Physician* **2021**, *24*, E511–E520. [PubMed]
36. Harms, P.G.; Ojeda, S.R. A rapid and simple procedure for chronic cannulation of the rat jugular vein. *J. Appl. Physiol.* **1974**, *36*, 391–392. [CrossRef]
37. US Food and Drug Administration. Guidance for industry: Bioanalytical method Validation. Available online: https://www.fda.gov/files/drugs/published/Bioanalytical-Method-Validation-Guidance-for-Industry.pdf (accessed on 10 May 2020).
38. Boroujerdi, M. Noncompartmental approach in pharmacokinetics based on statistical moments. In *Pharmacokinetics: Principles and Applications*; Mc Graw-Hill: New York, NY, USA, 2002; pp. 331–341.
39. Zhang, Y.; Huo, M.; Zhou, J.; Xie, S. PKSolver: An add-in program for pharmacokinetic and pharmacodynamic data analysis in Microsoft Excel. *Comput. Methods Programs Biomed.* **2010**, *99*, 306–314. [CrossRef]
40. Bennett, G.J.; Xie, Y.K. A peripheral mononeuropathy in rat that produces disorders of pain sensation like those seen in man. *Pain* **1988**, *33*, 87–107. [CrossRef]
41. Dixon, W.J. Efficient analysis of experimental observations. *Annu. Rev. Pharmacol. Toxicol.* **1980**, *20*, 441–462. [CrossRef] [PubMed]
42. Dunham, N.W.; Miya, T.S. A note on a simple apparatus for detecting neurological deficit in rats and mice. *J. Am. Pharm. Assoc.* **1957**, *46*, 208–209. [CrossRef]
43. Nassis, C.Z.; Lago, L.C.; Mory, S.B.; Raquel, M.K.S.; Figueiredo, C.R.; Lebre, A.T.; Giesbrecht, A.M. Estudo da ação depressora inespecífica do suco extraído das folhas de *Bryophyllum calycinum* Salisb. (*Crassulaceae*) sobre o sistema nervoso central. Comparação com efeitos da difenidramina. *Arq. Méd. ABC* **1991**, *14*, 64–68.
44. Mello, F.B.; Langeloh, A.; Mello, J.R.B. Study of the toxicity and efficacy in Wistar rats of a phytoterapic used as sedative and/or hypnotic. *Lat. Am. J. Pharm.* **2007**, *26*, 38–44.
45. European Patent Organization. Pregabalin Patent. Available online: www.escavador.com.br/patentes/383391/composicao-PG (accessed on 11 November 2016).
46. Tscharnuter, W. Photon Correlation Spectroscopy in Particle Sizing. In *Encyclopedia of Analytical Chemistry: Applications, Theory, and Instrumentation*, 1st ed.; Meyers, R.A., Ed.; John Wiley & Sons: Chichester, UK, 2006; pp. 5469–5485. [CrossRef]

47. Boyle, J.; Eriksson, M.E.; Gribble, L.; Gouni, R.; Johnsen, S.; Coppini, D.V.; Kerr, D. Randomized, placebo-controlled comparison of amitriptyline, duloxetine, and pregabalin in patients with chronic diabetic peripheral neuropathic pain: Impact on pain, polysomnographic sleep, daytime functioning, and quality of life. *Diabetes Care.* **2012**, *35*, 2451–2458. [CrossRef]
48. Cohen, S.P.; Mao, J. Neuropathic pain: Mechanisms and their clinical implications. *BMJ.* **2014**, *348*, f7656. [CrossRef]
49. Zilliox, L.A. Neuropathic Pain. *Continuum* **2017**, *23*, 512–532. [CrossRef]
50. Cavalli, E.; Mammana, S.; Nicoletti, F.; Bramanti, P.; Mazzon, E. The neuropathic pain: An overview of the current treatment and future therapeutic approaches. *Int. J. Immunopathol. Pharmacol.* **2019**, *33*, 1–10. [CrossRef]
51. Silva, H.S.R.C.; dos Santos, K.S.C.R.; Ferreira, E.I. Chitosan: Hydrossoluble derivatives, pharmaceutical applications, and recent advances. *Quim. Nova* **2006**, *29*, 776–785. [CrossRef]
52. Ergun, R.; Guo, J.; Huebner-Keese, B. *Encyclopedia of Food and Health*; Caballero, B., Finglas, P.M., Toldrá, F., Eds.; Elsevier: Waltham, MA, USA, 2016; pp. 694–702. [CrossRef]
53. Naves, V.M.L. Development of chitosan and hydroxypropylmethyl cellulose phthalate nanoparticles containing methotrexate for potential treatment of glioblastoma. Ph.D. Thesis, Universidade Federal de Alfenas, Alfenas, Brazil, 22 February 2018. Available online: https://bdtd.unifal-mg.edu.br:8443 (accessed on 30 October 2021).
54. Apolinário, A.C.; Salata, G.C.; Bianco, A.F.R.; Fukumori, C.; Lopes, L.B. Opening the pandora's box of nanomedicine: There is indeed "plenty of room at the bottom". *Quim. Nova* **2020**, *43*, 212–225.
55. Masserini, M. Nanoparticles for brain drug delivery. *ISRN Biochem.* **2013**, *2013*, 238428. [CrossRef] [PubMed]
56. Ibrahim, M.M.; Maria, D.N.; Mishra, S.R.; Guragain, D.; Wang, X.; Jablonski, M.M. Once daily pregabalin eye drops for management of glaucoma. *ACS Nano.* **2019**, *13*, 13728–13744. [CrossRef] [PubMed]
57. Külkamp, I.; Paese, K.; Guterres, S.S.; Pohlmann, A.R. Stabilization of lipoic acid by encapsulation in polymeric nanocapsules designed for cutaneous administration. *Quim. Nova* **2009**, *32*, 2078–2084. [CrossRef]
58. Liu, C.H.; Wu, C.T. Optimization of nanostructured lipid carriers for lutein delivery. *Colloid Surf. A Physicochem. Eng. Asp.* **2010**, *353*, 149–156. [CrossRef]
59. Ruela, A.L.; de Figueiredo, E.C.; de Araújo, M.B.; Carvalho, F.C.; Pereira, G.R. Molecularly imprinted microparticles in lipid-based formulations for sustained release of donepezil. *Eur. J. Pharm. Sci.* **2016**, *93*, 114–122. [CrossRef] [PubMed]
60. Gonsalves, A.A.; Araújo, C.R.M.; Soares, N.A.; Goulart, M.O.F.; de Abreu, F.C. Different strategies for crosslinking of chitosan. *Quim. Nova* **2011**, *34*, 1215–1223. [CrossRef]
61. Powers, K.W.; Brown, S.C.; Krishna, V.B.; Wasdo, S.C.; Moudgil, B.M.; Roberts, S.M. Research strategies for safety evaluation of nanomaterials. Part VI. Characterization of nanoscale particles for toxicological evaluation. *Toxicol. Sci.* **2006**, *90*, 296–303. [CrossRef] [PubMed]
62. Sipos, P.; Szucs, M.; Szabó, A.; Eros, I.; Szabó-Révész, P. An assessment of the interactions between diclofenac sodium and ammonia methacrylate copolymer using thermal analysis and Raman spectroscopy. *J. Pharm. Biomed. Anal.* **2008**, *46*, 288–294. [CrossRef] [PubMed]
63. Soares, G.A.; de Castro, A.D.; Cury, B.S.; Evangelista, R.C. Blends of cross-linked high amylose starch/pectin loaded with diclofenac. *Carbohydr. Polym.* **2013**, *91*, 135–142. [CrossRef]
64. Chew, M.L.; Plotka, A.; Alvey, C.W.; Pitman, V.W.; Alebic-Kolbah, T.; Scavone, J.M.; Bockbrader, H.N. Pharmacokinetics of pregabalin controlled-release in healthy volunteers: Effect of food in five single-dose, randomized, clinical pharmacology studies. *Clin. Drug Investig.* **2014**, *34*, 617–626. [CrossRef] [PubMed]
65. Matha, K.; Lollo, G.; Taurino, G.; Respaud, R.; Marigo, I.; Shariati, M.; Bussolati, O.; Vermeulen, A.; Remaut, K.; Benoit, J.P. Bioinspired hyaluronic acid and polyarginine nanoparticles for DACHPt delivery. *Eur. J. Pharm. Biopharm.* **2020**, *150*, 1–13. [CrossRef] [PubMed]
66. DiPiro, J.T.; Spruill, W.J.; Wade, W.E.; Blouin, R.A.; Pruemer, J.M. *Concepts in Clinical Pharmacokinetics*, 4th ed.; American Society of Health-System Pharmacists: Bethesda, MD, USA, 2005; pp. 29–44.
67. Bee, L.A.; Dickenson, A.H. Descending facilitation from the brainstem determines behavioral and neuronal hypersensitivity following nerve injury and efficacy of pregabalin. *Pain* **2008**, *140*, 209–223. [CrossRef] [PubMed]
68. Field, M.J.; McCleary, S.; Hughes, J.; Singh, L. Gabapentin and pregabalin, but not morphine and amitriptyline, block both static and dynamic components of mechanical allodynia induced by streptozocin in the rat. *Pain* **1999**, *80*, 391–398. [CrossRef]
69. Nakai, K.; Nakae, A.; Hashimoto, R.; Mashimo, T.; Hosokawa, K. Antinociceptive effects of mirtazapine, pregabalin, and gabapentin after chronic constriction injury of the infraorbital nerve in rats. *J. Oral Facial Pain Headache* **2014**, *28*, 61–67. [CrossRef] [PubMed]
70. Pineda-Farias, J.B.; Caram-Salas, N.L.; Salinas-Abarca, A.B.; Ocampo, J.; Granados-Soto, V. Ultra-low doses of naltrexone enhance the antiallodynic effect of pregabalin or gabapentin in neuropathic rats. *Drug Dev. Res.* **2017**, *78*, 371–380. [CrossRef]
71. Okamoto, Y.; Kawakami, K.; Miyatake, K.; Morimoto, M.; Shigemasa, Y.; Minami, S. Analgesic effects of chitin and chitosan. *Carbohydr. Polym.* **2002**, *49*, 249–252. [CrossRef]
72. Organization for Economic Co-operation and Development (OECD). Test No. 423: Acute Oral toxicity—Acute Toxic Class Method, in OECD Guidelines for the Testing of Chemicals; Section 4. Available online: https://doi.org/10.1787/9789264071001-en (accessed on 22 October 2019).
73. Khan, J.; Noboru, N.; Imamura, Y.; Eliav, E. Effect of pregabalin and diclofenac on tactile allodynia, mechanical hyperalgesia and pro inflammatory cytokine levels (IL-6, IL-1β) induced by chronic constriction injury of the infraorbital nerve in rats. *Cytokine* **2018**, *104*, 124–129. [CrossRef] [PubMed]

Article

Ciprofibrate-Loaded Nanoparticles Prepared by Nanoprecipitation: Synthesis, Characterization, and Drug Release

Raissa Lohanna Gomes Quintino Corrêa [1,*], Renan dos Santos [1], Lindomar José Calumby Albuquerque [1,2], Gabriel Lima Barros de Araujo [3], Charlotte Jennifer Chante Edwards-Gayle [4], Fabio Furlan Ferreira [1,5] and Fanny Nascimento Costa [1,4,*]

1. Center for Natural and Human Sciences (CCNH), Federal University of ABC (UFABC), Santo André 09210-580, Brazil; reeh_sants@hotmail.com (R.d.S.); lindomaralbuquerque@gmail.com (L.J.C.A.); fabio.furlan@ufabc.edu.br (F.F.F.)
2. Brazilian Synchrotron Light Laboratory (LNLS), Brazilian Center for Research in Energy and Materials (CNPEM), Campinas 13083-170, Brazil
3. Department of Pharmacy, Faculty of Pharmaceutical Sciences, University of Sao Paulo, Sao Paulo 05508-900, Brazil; gabriel.araujo@usp.br
4. Diamond Light Source, Harwell Science & Innovation Campus, Didcot, Oxfordshire OX11 0DE, UK; charlotte.edwards-gayle@diamond.ac.uk
5. Nanomedicine Research Unit (NANOMED), Federal University of ABC (UFABC), Santo André 09210-580, Brazil
* Correspondence: raissa.lohanna@ufabc.edu.br (R.L.G.Q.C.); fanny.nascimento.costa@diamond.ac.uk (F.N.C.)

Abstract: Ciprofibrate (CIP) is a highly lipophilic and poorly water-soluble drug, typically used for dyslipidemia treatment. Although it is already commercialized in capsules, no previous studies report its solid-state structure; thus, information about the correlation with its physicochemical properties is lacking. In parallel, recent studies have led to the improvement of drug administration, including encapsulation in polymeric nanoparticles (NPs). Here, we present CIP's crystal structure determined by PXRD data. We also propose an encapsulation method for CIP in micelles produced from Pluronic P123/F127 and PEO-b-PCL, aiming to improve its solubility, hydrophilicity, and delivery. We determined the NPs' physicochemical properties by DLS, SLS, ELS, SAXS and the loaded drug amount by UV-Vis spectroscopy. Micelles showed sizes around 10–20 nm for Pluronic and 35–45 nm for the PEO-b-PCL NPs with slightly negative surface charge and successful CIP loading, especially for the latter; a substantial reduction in ζ-potential may be evidenced. For Pluronic nanoparticles, we scanned different conditions for the CIP loading, and its encapsulation efficiency was reduced while the drug content increased in the nanoprecipitation protocol. We also performed in vitro release experiments; results demonstrate that probe release is driven by Fickian diffusion for the Pluronic NPs and a zero-order model for PEO-b-PCL NPs.

Keywords: ciprofibrate; drug delivery; Rietveld method; crystallography; nanotechnology

1. Introduction

Ciprofibrate (CIP), chemical formula $C_{13}H_{14}Cl_2O_3$, is classified as a synthetic active pharmaceutical ingredient (API), which belongs to the fibrate class of drugs, generally used against dyslipidemia, a condition characterized by abnormal lipid levels in the blood system [1]. Dyslipidemia is a risk factor for developing cardiac diseases such as atherosclerosis, an inflammation characterized by the formation of fat, calcium, and other elements' plates in the walls of the heart's arteries and vascular system in general. Atherosclerosis can lead to different cardiac diseases responsible for more than 17.3 million deaths annually worldwide [2]. Nowadays, CIP is commercially available for oral formulations as tablets and capsules (100 mg) [3], and patients report several side effects such as headaches, nausea, and diarrhea [4].

In pharmaceutical development, there is considerable interest in crystal structure investigations, as a way to understand how their structures (either amorphous or crystalline) and properties such as density, size, and particle shape correlate to other physicochemical properties, aiming the optimization of these drugs for solid dosage [5].

A relevant characteristic of ciprofibrate is its hydrophobic behavior. Designated by the Biopharmaceutics Classification System (BCS) as a class II drug, CIP presents low solubility and high permeability [6]. The high permeability allows a complete absorption of the drug by the small intestine; otherwise, its poor solubility limits its application to treatments. One possible way to overcome this problem is the use of micro- or nanostructures for improved drug delivery.

Nanomedicine—an interdisciplinary area that merges nanotechnology and medicine—has investigated several solutions that reach more efficient treatments with minimum side effects for various diseases. Several types of nanocarriers have been developed; structures such as polymeric micelles, liposomes, and NP that conjugate with drugs by diverse mechanisms (encapsulation, surface adsorption, and others) are used to deliver controlled and localized drug dosages to the body, implementing the drug delivery concept [7–10]. Each of these techniques presents a myriad of controllable features that can influence important aspects of the nanostructures. Encapsulation methods, in particular, may vary among different combinations of hydrophilic and lipophilic block copolymers and steps for synthesis, changing size, loading capacity, and even drug release behavior [11,12].

The poloxamer surfactant (or Pluronic, known by its trade name) is one of the most extensively investigated biomaterials used to build nanocarriers. These are typical tri-block copolymers comprising two unities of poly (ethylene oxide) (PEO) and one of poly (propylene oxide) (PPO) in an alternated linear structure. Although not classified as biodegradable, PEO and derivatives present many advantages for biomedical usage such as relative safety profile (lethal dosage being LD50s > 5 g kg^{-1}), FDA approval, lack of immunogenicity, and particularly, the ease of excretion from living organisms [13]. The remaining parts of the molecules are found in living systems since they are aliphatic chains (derived from fatty acid) and sugar molecules that are dietary sources of fuel and important structural components of cells. Additionally, taking into account the molecular weight of the amphiphilic chains (PCL Mw 18,500 g mol^{-1}, P123 Mw 5750 g mol^{-1}, and F127 12,600 g mol^{-1}), the materials can undergo renal clearance (Mw < 40,000 g mol^{-1}) [13,14]. Their block copolymer structures can self-assemble in different structural forms such as micelles, worm-like micelles, or vesicles in the aqueous medium. This approach allows the encapsulation of hydrophobic drugs such as CIP, increasing their solubility, improving circulation in vivo, and avoiding aggregation problems [15].

Different types of Pluronic are used together as they present better properties such as colloidal stability and better drug-carrying efficiency [15], especially for the self-assembly into micelle morphology [16]. In this work, P123 and F127 block copolymers were used together (PEO$_{20}$-b-PPO$_{70}$-b-PEO$_{20}$ and PEO$_{100}$-b-PPO$_{65}$-b-PEO$_{100}$, respectively) to produce ciprofibrate-loaded micelles. F127 is a widely known and investigated copolymer for its thermoreversible gelation property at body temperature [17]. Another block copolymer was considered for preparing the CIP-loaded micelles: the PEO$_{113}$-b-PCL$_{118}$ was chosen due to its biocompatibility, biodegradability, and proven record in soft-based nanocarrier platforms for therapeutic applications [18,19]. The nanoprecipitation process was selected to prepare the CIP-loaded micelles, and an illustration of the method is provided in Figure 1.

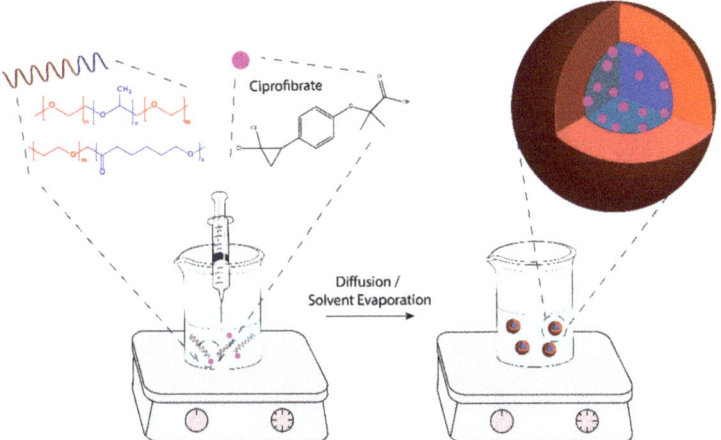

Figure 1. Scheme of preparation by nanoprecipitation of ciprofibrate-loaded nanoparticles.

Recent results suggest that fenofibrate, which belongs to the same drug class as CIP, may have an essential role in controlling SARS-CoV-2 infection in in vitro models [20]. Considering that the reduction in serum triglycerides and LDL cholesterol has a positive impact in the fighting against numerous age-related diseases and also against the recent COVID-19 infection, the relevance of more in-depth insight into ciprofibrate's crystal structure to determine its fundamental properties and the possibility to use copolymer encapsulation as a strategy to improve its biodistribution becomes clear.

Herein, we present the crystal structure of CIP in solid state, solved using powder X-ray diffraction data. The drug was also encapsulated in both P123/F127 Pluronic and PEO$_{113}$-b-PCL$_{118}$ micelles for comparison via nanoprecipitation technique in water and evaluated in different solvent concentrations. We also propose the best condition for encapsulation for this system and demonstrate the fundamental physicochemical properties of the NPs (size distribution, surface charge, surface morphology). Furthermore, in vitro results and mathematical models for the drug release in buffer solution are presented.

2. Materials and Methods

CIP was obtained as a courtesy of DEINFAR (Laboratório de Desenvolvimento e Inovação Farmacotécnica) from the Faculty of Pharmaceutical Sciences (University of São Paulo, São Paulo, Brazil). PEO$_{113}$-b-PCL$_{118}$ blocks (Mw = 18,500 g mol^{-1}) were purchased from Polymer Source, Inc. (Dorval, QC, Canada). Pluronic F127 (Mw = 12,600 g mol^{-1}), Pluronic P123 (Mw = 5750 g mol^{-1}), ethanol, acetone, PBS and Tween 20 were purchased from Sigma-Aldrich (São Paulo, Brazil).

2.1. Preparation of CIP-Loaded Nanoparticles

The polymeric micelles were produced by nanoprecipitation (as schematized in Figure 1) from stock polymer/drug organic solutions prepared in ethanol. For Pluronic nanoparticles, first, P123 and F127 blocks were weighed and dissolved in ethanol (EtOH) (maintaining a 2:1 w/w fixed molar ratio of P123 and F127). The polymer concentration was fixed at 10 mg mL^{-1} as proposed by Sortini et al. [21]. This solution was stabilized via sonication for approximately 10 min until it became visually transparent. The organic solution was transferred to a syringe and afterward added dropwise into 5 mL of water solution and then removed by evaporation. For the PEO$_{113}$-b-PCL$_{118}$ nanoparticles, acetone was used to completely dissolve the polymeric chains [6,22]. For the CIP-loaded micelles, CIP was weighed and solubilized in ethanol according to the desired feeding and was mixed with the polymeric organic solution before the addition to the aqueous phase.

2.2. Particle Size and Morphology of CIP-Loaded Nanoparticles

The average diameter and size distribution (polydispersity) of the NPs were determined via Dynamic Light Scattering (DLS) and Static Light Scattering (SLS). Samples were loaded into test tubes (10 μL) and diluted in 1 mL of distilled water. Measurements were performed using ALV/CGS-3 platform-based goniometer system (ALV GmbH, Langen, Germany) consisting of a polarized HeNe laser (22 mW) operating at a wavelength λ = 633 nm, an ALV 7004 digital correlator, and a pair of pseudocorrelation APD detectors operating in a crusade mode. The data were collected and further averaged using ALV Correlator Control software. The polydispersity was estimated using the cumulant analysis of the autocorrelation functions measured at 90°. The temporal correlation functions were analyzed using the REPES algorithm (incorporated into the ALV Correlator program) to confirm the monomodal distribution of NPs. The autocorrelation functions reported are based on three independent runs of 60 s counting time for each sample.

The NPs' surface charges were obtained via Electrophoretic Light Scattering (ELS) tests. Samples were added into cuvettes (10 μL), placed into the apparatus, and exposed to the laser beam. Experiments were carried out using a Zetasizer Nano-ZS ZEN3600 instrument (Malvern Instruments, Worcestershire, UK). The electrophoretic mobility (μ_e) was calculated through the Smoluchowski approximation. Each zeta-potential value reported is an average of 3 independent measurements with repeatability of ±2%.

2.3. X-ray Diffraction Analysis

CIP's crystal structure was determined using Powder X-ray Diffraction (PXRD) data. Moreover, to the best of our knowledge, this is the first time it is reported. The method employed to solve the CIP's crystal structure is well described in the literature [6,23]. The sample was hand-ground in an agate mortar and loaded between two cellulose acetate foils (0.014 mm) in a spinning sample holder. Powder X-ray diffraction data were collected utilizing a transmission mode copper source, filtered by a germanium monochromator (111). Diffraction intensities were collected by a linear detector Dectris Mythen 1K (Baden-Daettwil, Switzerland) with 0.015° step and integration time of 60 s at every 1.05°. The experiment used a STADI-P (Stoe, Darmstadt, Germany) powder diffractometer available at the Laboratory of Crystallography and Structural Characterization of Materials (LCCEM).

2.4. In Vitro Drug Release Characteristics of CIP-Loaded Nanoparticles

To measure the CIP release from the NPs and to compare the stabilities of both Pluronic and PEO_{113}-b-PCL_{118}, samples with 5 mg mL^{-1} containing 10% (w/w) and 20% (w/w) CIP were diluted in PBS pH 7.4 and placed in a dialysis bag (MWCO: 3.500–5.000 Da, Spectra/Por), which was dialyzed against 500 mL of PBS pH 7.4 containing 0.4% (w/v) Tween 20 at 37 °C for 48 h, under constant magnetic stirring. Aliquots of 50 μL were taken from the dialysis bag at increasing time intervals and afterward diluted 10 times in ethanol and measured through UV-Vis spectroscopy technique using a Cary 50 UV-Vis spectrophotometer (Varian, Inc., Crawley, UK). First, CIP's analytical calibration curve in EtOH with a linear response in the range 0.0001–0.05 mg mL^{-1} was recorded and used to determine CIP contents. A sample containing empty NPs in EtOH was also measured as a blank sample for comparison. Then, samples were added into cuvettes (10 μL), diluted in EtOH, and placed into the equipment. Dilution proportions varied according to the different investigations, as some samples were highly concentrated and visually turbid, which may affect the results.

To describe the drug dissolution as a function of time, the drug release profile data were submitted to quantitative analysis, fitted to several kinetic release models. Statistical analysis was performed and indicated the models that best demonstrate the CIP's release mechanism for both polymeric matrices.

2.5. Small-Angle X-ray Scattering

Small-Angle X-ray Scattering (SAXS) experiments were performed at the beamline B21 of the Diamond Light Source (Didcot, UK) [23,24]. Samples were loaded into quartz capillaries by the Arinax liquid-handling robot and exposed for 1 s, acquiring 20 frames. The wavelength was 0.95 Å, and the camera length was 3.71 m. Modeling to a spherical shell model was done using SASfit. It is worth noting that the samples were stored in a fridge at a range of 4–8 °C for three months before the SAXS experiment.

3. Results

3.1. Characterization of CIP-Loaded Nanoparticles

Table 1 presents general physicochemical aspects of the NPs: their hydrodynamic and gyration radius, polydispersity, and charge, for samples prepared under different concentrations (20–2.5 mg mL^{-1}). Figure 2 is a plot obtained from DLS measurements and shows different size distribution profiles for each concentration.

Table 1. Results for DLS and ELS evaluation for different CIP-loaded Pluronic micelle concentrations [†] in water.

Pluronic Concentration (mg mL^{-1})	R_H (nm)	PDI	ζ-Potential (mV)	R_G (nm)	$\rho = \frac{R_G}{R_H}$
20.0	13 ± 3	0.22	−4 ± 1	-	-
10.0	11 ± 2	0.19	−4.3 ± 0.6	14.9 ± 0.5	1.3
5.0	12 ± 2	0.18	−5.9 ± 0.5	16.4 ± 0.5	1.3
2.5	52 ± 12	0.33	−9 ± 3	-	-

[†] The amount of drug remained fixed as 10% of the total amount of Pluronic (w/w); PDI stands for the polydispersity of NPs; R_H is the hydrodynamic radii of each sample in terms of mass distribution; R_G is the gyration radii of each sample; ρ is the structure-sensitive parameter.

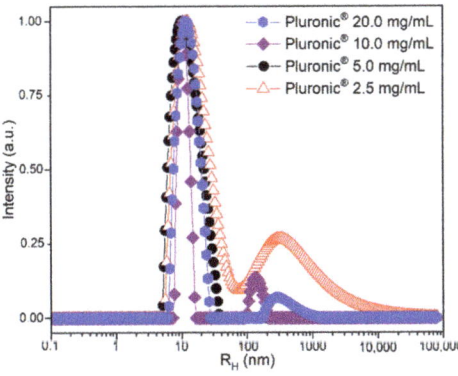

Figure 2. Size distribution of Pluronic micelles measured at 90° by Dynamic Light Scattering.

Subsequently, samples were prepared with different CIP amounts and characterized by UV-Vis to determine drug loading content (DLC) and encapsulation efficiency (EE). Results shown in Table 2 indicate a remarkable high efficiency (96%) for the Pluronic micelles prepared with 10% (w/w) CIP.

Considering the best values for drug encapsulation in Table 2 (samples that presented more efficient CIP uptake), a new set of NPs was synthesized using the block copolymer PEO$_{113}$-b-PCL$_{118}$. Table 3 summarizes aspects such as size, charge, and drug loading for these systems. Table 4 shows the DLC and EE indexes for PEO$_{113}$-b-PCL$_{118}$ micelles.

Table 2. Results for drug content evaluation through UV-Vis spectroscopy for different micelle concentrations in water.

Pluronic Conc. (mg mL^{-1})	Drug Amount (w/w)	R_H (nm)	PDI	ζ-Potential (mV)	Amount of Drug by UV (mg mL^{-1})	EE (%)	DLC (%)
5.0	-	12 ± 2	0.21	−4.7 ± 0.9	-	-	-
5.0	10%	10 ± 2	0.20	−7 ± 1	0.48 ± 0.03	96	8.8
5.0	20%	13 ± 2	0.16	−7.3 ± 0.8	0.51 ± 0.02	51	9.3
5.0	30%	14 ± 2	0.18	−5.9 ± 0.7	0.49 ± 0.04	32.6	8.9
5.0	40%	13 ± 2	0.22	−5 ± 1	0.24 ± 0.02	12	4.6

Table 3. Results for DLS, ELS, and drug content evaluation for PEO$_{113}$-PCL$_{118}$ micelles with the best CIP concentrations.

PEO/PCL Conc. (mg mL^{-1})	Drug Amount (w/w)	R_H (nm)	PDI	ζ-Potential (mV)	R_G (nm)	$\rho = \frac{R_G}{R_H}$
2.5	-	34 ± 1	0.38	−22 ± 2	-	-
2.5	10%	41 ± 1	0.12	−12 ± 1	33.5 ± 0.5	0.82
2.5	20%	41.8 ± 0.9	0.08	−9.5 ± 0.7	32.6 ± 0.5	0.78

Table 4. Results for drug content evaluation using UV-Vis spectroscopy for PEO$_{113}$-b-PCL$_{118}$ micelles with the best CIP concentrations.

Total Polymer (mg)	Total CIP (mg)	CIP Amount (w/w)	Theoretical CIP Final Conc. (mg mL^{-1})	Amount of Drug by UV (mg mL^{-1})	EE (%)	DLC (%)
5.0 *	-	-	-	-	-	-
5.0	0.5	10%	0.25	0.24 ± 0.01	98	8.7
5.0	1.0	20%	0.50	0.23 ± 0.02	96	8.4

* Control sample.

3.2. X-ray Diffraction Analysis of CIP

CIP powder diffraction data were collected to characterize this active pharmaceutical ingredient in the solid state. Figure 3 shows the PXRD diffraction pattern of the pure CIP sample and its final Rietveld refinement, and Table 5 summarizes the complete crystallographic information, deposited in the Cambridge Crystallographic Data Centre under the ID 2097980. Figure 4 shows CIP's crystal structure. Further details on atom coordinates, bond lengths, and angles can be found in the Supplementary Materials in Tables S1–S4.

Table 5. Crystal structure of CIP determined via PXRD and statistical information of the Rietveld refinement.

Crystal System	Monoclinic
Space group	$P2_1/c$
a; b; c (Å)	10.7646(3); 10.2368(3); 12.8079(4)
β (°)	102.933(2)
Volume (Å^3)	1375.56(7)
Z; Z'	4; 1
R_{exp} (%)	2.637
R_{wp} (%)	5.746
R_{Bragg} (%)	2.984
χ^2	2.179

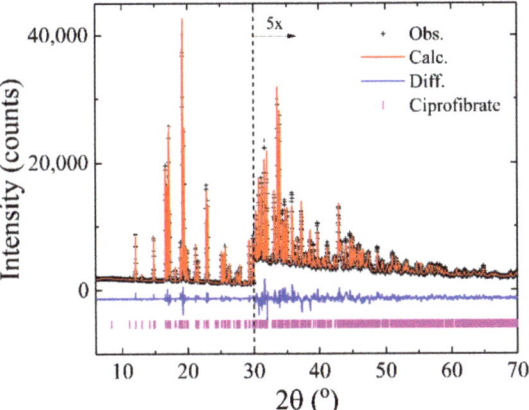

Figure 3. Rietveld plot for CIP showing the excellent agreement between the experimental data (black crosses) and the calculated profile (red line). The blue line displays the difference between observed and calculated data. The magenta vertical bars at the bottom represent the Bragg peaks' positions. The region from 30 to 70° in 2θ is magnified 5 times to clarify the good agreement between observed and calculated data in higher angles.

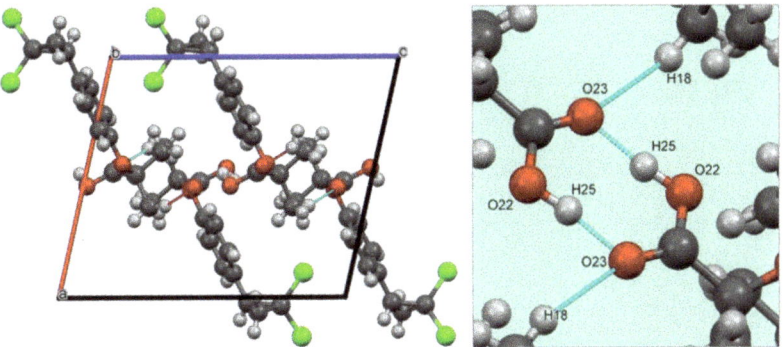

Figure 4. CIP's crystal structure forms a network of molecular aggregates along the b-axis, with four formula units per unit cell (Z = 4). Hydrogen bonds (cyan lines) are shown on the right side in an enlarged region of the unit cell. Atom color code: red = oxygen (O); green = chloride (Cl); grey = carbon (C); light grey = hydrogen (H).

3.3. Small-Angle X-ray Scattering Analysis of Pluronic and PEO_{113}-b-PCL_{118} Nanoparticles' Structure

Figure 5 shows the data for PEO_{113}-b-PCL_{118} (top) and Pluronic (bottom) nanoparticles for unloaded and loaded (10 and 20 wt% CIP) samples, fitted to a spherical shell model using SASfit. An exception was made for the Pluronic 20 wt% CIP sample, which was better fitted to a long cylinder. Further details of fits can be found in the Supplementary Materials (Tables S5–S7).

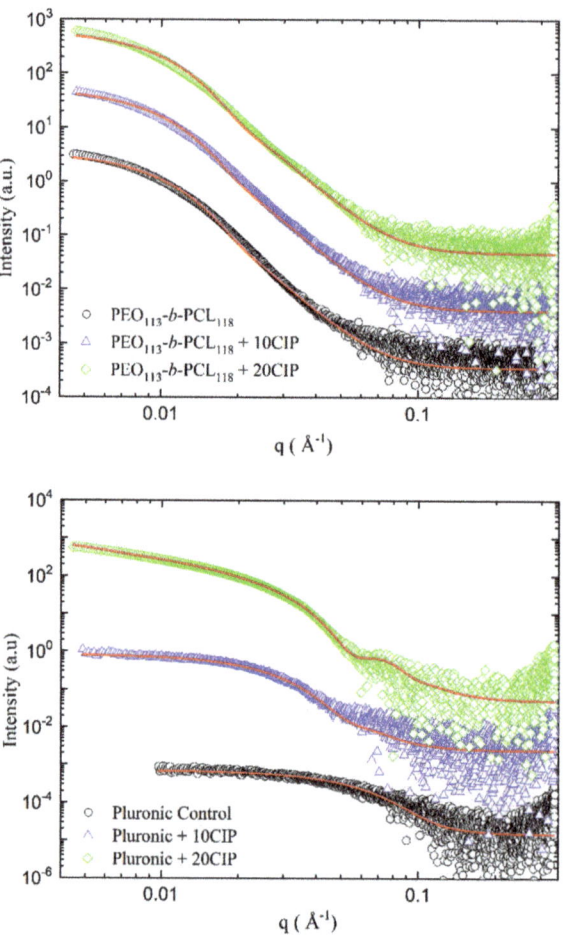

Figure 5. SAXS curves for PEO_{113}-b-PCL_{118} and Pluronic nanoparticles with different concentrations of CIP in PEO_{113}-b-PCL_{118} samples (**top**) and Pluronic samples (**bottom**). The fits of data are displayed in red. Details of fits are found in the Supplementary Materials (Tables S5 and S6). Scattering curves have been scaled for the sake of clarity.

3.4. In Vitro Release Profile of CIP from Pluronic and PEO_{113}-b-PCL_{118} Nanoparticles

In vitro release was performed for best encapsulation conditions (10% and 20% CIP) using the two polymeric matrices to analyze their influence on the drug release profile. Samples were collected during a 48 h experiment and analyzed via UV-Vis spectrophotometry. Figure 6 shows the release profiles over time obtained for Pluronic (top) and PEO_{113}-b-PCL_{118} (bottom).

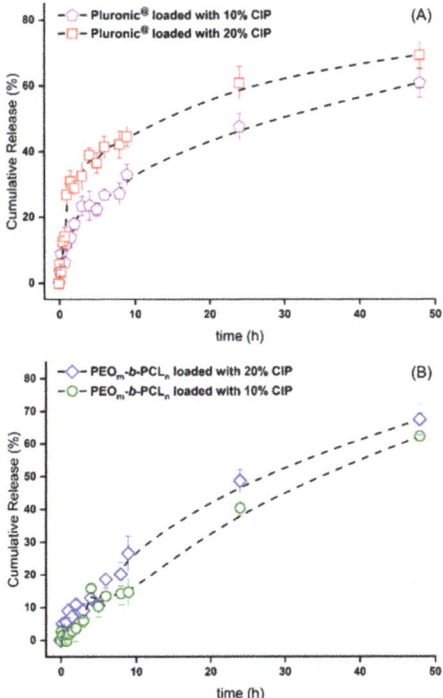

Figure 6. CIP release profile in best encapsulation conditions with (**A**) Pluronic P123/F127 and (**B**) PEO$_{113}$-*b*-PCL$_{118}$ micelles. Error bars indicate the standard deviation of three replicates.

3.5. Release Mechanisms for Pluronic and PEO$_{113}$-b-PCL$_{118}$ Nanoparticles

Table 6 presents the result of a statistical analysis of the last release profile data of Pluronic P123/F127 and PEO$_{113}$-*b*-PCL$_{118}$ CIP-loaded nanoparticles, utilizing the linear regression model—except for the Korsmeyer–Peppas model, which was fitted using a polymeric regression—and involving the correlation coefficient (R). R-values closest to 1 describe a better release mechanism [25]. Figure 7 shows plots of cumulative release data of each of the loaded samples of both polymers over time, fitted to their respective best models. The kinetic constant values of these models are presented in Table 7.

Table 6. R-values for different drug release model fittings.

Polymeric Matrix	Drug/Polymer Ratio	Zero Order	First Order	Higuchi	Hixson–Crowell	Korsmeyer–Peppas
P123/F127	10%	0.91247	0.70178	**0.98544**	0.78473	**0.96861**
P123/F127	20%	0.80424	0.55803	**0.93048**	0.64722	**0.91804**
PEO/PCL	10%	**0.98031**	0.70064	**0.97411**	0.76204	0.82982
PEO/PCL	20%	**0.97149**	0.69793	**0.98321**	0.84923	0.95231

Bold values indicate R-coefficients closest to 1.

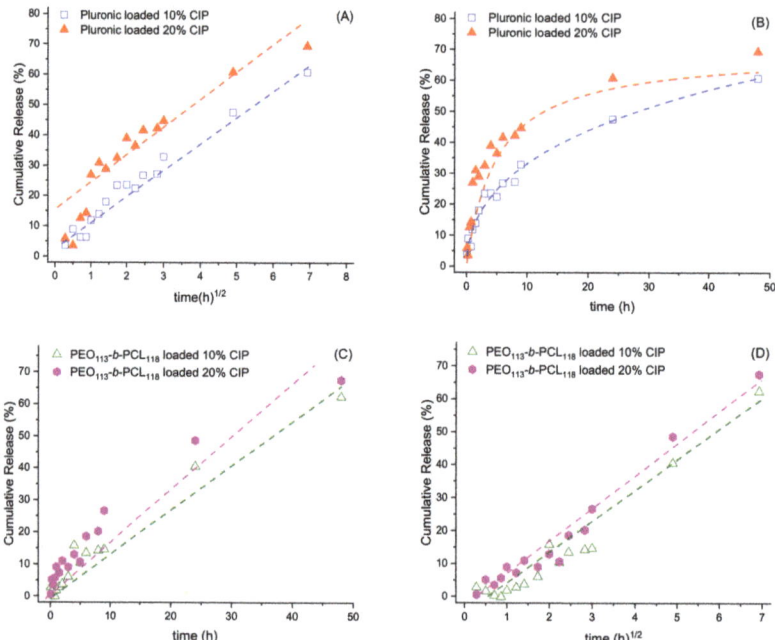

Figure 7. Model fitting using Higuchi (**A**,**D**), Korsmeyer–Peppas (**B**), and zero-order (**C**) approaches based on data given in Figure 6 for Pluronic and PEO$_{113}$-b-PCL$_{118}$ nanoparticles loaded with ciprofibrate.

Table 7. Release kinetic constants obtained from statistical analysis.

Polymeric Matrix	Drug/Polymer Ratio	Zero Order	Higuchi	Korsmeyer–Peppas	
		K_0 (%h^{-1})	K_H (%h$^{-1/2}$)	K_{KP} (%h^{-n})	N
P123/F127	10%	-	8.76	0.12	0.43
P123/F127	20%	-	9.73	0.20	0.38
PEO/PCL	10%	1.32	9.39	-	-
PEO/PCL	20%	1.39	10.09	-	-

4. Discussion

4.1. Physicochemical Aspects of CIP-Loaded Nanoparticles

Table 1 summarizes the influence of the dispersed system's concentration (polymeric solution with drug) in the NP aggregation process. No significant differences were observed for the NPs' diameter and ζ-potential, responsible for their dispersive media stability. It remains valid except for the 2.5 mg mL^{-1} of Pluronic-containing sample, which presented a much larger diameter, higher polydispersity, and a slightly more negative surface charge. It is possible to see in Figure 2 that when the polymeric concentration drops severely, the dilution effect leads to aggregates, increasing size and PDI.

Table 1 also shows the structure-sensitive parameter (ρ) values, obtained from DLS and SLS measurements, which could indicate the spherical shape objects. This parameter provides information on the shape, inner structure, and conformation of scattering objects. For this type of assembling, ρ is dependent on the inner structure and compactness [26], being close to 0.775 for compact spheres, ~0.8–0.9 for block copolymer micelles due to solvation phenomena, and ~1.0 for hollow spheres and vesicles [27]. Thus, the reported values in Table 1 suggest that the micelles are spherical but highly swollen by water. The 5 mg mL^{-1} polymeric final concentration was selected for the CIP encapsulation study as this condition provides nanoparticles with good features and a concentration of F127 above the critical micelle concentration (CMC) at room temperature (0.357 mM) [28]. Hence, this was the chosen condition for the CIP encapsulation study.

Subsequently, considering the final polymeric concentration in water, it was observed that samples with higher drug concentrations induced sedimentation of drug crystals on the recipient's bottom after the synthesis. Values shown in Table 2 demonstrate that features such as size and ζ-potential remained the same average from the previous samples (Table 1). However, it is possible to see that for increasing CIP-loading values, there is a decrease in ζ-potential (except for the blank sample). Thus, samples with more efficient CIP uptake corresponded to 10% and 20% of drug amount.

When considered these selected systems produced with PEO_{113}-b-PCL_{118} micelles (Table 3), it is worth noting that the average size (34 nm) was more significant than for the same Pluronic system (Table 1). For the concentration of 2.5 mg mL^{-1}, PEO_{113}-b-PCL_{118} samples did not show the same high PDI, although this may be observed for the blank sample. The lowest PDI value was obtained for the 20% (w/w) CIP condition. These polymers present different molar weight distributions and different CMCs; therefore, these distinct aggregation behaviors are expected.

There is also a reduction in the negative surface charge with increasing CIP loading, probably due to the inversely proportional interaction between PCL chains and the drug. Previous simulations indicate the formation of H-bonds between PCL blocks and hydrophobic drugs [29]; thus, as the CIP amount increases inside the micelle, there is more interaction with PCL chains and consequently less PCL available for interaction with the outer media, reducing the ζ-potential. Additionally, ρ factors reached close values compared to the theoretical reference of 0.775 [27], indicating the presence of hard spheres, showing a different hydration profile than Pluronic micelles.

It is essential to highlight that higher CIP ratios were tested; however, CIP's solubility was limited in acetone. According to data in Table 4, the encapsulation efficiency reaches values closer to 100% compared to previous best Pluronic formulations (Table 2), indicating higher efficiency of the PEO_{113}-b-PCL_{118} spheres. The DLC values were remarkably similar to those obtained for Pluronic NPs, showing that the produced particles' size could be a limiting feature related to the final drug amount in the formulation.

4.2. CIP's Crystallographic Characterization

The structural characterization of small drugs such as ciprofibrate is essential to understand how biological activity relates to the physicochemical features. Several drugs are manufactured as crystalline powders (including CIP). Depending on the chemical structure's degrees of freedom, it might form polymorphs—the ability of a material to exist in two or more crystalline forms with different arrangements or conformations in the crystal lattice—thus changing the crystal structure and, consequently, its properties.

As reported in Table 5, CIP crystallized under a monoclinic crystal system with space group $P2_{1/c}$ and unit cell parameters a = 10.7646(3) Å, b = 10.2368(3) Å, c = 12.8079(4) Å, β = 102.933(2)°, and V = 1375.56(7) Å3. Its structure comprises four formula units per unit cell (Z = 4), accommodating one molecule in the asymmetric unit (Z' = 1), as shown in Figure 4. Within the unit cell, the molecules are held together by weak hydrogen bonds (or nonclassical) between atoms C(10)–H(18)$\cdots$O(23), the distances of which are D–H = 0.97(2) Å, H$\cdots$A = 2.49(2) Å, D$\cdots$A = 3.41(2) Å, and D–H$\cdots$A = 159(1)°, and by an intramolecular classical hydrogen bond O(22)–H(25)$\cdots$O(23) (D–H = 0.993(5) Å, H$\cdots$A = 1.584(5) Å, D$\cdots$A = 2.571(5) Å, and D–H$\cdots$A = 171.8(4)), where "D" and "A" are hydrogen donor and acceptor, respectively, as represented by cyan lines in Figure 4.

4.3. Structure of Pluronic and PEO_{113}-b-PCL_{118} Nanoparticles

Small-Angle X-ray Scattering (SAXS) is a powerful technique to determine the structural properties of nanoparticles in solution and give their average shape. This gives information on how these nanoparticles may behave in vivo as drug delivery molecules.

For unloaded PEO$_{113}$-b-PCL$_{118}$, SAXS data show polydisperse spheres with an average radius of 12.2 nm (Figure 5, top). This finding agrees with the DLS data, which show a polydisperse distribution of radii. The radii increase to 12.5 nm at 10 wt% CIP and 12.6 nm at 20 wt% CIP upon addition of CIP. Moreover, there is a decrease in the scattering intensity of the core, indicating that CIP is loaded into it.

For the Pluronic nanoparticles (Figure 5, bottom), a spherical shell was fitted to the unloaded sample and the sample with 10% CIP. Empty Pluronic shows polydisperse spheres with a radius of 2.3 nm, which swells to 5.7 nm upon the addition of 10 wt% CIP. Moreover, the scattering density of the core decreases, implying the loading of the drug. For 20 wt% CIP, the decay of q^{-1} in the low-q data region implied cylindrical particles. These data were fitted to a long cylinder with a radius of 6.5 nm, implying that the spherical nanoparticles are less stable and may aggregate at this concentration of CIP. It may be inferred that, as these samples were submitted to low-temperature long-term storage, this may have influenced their morphological structure.

4.4. In Vitro Release Profile of CIP from Pluronic and PEO$_{113}$-b-PCL$_{118}$ Nanoparticles

The data displayed in Figure 6 indicate a faster release for the mixed Pluronic NPs than for the PEO$_{113}$-b-PCL$_{118}$ system. For CIP 20% condition (red squares), nearly 45% of the drug content was released in the first 9 h, while 33% was released from CIP 10% condition (purple pentagons).

For the PEO$_{113}$-b-PCL$_{118}$ system, a more consistent release was observed; for the CIP 20% condition, only 25% of the drug was released after 9 h, and for the CIP 10% sample, 16% of the drug was released after 9 h. The data evidenced that nearly 60–70% of the loaded CIP is released within two days and—based on previous studies of our research group with PCL copolymers—the remaining is slowly released within the next day, or it is adsorbed into the dialysis bag [18].

This quick drug release in the first hours of in vitro experiments agrees with previous studies concerning mixed Pluronic formulations [21,30,31], although, in this study, a slower profile was obtained. This kinetic may be altered due to several conditions. First, the experiment was conducted at a temperature of 37 °C, in which the F127 is most probably in its gelation form due to its sensitive CMT [32], which tends to release the encapsulated drugs faster. As previously demonstrated [30], mixed formulations of F127 and more hydrophobic Pluronic chains can control the hydrogel formation and increase the release interval, contributing to a more stable composition. This aspect is mainly observed for hydrophobic drugs. The addition of hydrophilic F127 to the composition increases the PEO chain lengths and, consequently, the amount of water in the micelle, leading to so-called hydrophilic channels [21]. Finally, the encapsulated drug solubility may also alter the hydrogel state and, as previously reported, lower the gelification temperature in case of hydrophobic behavior. The release time increment is probably due to the presence of P123 in a higher proportion than F127 in the micelles. However, the temperature and hydrophobicity of the CIP may still contribute to a rapid release.

4.5. Release Mechanisms for Pluronic and PEO$_{113}$-b-PCL$_{118}$ Nanoparticles

Considering the obtained release profiles (Figure 6) and the different kinetic (Table 7) and physicochemical features of the block copolymers (Tables 1–4), it is essential to dig out the associated release mechanism for the produced NPs. In the literature, four different drug release mechanisms are known: (1) drug desorption, (2) drug diffusion, (3) erosion, and (4) combined erosion–diffusion. Accordingly, the CIP concentration values obtained via UV-Vis spectroscopy were later fitted to selected mathematical models to comprehend the release mechanisms involved in both polymeric systems.

It is worth noting in Table 6 that, for Pluronic systems, the Higuchi model presents the highest R-values, indicating the release mechanism is diffusional through the micelle, which predicts a rapid release of the drug [25], in agreement with previous studies [30]. This model is a time-dependent linear square root that obeys Fick's diffusion law; therefore, the release is directly proportional to the surface area and the concentration difference and inversely proportional to the membrane's thickness. Moreover, it is possible to notice that the Korsmeyer–Peppas R-values are also higher for Pluronic systems, reinforcing the diffusion as a suitable proposal for the release mechanism.

According to Bruschi M. [33], for spherical particles, the release exponent n is expected to be 0.43—precisely what was achieved for Pluronic micelles with 10% CIP (Table 7). It is a good approximation for the Pluronic sample with 20% CIP, confirming previous results regarding NPs' morphology. Additionally, this n value also indicates the dominating release mechanism: for $n < 0.45$, liberation occurs by Fickian diffusion [25], suggesting that Higuchi's model is the preferential form of drug release. However, it is important to highlight that this model is suitable for these drug concentrations. The interaction between drug and polymer and the total volume of the encapsulated drug may alter the release mechanism.

As for the PEO_{113}-b-PCL_{118} micelles, both zero-order and Higuchi models describe the release mechanism (Table 6 and Figure 7C,D). According to Costa et al. [25], the zero-order model describes systems that do not disaggregate with time and promote a slower, linear dissolution that does not depend on the drug's concentration. These are ideal drug delivery systems, as they sustain a prolonged action.

5. Conclusions

It was possible to characterize CIP's crystal structure through this study, which has never been reported in the specific literature before. It crystallizes in a monoclinic crystal system in its solid-state, a pattern maintained due to strong H bonds between the –OH terminations of the molecules. We have also proposed a synthesis method and two polymeric matrices to encapsulate CIP and improve its solubility—a mixed proportion of Pluronic P123/F127 and a matrix composed of PEO_{113}-b-PCL_{118}. These polymers are already well established as biopolymers but have not been tested yet with CIP mainly. Results indicate that both systems produced micelles with suitable physicochemical characteristics, such as small size and relatively neutral zeta potential (from -10 to $+10$ mV), with great potential to enhance delivery efficiency in the human body [34]. The synthesis method was also demonstrated to be suitable, as repeated productions led to similar samples with minimum deviations. Besides, polydispersity values for the samples were very low, ideal for a stable solution.

As for the shape of the NPs, morphological characteristics demonstrated good agreement among diverse techniques such as DLS, SLS, SAXS, and statistical analysis using in vitro tests results after UV-Vis spectrophotometry for both systems. The SAXS profiles corroborate the spherical shape expected by combining the DLS/SLS measurements for both polymeric systems, except for the Pluronic micelles with 20 wt% of CIP loading, which presented a morphological change, as discussed adequately in Section 4.3. The small values of the gyration radius also agree with the hydrodynamic radius. Moreover, comparing the scattering densities, results indicate that the drug was incorporated in the nucleus of the polymeric nanoparticles, as we expected.

In vitro release experiments also have indicated that the drug remains in the system for up to 48 h. Although both polymeric systems have shown very similar physicochemical characteristics, PEO_{113}-b-PCL_{118} micelles showed zero-order release kinetics in in vitro drug release tests. Statistical analyses also indicated that the proposed CIP-loaded mixed Pluronic system has a release mechanism based on Fickian diffusion, which is independent of the amount of CIP encapsulated. In conclusion, concerning the NPs' physicochemical stability over time, this study proposes that CIP quantities up to 20% (w/w) can be encapsulated preferentially in PEO_{113}-b-PCL_{118} micelles as an alternative to increase its hydrophilicity and, therefore, its release time. Considering the recent advances in nanomedicine and the continued efforts to improve drug delivery techniques in order to provide targeted release of drugs, hydrophobicity, and bioavailability [7,35,36], this study represents an initial step to understand both CIP's physicochemical properties and the proposed nanoparticle system of encapsulation. The present work may also contribute to studies aiming to better comprehend the mechanism of action of the fibrate class drugs in diminishing lipid levels in the human body and, beyond that, to the rising number of studies correlating the presence of dyslipidemia condition with more severe forms of COVID-19 infection [37–39]. Therefore, further studies can be conducted for possible improvements in ciprofibrate's pharmaceutical applications.

Supplementary Materials: The following are available online at https://www.mdpi.com/article/10.3390/polym13183158/s1, Table S1: Final atomic coordinates of all atoms for ciprofibrate crystal structure; Table S2: Bond lengths; Table S3: All angles; Table S4: Torsion angle data; Table S5: Spherical shell data fits; Table S6: Long cylinder fit for Pluronic + 20CIP; Table S7: Radius of gyration and D_{max} (maximum particle dimensions) calculated using the ATSAS software.

Author Contributions: Conceptualization, F.N.C. and G.L.B.d.A.; methodology, F.N.C., F.F.F. and L.J.C.A.; validation, F.F.F. and G.L.B.d.A.; formal analysis, L.J.C.A. and R.L.G.Q.C.; investigation, C.J.C.E.-G., L.J.C.A., R.L.G.Q.C. and R.d.S.; resources, F.N.C., F.F.F. and G.L.B.d.A.; data curation, F.N.C., L.J.C.A. and R.L.G.Q.C.; writing—original draft preparation, C.J.C.E.-G., F.N.C., L.J.C.A. and R.L.G.Q.C.; writing—review and editing, R.L.G.Q.C., L.J.C.A., F.N.C., G.L.B.d.A. and F.F.F.; visualization, F.N.C., L.J.C.A. and R.L.G.Q.C.; supervision, project administration, and funding acquisition, F.N.C. and F.F.F. All authors have read and agreed to the published version of the manuscript.

Funding: This research was funded by CNPq, grant number 305601/2019-9. Additionally, L.J.C.A. acknowledges the fellowship granted by FAPESP (grant 2016/23844-8).

Institutional Review Board Statement: Not applicable.

Informed Consent Statement: Not applicable.

Data Availability Statement: Ciprofibrate's crystallographic information framework file is available on Cambridge Crystallographic Data Centre under the ID 2097980.

Acknowledgments: The group would like to thank the funders, the University of ABC (UFABC) for the technical support, Daniele Ribeiro de Araujo (UFABC), Amedea Barozzi Seabra (UFABC), and DEINFAR (Laboratório de Desenvolvimento e Inovação Farmacotécnica) for the courtesy with materials and appointments for this study. The authors also acknowledge Diamond Light Source for access to beamline B21 under proposal SM26698-5.

Conflicts of Interest: The authors declare no conflict of interest. The funders had no role in the study's design; in the collection, analyses, or interpretation of data; in the writing of the manuscript; or in the decision to publish the results.

References

1. Staels, B.; Dallongeville, J.; Auwerx, J.; Schoonjans, K.; Leitersdorf, E.; Fruchart, J.-C. Mechanism of Action of Fibrates on Lipid and Lipoprotein Metabolism. *Circulation* **1998**, *98*, 2088–2093. [CrossRef] [PubMed]
2. World Health Organization. *A Global Brief on Hypertension: Silent Killer, Global Public Health Crisis*; World Health Organization (WHO): Geneva, Switzerland, 2013.
3. Benet, L.Z. The Role of BCS (Biopharmaceutics Classification System) and BDDCS (Biopharmaceutics Drug Disposition Classification System) in Drug Development. *J. Pharm. Sci.* **2013**, *102*, 34–42. [CrossRef]
4. Costa, F.N.; da Silva, T.F.; Silva EM, B.; Barroso, R.C.; Braz, D.; Barreiro, E.J.; Lima, L.M.; Punzo, F.; Ferreira, F.F. Structural feature evolution from fluids to the solid phase-and crystal morphology study of LASSBio 1601: A cyclohexyl-N-acylhydrazone derivative. *RSC Adv.* **2015**, *5*, 39889–39898. [CrossRef]
5. de Oliveira, M.A.; da Silva, G.D.; Campos, M.S.T. Chemical degradation kinetics of fibrates: Bezafibrate, ciprofibrate and fenofibrate. *Braz. J. Pharm. Sci.* **2016**, *52*, 545–553. [CrossRef]
6. de Oliveira, A.M.; Jäger, E.; Jäger, A.; Stepánek, P.; Giacomelli, F.C. Physicochemical aspects behind the size of biodegradable polymeric nanoparticles: A step forward. *Colloids Surf. A Physicochem. Eng. Asp.* **2013**, *436*, 1092–1102. [CrossRef]
7. Shi, J.; Kantoff, P.W.; Wooster, R.; Farokhzad, O.C. Cancer nanomedicine: Progress, challenges and opportunities. *Nat. Rev. Cancer* **2017**, *17*, 20–37. [CrossRef] [PubMed]
8. Moghimi, S.M.; Hunter, A.C.; Murray, J.C. Nanomedicine: Current status and future prospects. *FASEB J.* **2005**, *19*, 311–330. [CrossRef] [PubMed]
9. Sercombe, L.; Veerati, T.; Moheimani, F.; Wu, S.Y.; Sood, A.K.; Hua, S. Advances and Challenges of Liposome Assisted Drug Delivery. *Front. Pharmacol.* **2015**, *6*, 1–13. [CrossRef]

10. Bobo, D.; Robinson, K.J.; Islam, J.; Thurecht, K.J.; Corrie, S.R. Nanoparticle-Based Medicines: A Review of FDA-Approved Materials and Clinical Trials to Date. *Pharm. Res.* **2016**, *33*, 2373–2387. [CrossRef]
11. Jiang, Z.; Liu, H.; He, H.; Ribbe, A.E.; Thayumanavan, S. Blended Assemblies of Amphiphilic Random and Block Copolymers for Tunable Encapsulation and Release of Hydrophobic Guest Molecules. *Macromolecules* **2020**, *53*, 2713–2723. [CrossRef]
12. Jia, L.; Wang, R.; Fan, Y. Encapsulation and release of drug nanoparticles in functional polymeric vesicles. *Soft Matter* **2020**, *16*, 3088–3095. [CrossRef] [PubMed]
13. Croy, S.; Kwon, G. Polymeric Micelles for Drug Delivery. *Curr. Pharm. Des.* **2006**, *12*, 4669–4684. [CrossRef] [PubMed]
14. de Castro, C.E.; Ribeiro, C.A.S.; da Silva, M.C.C.; Dal-Bó, A.G.; FGiacomelli, C. Sweetness Reduces Cytotoxicity and Enables Faster Cellular Uptake of Sub-30 nm Amphiphilic Nanoparticles. *Langmuir* **2019**, *35*, 8060–8067. [CrossRef] [PubMed]
15. Gref, R.; Lück, M.; Quellec, P.; Marchand, M.; Dellacherie, E.; Harnisch, S.; Blunk, T.; Müller, R.H. Stealth' corona-core nanoparticles surface modified by polyethylene glycol (PEG): Influences of the corona (PEG chain length and surface density) and of the core composition on phagocytic uptake and plasma protein adsorption. *Colloids Surf. B Biointerfaces* **2000**, *18*, 301–313. [CrossRef]
16. Oh, K.S.; Song, J.Y.; Cho, S.H.; Lee, B.S.; Kim, S.Y.; Kim, K.; Jeon, H.; Kwon, I.C.; Yuk, S.H. Paclitaxel-loaded Pluronic nanoparticles formed by a temperature-induced phase transition for cancer therapy. *J. Control. Release* **2010**, *148*, 344–350. [CrossRef]
17. Wood, I.; Martini, M.F.; Albano, J.M.R.; Cuestas, M.L.; VMathet, L.; Pickholz, M. Coarse grained study of pluronic F127: Comparison with shorter co-polymers in its interaction with lipid bilayers and self-aggregation in water. *J. Mol. Struct.* **2016**, *1109*, 106–113. [CrossRef]
18. Ribeiro, C.A.S.; de Castro, C.E.; Albuquerque, L.J.C.; Batista, C.C.S.; Giacomelli, F.C. Biodegradable nanoparticles as nanomedicines: Are drug-loading content and release mechanism dictated by particle density? *Colloid Polym. Sci.* **2017**, *295*, 1271–1280. [CrossRef]
19. Šachl, R.; Uchman, M.; Matějíček, P.; Procházka, K.; Štěpánek, M.; Špírková, M. Preparation and characterization of self-assembled nanoparticles formed by poly(ethylene oxide)-block-poly(ε-caprolactone) copolymers with long poly(ε-caprolactone) blocks in aqueous solutions. *Langmuir* **2007**, *23*, 3395–3400. [CrossRef]
20. Davies, S.P.; Mycroft-West, C.J.; Pagani, I.; Hill, H.J.; Chen, Y.H.; Karlsson, R.T.; Bagdonaite, I.; Guimond, S.E.; Stamataki, Z.; de Lima, M.A.; et al. The hyperlipidaemic drug fenofibrate significantly reduces infection by SARS-CoV-2 in cell culture models. *bioRxiv* **2021**. [CrossRef]
21. Pellosi, D.S.; Moret, F.; Fraix, A.; Marino, N.; Maiolino, S.; Gaio, E.; Hioka, N.; Reddi, E.; Sortino, S.; Quaglia, F. Pluronic® P123/F127 mixed micelles delivering sorafenib and its combination with verteporfin in cancer cells. *Int. J. Nanomed.* **2016**, *11*, 4479–4494. [CrossRef]
22. de Castro, C.E.; Bonvent, J.-J.; da Silva, M.C.C.; FCastro, L.F.; Giacomelli, F.C. Influence of Structural Features on the Cellular Uptake Behavior of Non-Targeted Polyester-Based Nanocarriers. *Macromol. Biosci.* **2016**, *16*, 1643–1652. [CrossRef]
23. Svergun, D.I. Determination of the regularization parameter in indirect-transform methods using perceptual criteria. *J. Appl. Crystallogr.* **1992**, *25*, 495–503. [CrossRef]
24. Cowieson, N.P.; Edwards-Gayle, C.J.; Inoue, K.; Khunti, N.S.; Doutch, J.; Williams, E.; Daniels, S.; Preece, G.; Krumpa, N.A.; Sutter, J.P.; et al. Beamline B21: High-throughput small-angle X-ray scattering at Diamond Light Source. *J. Synchrotron Radiat.* **2020**, *27*, 1438–1446. [CrossRef]
25. Costa, P.; Lobo, J.M.S. Modeling and comparison of dissolution profiles. *Eur. J. Pharm. Sci.* **2001**, *13*, 123–133. [CrossRef]
26. Li, M.; Jiang, M.; Zhu, L.; Wu, C. Novel Surfactant-Free Stable Colloidal Nanoparticles Made of Randomly Carboxylated Polystyrene Ionomers. *Macromolecules* **1997**, *30*, 2201–2203. [CrossRef]
27. Sedlák, M.; Konňák, C. A New Approach to Polymer Self-assembly into Stable Nanoparticles: Poly(ethylacrylic acid) Homopolymers. *Macromolecules* **2009**, *42*, 7430–7438. [CrossRef]
28. Dutra LM, U.; Ribeiro ME, N.P.; Cavalcante, I.M.; Brito DH, A.D.; Semião LD, M.; Silva RF, D.; Fechine, P.B.A.; Yeates, S.G.; Ricardo, N.M.P.S. Binary mixture micellar systems of F127 and P123 for griseofulvin solubilization. *Polímeros* **2015**, *25*, 433–439. [CrossRef]
29. Patel, S.K.; Lavasanifar, A.; Choi, P. Roles of Nonpolar and Polar Intermolecular Interactions in the Improvement of the Drug Loading Capacity of PEO- b -PCL with Increasing PCL Content for Two Hydrophobic Cucurbitacin Drugs. *Biomacromolecules* **2009**, *10*, 2584–2591. [CrossRef] [PubMed]
30. Oshiro, A.; da Silva, D.C.; de Mello, J.C.; de Moraes, V.W.; Cavalcanti, L.P.; Franco, M.K.; Alkschbirs, M.I.; Fraceto, L.F.; Yokaichiya, F.; Rodrigues, T.; et al. Pluronics F-127/L-81 Binary Hydrogels as Drug-Delivery Systems: Influence of Physicochemical Aspects on Release Kinetics and Cytotoxicity. *Langmuir* **2014**, *30*, 13689–13698. [CrossRef] [PubMed]
31. Zhao, M.; Hu, B.; Gu, Z.; Joo, K.I.; Wang, P.; Tang, Y. Degradable polymeric nanocapsule for efficient intracellular delivery of a high molecular weight tumor-selective protein complex. *Nano Today* **2013**, *8*, 11–20. [CrossRef]
32. Bohorquez, M.; Koch, C.; Trygstad, T.; Pandit, N. A study of the temperature-dependent micellization of pluronic F127. *J. Colloid Interface Sci.* **1999**, *216*, 34–40. [CrossRef] [PubMed]
33. Bruschi, M.L. *Strategies to Modify the Drug Release from Pharmaceutical Systems*; Elsevier: Amsterdam, The Netherlands, 2015.
34. Wilhelm, S.; Tavares, A.J.; Dai, Q.; Ohta, S.; Audet, J.; Dvorak, H.F.; Chan, W.C. Analysis of nanoparticle delivery to tumours. *Nat. Rev. Mater.* **2016**, *1*, 16014. [CrossRef]
35. Sharma, P.; Negi, P.; Mahindroo, N. Recent advances in polymeric drug delivery carrier systems. *Adv. Polym. Biomed. Appl.* **2018**, *10*, 369–388.

36. Sanna, V.; Pala, N.; Sechi, M. Targeted therapy using nanotechnology: Focus on cancer. *Int. J. Nanomed.* **2014**, *9*, 467–483. [CrossRef]
37. Hariyanto, T.I.; Kurniawan, A. Dyslipidemia is associated with severe coronavirus disease 2019 (COVID-19) infection. *Diabetes Metab. Syndr. Clin. Res. Rev.* **2020**, *14*, 1463–1465. [CrossRef]
38. Atmosudigdo, I.S.; Lim, M.A.; Radi, B.; Henrina, J.; Yonas, E.; Vania, R.; Pranata, R. Dyslipidemia Increases the Risk of Severe COVID-19: A Systematic Review, Meta-analysis, Meta-regression. *Clin. Med. Insights Endocrinol. Diabetes* **2021**, *14*. [CrossRef]
39. Ballavenuto, J.M.A.; de Oliveira, J.D.D.; Alves, R.J. Glicogenose Tipo I (Doença de Von Gierke): Relato de Dois Casos com Grave Dislipidemia. *Arq. Bras. Cardiol.* **2020**, *114* (Suppl. 1), 23–26. [CrossRef]

Review

Antimicrobial Polymeric Structures Assembled on Surfaces

Iulia Babutan [1,2], Alexandra-Delia Lucaci [3] and Ioan Botiz [1,*]

[1] Interdisciplinary Research Institute on Bio-Nano-Sciences, Babeș-Bolyai University, 42 Treboniu Laurian Str., 400271 Cluj-Napoca, Romania; iulia.babutan@ubbcluj.ro
[2] Faculty of Physics, Babeș-Bolyai University, 1 M. Kogălniceanu Str., 400084 Cluj-Napoca, Romania
[3] George Emil Palade University of Medicine, Pharmacy, Science, and Technology of Târgu Mureș, 38 Gheorghe Marinescu Str., 540142 Târgu Mureș, Romania; lucaci.alexandra-delia@stud16.umftgm.ro
* Correspondence: ioan.botiz@ubbcluj.ro

Abstract: Pathogenic microbes are the main cause of various undesired infections in living organisms, including humans. Most of these infections are favored in hospital environments where humans are being treated with antibiotics and where some microbes succeed in developing resistance to such drugs. As a consequence, our society is currently researching for alternative, yet more efficient antimicrobial solutions. Certain natural and synthetic polymers are versatile materials that have already proved themselves to be highly suitable for the development of the next-generation of antimicrobial systems that can efficiently prevent and kill microbes in various environments. Here, we discuss the latest developments of polymeric structures, exhibiting (reinforced) antimicrobial attributes that can be assembled on surfaces and coatings either from synthetic polymers displaying antiadhesive and/or antimicrobial properties or from blends and nanocomposites based on such polymers.

Keywords: synthetic antimicrobial polymers; assembled nanostructures; surfaces and coatings; antimicrobial properties

1. Introduction

The risk of microbial infection associated with multidrug-resistant microbes, i.e., certain microbes that have developed resistance to at least one drug of three different antimicrobial drug categories, has become an increasingly important problem for human health [1–4]. An efficient way to prevent and control possible microbial infections is the administration of antibiotics [5]. Nonetheless, excessive consumption of antibiotics leads, for instance, to developing new strains of antibiotic-resistant bacteria [6]. This happens inclusively in hospitals where patients with compromised immunity are treated for chronic diseases. Moreover, in recent years, the development of new types of antibiotics has been hindered by significant scientific challenges, regulatory uncertainties and industrial difficulties [7]. As a result, millions of people are infected yearly with multidrug-resistant bacteria [8], and the number of deaths associated with such infections is significant [9]. Furthermore, considerable developments of biomaterials led to the appearance of vital medical devices, such as catheters, joint implants or contact lenses, just to name a few. Unfortunately, while the surfaces of these devices are constantly exposed to microbial adhesion [10], treating the eventual infections requires significant efforts [11] and therapeutic targets [12].

Antimicrobial polymers are (biocidal) materials that can prevent and suppress the growth of various undesired microorganisms, including bacteria. Moreover, they can combat the bacterial resistance to antibiotics because, unlike conventional antibiotics, polymers exhibit antimicrobial mechanisms that cannot be outwitted by pathogens [13]. Furthermore, antimicrobial polymers can be easily adapted to applications, such as coatings and used to sterilize various surfaces, inclusively those of medical instruments. Thus, such polymers could become a good alternative to antibiotics and disinfectants and, why not, could eventually replace them in the future. This could be possible, especially because polymers come with important advantages, as they can adopt more or less complex chemical structures [14,15] that can favor their assembly and crystallization processes. These processes actually dictate the final properties of polymers in bulk, solutions, or thin-films [16–19]. Due to the potentiality to precisely control other processing parameters, such as melting, crystallization or glass-transition temperature, a polymer can display a highly tunable molecular ordering on multiple-length scales, ranging from nanometers to macroscopic dimensions that can generate a diverse landscape of nanostructures [16,20–22]. Further expansion of this landscape on the molecular, microscopic and macroscopic scales can be induced by favoring physical and chemical interactions of specific chain segments with their neighbors [15,16,23] or by degrading the phase purity through the addition of other (polymeric) components [24–27].

Antimicrobial polymeric systems include biopolymers and synthetic polymers (hereafter simply "polymers"). While biopolymers, such as polypeptides, polysaccharides or polynucleotides, are natural chains produced by the cells of various living organisms, polymers are human-made from precursors, such as petroleum derivatives or even biological components like peptides [28], lysine or arginine [29]. Antimicrobial polymers are designed to imitate the antimicrobial structures produced by the immune systems of various living organisms to kill microbes and can contain in their backbone or their side-chains various moieties with biocidal properties.

In this work, we emphasize the antimicrobial properties of recent structures (self-)assembled from polymers on various surfaces, thin films and coatings. We highlight these antimicrobial structures and the corresponding antimicrobial mechanisms developed not only in pure polymeric materials but also in polymer-based nanocomposites and polymer–polymer or polymer–biopolymer blends, respectively.

2. Main Antimicrobial Mechanisms Associated with Polymeric Structures

To depict how significant is the role of antimicrobial polymers in the war against microbes, we start our review with a brief classification of the polymer-based antimicrobial mechanisms (Table 1). These mechanisms of action are mainly associated to polymeric structures loaded with drugs or formed in hydrogels, or bound to surfaces.

Table 1. Summary of the main antimicrobial structures used to develop various mechanisms to repel and destroy microbes.

Antimicrobial Polymer Devices	Employed Polymeric Structures	Antimicrobial Principle	Ref.
Drug-loaded polymers	Nanoparticles, micelles, vesicles, dendritic structures	Delivery and release of drugs or other biocidal components	[30–34]
Polymeric hydrogels	Gel-like microstructures	Employment of drugs/biocides to kill microbes	[35,36]
Surface-bound polymers	Various structures: (bottle)brushes, spherical nanoparticles (micelles, vesicles, rods, fibers, worms, multilayers, etc.	Neutral polymer-based surfaces (steric repulsion) Anionic polymer-based surfaces (electrostatic repulsion) Ultrahydrophobic (low-energy) polymer-based surfaces Contact killing surfaces (cationic, use of biocidal moieties) Biocide releasing surfaces (use of biocides) Stimuli-responsive surfaces (temperature, pH, etc.) Adaptive bactericidal surfaces	[37–40]

2.1. Drug-Loaded Polymers and the Associated Antimicrobial Mechanisms

Polymeric nanoparticles (NPs) are self-assembled (hierarchical) structures that can be rapidly adapted to achieve controlled and targeted drug load and release at the site of microbial infections by regulating the characteristics of polymers and the surface chemistry of resulting structures. Polymeric structures loaded with biocides can be concentrated preferentially at the infected site and can act as a depot that provides a continuous supply of encapsulated therapeutics over days or even weeks (see the mechanism of action of antibiotics schematically depicted in Figure 1a). A significant advantage of a drug delivery system based on polymeric NPs is that it can protect therapeutic agents against enzymatic degradation, while the dose required for the drug to be therapeutically efficient is significantly lower. Consequently, its systemic toxicity is reduced. Other advantages of polymeric NPs include the possibility to augment the bacterial sensitivity to drugs by overcoming two resistance mechanisms, namely increased efflux and decreased antibiotic uptake (Figure 1b), and to achieve curative effects more easily by packaging multiple drugs within the same NP, or by combining several types of NPs. This strategy enhances the antibacterial effects and helps preventing bacterial resistance [41].

Some polymeric self-assembled NPs have their antibacterial capacity from their multimodal mechanism of action: outer membrane destabilization, inner membrane perturbation and unregulated ionic movements, leading to apoptotic-like death followed by bacterial cell lysis [42,43]. Star polymer NPs [42] can encapsulate guest drug "cargos" with fewer undesired bursts [30] and can maintain solubility and viscosity similar to that of low molecular weight linear or branched polymers [44]. Although some star polymer NPs may be toxic and can present off-target side-effects against probiotic bacteria [42], the unwanted effects can be eliminated by adding sugar-based moieties on star polymers to target the receptors expressed on macrophages and to render polymers antimicrobial [45]. Similarly, glucosamine-functionalized star polymers can be employed to penetrate the peptidoglycan layer of bacterial cells more easily [46].

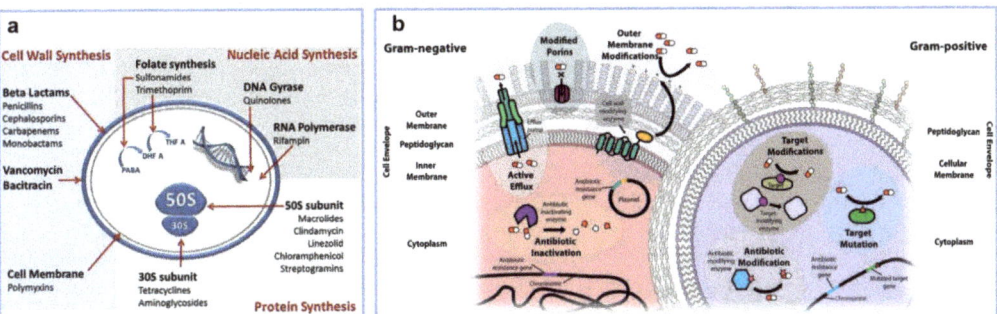

Figure 1. (a) Schematics depict the mechanism of action of antibiotics. (b) Illustration of mechanisms of antibiotic resistance corresponding to Gram-negative and Gram-positive bacteria. In comparison to Gram-positive bacteria, Gram-negative bacteria display an additional outer lipid membrane layer. This layer has the role of obstructing the entry of drugs into the bacterial cell, leading to antibiotic resistance. Several components contribute to this process: efflux pumps, porins with modified or reduced expression and cell wall modifying enzymes. On the other hand, Gram-positive bacteria present a thicker peptidoglycan layer whose structure can be altered to decrease antibiotic uptake. Enzymes encoded in Gram-positive and Gram-negative bacterial genetic material can inactivate or modify antibiotics. Furthermore, certain enzymes can protect the target of antibiotics by modifying their structure and/or number. Adapted with permission from ref. [47] (a) and ref. [48] (b).

Polymeric micelles are core-shell NPs, typically formed through the self-assembly of amphiphilic block copolymers, in which the core can accommodate hydrophobic drugs, while the shell makes the micelle water-soluble and promotes delivering poorly soluble agents. They can also act as antimicrobial polymers by themselves without incorporating other biocidal agents [49]. Early biodegradable antimicrobial polymeric micelles, self-assembled from amphiphilic polycarbonates (PCs), exerted biocidal action by a cell wall and membrane disruption, which led to bacterial lysis [50]. The most important advantages presented by this system are the low hemolytic activity and low toxicity observed on mice. However, these micelles are ineffective against Gram-negative bacteria. To overcome this major drawback, a random copolymer structure can be used instead of a block copolymer, making the hydrophobic moieties more accessible, therefore, increasing the likelihood of membrane insertion and disruption [51]. Furthermore, multifunctional micelles, capable of both detection and inhibition of bacteria [52], as well as NPs exhibiting inherent antimicrobial activity and drug-loading capability, can be produced [53]. Other types of micelles include antimicrobial rod-shaped micelles that can be used efficiently against *C. albicans*, fungal species that are generally more difficult to kill due to their multilayered, thicker, and less negatively charged cell walls [31,32]. Although there were numerous attempts to create effective, biocompatible, and nontoxic polymeric micelles, the role of self-assembly on biological activity was not completely defined, with multiple studies declining the positive contributory role of micellization [54–56]. Moreover, it was postulated that the polymer chains might be more active as free molecules in solution. In contrast, the supramolecular structures could reduce the polymer chain mobility [54], hinder the exposure of cationic groups, which are essential for electrostatic interaction with bacterial cell walls [42] or shield the hydrophobic components required for the interaction with bacterial membrane within the core [57].

Polymeric vesicles (polymersomes) are other spherical polymeric capsules with a hollow inner compartment confined by a bilayered membrane composed of amphiphilic block copolymers that can be loaded with both hydrophobic and hydrophilic drugs. A distinctive characteristic of polymersomes is that they offer three different regions available for functionalization, more precisely the inner hydrophobic cavity, the polymer shell, and the periphery [58]. Stronger polymer–bacteria affinity exists when the polymers are self-assembled as vesicles, rather than in cases where polymer chains remained unassembled, probably due to increased local concentration of positive charges [33]. This is the case of perforated high-genus polymer vesicles that can encapsulate doxorubicin and manifest an acid-accelerated drug release profile that can be used for targeted therapeutic delivery when simultaneous antibacterial and anti-cancer medication is needed [59].

Dendrimers, i.e., highly branched symmetrical 3D macromolecules with core–shell architecture and nanometer-scale dimensions, consist of a (hydrophobic) core, layers of branched repeating units and an outer (hydrophilic) layer of functional end groups [60,61]. Although the diffusion of dendrimers is limited, initial adsorption and binding to cell membranes are stronger than for linear polymers [34]. In this case, biocidal hydrophobic and hydrophilic agents can be loaded inside the dendritic structure by noncovalent encapsulation or can be attached to surface moieties by covalent binding [61,62] followed by a targeted release that can increase the duration of action of the antimicrobials [63–65]. This is, inclusively, the case of cationic dendrimers that can eventually be loaded with antiproliferative constituents and can be employed as novel antibacterial agents against both Gram-positive and Gram-negative bacteria [66–68]. Usually, the bacteria-killing process is following these steps: adsorption onto negatively charged bacterial cell wall, diffusion through the cell wall, binding to the cell membrane, disruption and disintegration of the membrane followed by the release of electrolytes and nucleic materials from the cell, leading to death [62].

2.2. Antimicrobial Mechanisms Specific to Polymeric Hydrogels

Hydrogels are hydrophilic, crosslinked polymer networks with a unique 3D structure, capable of absorbing more than 20% of their weight of water while maintaining their structure and ability to control the release of therapeutics. They can be manufactured of polymers that are capable of converting to gels in the presence of different stimuli, such as temperature, pH, UV irradiation, etc. Hydrogels can be manufactured in a manner that confers mechanical features and structures similar to natural tissues. Because of their high biocompatibility, mucoadhesion and ability to respond to microenvironmental changes, hydrogels are attractive drug-delivery systems, especially for local antimicrobial applications. They can be successfully used as coatings for catheters, wound dressings, contact lenses, etc. The use of hydrogels loaded with antimicrobial therapeutics promotes delivering an adequate dose directly to the infected site and offers a stable and prolonged release, without undesired bursts, thus hindering the development of new drug-resistant strains. They offer a high surface-to-volume ratio and, most importantly, their hydrophilic nature provides an excellent solubilizing environment for numerous antibiotics like ciprofloxacin, amoxicillin, gentamicin [35].

Another advantage of antimicrobial hydrogels consists in the possibility to either construct hydrogel networks with inherent antibacterial activity by utilizing polymer molecules that exert antimicrobial properties by themselves [69,70] or to further employ at least two components exhibiting antimicrobial properties to generate hydrogels with synergistic effects that can further enhance the overall antimicrobial efficiency [71–73]. For instance, hydrogels are often loaded with inorganic NPs with great potential in biomedical applications [35,36]. Silver (Ag) is a popular antimicrobial agent that expresses biocidal properties against a wide range of bacteria, fungi and viruses due to its multiple mechanisms of action (including the release of Ag^+, the intrinsic antibacterial properties of Ag-based on its penetration into the bacterial membrane and cell walls, the antibacterial effects caused by reactive oxygen species/ROS generated by Ag^+ that can induce oxidative stress in bacterial cells) [74]. For example, exposure of microbes to Ag^+ activates the interaction with proteins of the microbial cell wall and membrane, leading to the impairment of cellular transport systems, electrolytes imbalance, membrane perforations, loss of cytoplasmic organelles, alteration of bacterial cell division and finally, to cell death [75]. Thus, loading polymeric hydrogel matrices with Ag NPs leads to highly antimicrobial systems capable to efficiently neutralize Gram-positive and Gram-negative bacteria over longer periods [35,36]. Gold (Au) NPs also possess antibacterial properties and can be embedded within hydrogel networks. Although Au NPs are not as efficient as Ag NPs, they can cause bacterial membrane disruption, inducing cell death, have a large antibacterial spectrum, including methicillin-resistant *S. aureus* and *P. aeruginosa* and are nontoxic to osteoblastic cells [35,36]. Furthermore, ZnO shows antibacterial activity and low toxicity against mammalian cells. Its mechanism of action is based on damaging the lipids and proteins of the bacterial cell membrane and ROS generation. Significant peculiarities of ZnO NPs include their activity against high-temperature-resistant and high-pressure resistant bacterial spores and beneficial influence on bone regeneration [35,36]. Beside hydrogels, Ag, Au, ZnO and other types of inorganic NPs can also be incorporated in antimicrobial polymer composites in order to boost the antimicrobial properties of the latter, as we will see in Section 5.

2.3. Surface-Bounded Polymers and the Corresponding Antimicrobial Mechanisms

As it is schematically depicted in Figure 2 (top), there are two ways to keep various surfaces microbial-free: to prevent the attachment of microbes, including bacteria, on surfaces before any contact (antiadhesive mechanisms) and to kill microbes on surfaces after contact (biocidal mechanisms). For better clarity of the text, we further discuss the above mechanisms in the following two subsections.

2.3.1. Mechanisms Employed to Repel Microbes from Surfaces

Protein and microbial repellent/antifouling coatings prevent the attachment of proteins and microbes on surfaces but do not actively interact with or kill them. A typical repelling mechanism is corresponding to hydrophilic polymers that are coated or assembled on various surfaces, where they produce an aqueous interface, which repels microbes/foulants (when proteins or microorganisms are in proximity of the surface, water molecules are released from the interface, while polymer chains adopt more compressed conformations [37]). Therefore, the microbial adhesion strongly relies on the precise arrangements of the polymer chains into well-defined, often not easy to fabricate surface structures [76] that dictate the surface charge and chemistry, hydrophobicity, topography, roughness, stiffness (Figure 2 bottom) [77]. The repelling mechanism is based on various interactions of microorganisms with the surface they try to colonize. These interactions include the steric and the electrostatic repulsion (Figure 2 top).

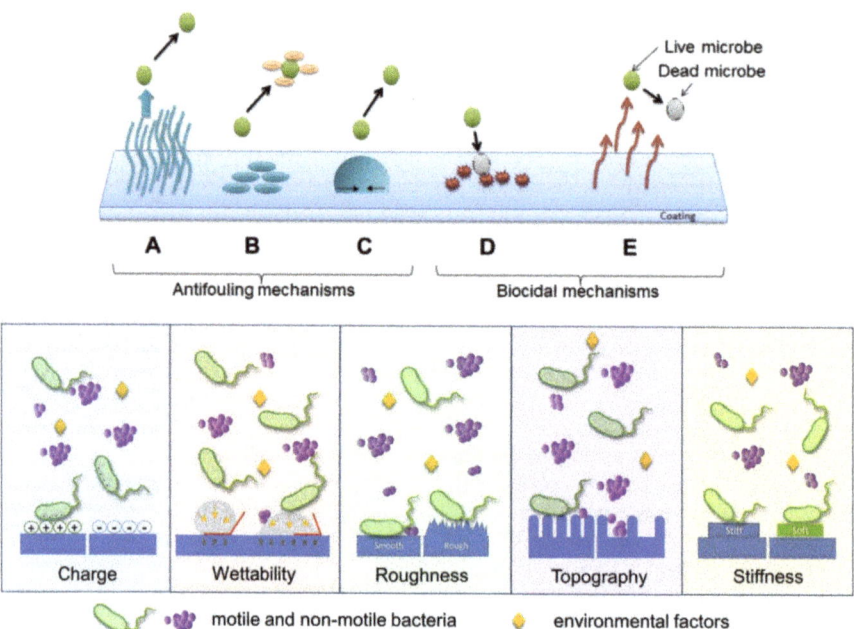

Figure 2. Examples of antimicrobial mechanisms of polymers: (**A**) steric repulsion, (**B**) electrostatic repulsion, (**C**) low surface energy (i.e., surfaces exhibiting high contact angle), (**D**) biocide releasing, (**E**) contact-killing of microbes (top). Schematic illustration depicting various surface parameters that exert significant influence on bacterial adhesion (bottom). Reprinted with permission from ref. [37] (top) and ref. [38] (bottom).

Hydrophilic oligomers and polymers, such as oligoethylene glycol (OEG), polyethylene glycol (PEG), poly(poly(ethylene glycol) methacrylate) (PPEGMA) [78–80], take advantage of steric repulsion to repel microorganisms. For instance, PEG functionalized coatings form the hydration layer via hydrogen bonds and can prevent direct contact between microorganisms or proteins and surfaces with a certain efficiency. This repelling efficiency is low on surfaces of linear PEGs, in which molecules do not overlap, and higher on (star-shaped) PEG surfaces where molecules do interpenetrate and form a denser microstructure [81] (generally, a denser PEG microstructure can be obtained with shorter chains [82,83]). Moreover, the repelling efficiency is higher for microstructures made of PEG chains covalently bonded rather than less stable physioadsorbed chains [84].

The main disadvantage of the repelling mechanism of PEG-based polymers is its dependence on temperature [85] and time [76]. To overcome such drawbacks, zwitterionic polymers, i.e., neutral molecules that comprise both positive and negative charges, can be employed. The repelling mechanism based on zwitterionic polymers, such as poly(2-methacryloyloxyethyl phosphorylcholine) (PMPC) [86], poly(sulfobetaine methacrylate) (PSBMA) [87,88], poly(carboxybetaine methacrylate) (PCBMA) [87] or polybetaines [89] mainly operates by electrostatic repulsion, as multilayered surface structures are obtained via electrostatic interactions instead of hydrogen bonding. Owing to their great potential in withstanding protein adsorption [90] and to their specific molecular arrangements that allow a significant amount of water molecules to be bound on their surface, polymeric zwitterions are considered very attractive candidates for developing highly antiadhesive coatings [37,85]. Electrostatic repulsion can also be exploited to create antibacterial surfaces by employing anionic polymers. This is possible because bacterial cells are negatively charged (with only a few exceptions), and thus when approaching a negatively charged surface, electrostatic repulsion intervenes and prevents the bacterial adhesion [91,92].

At the end of this section, we briefly mention that polymers, such as polystyrene (PS), PC and polyethylene (PE), can also be used to fabricate extrinsic antimicrobial surfaces [93]. Such polymer chains can be directed to assemble into well-defined superhydrophobic micro-/nanostructures exhibiting low surface energy (Figure 2 top). It was demonstrated that only 2% of bacteria from a droplet adhered on such surfaces; less than 0.1% of bacteria remained attached on the surface after it was rinsed [93]. Bacterial cells can colonize a superhydrophobic structure only when the latter become fully wet and the air entrapped between the water droplets, and the structure is excluded [91].

2.3.2. Mechanisms Employed to Kill Microbes on Surfaces

Contact-killing surfaces express their biocidal properties when in contact with various microorganisms, including bacteria (Figure 3). Their major advantage is the possibility to hinder the rise of bacterial resistance, considering their non-specific mechanisms of action: physical damage of bacterial cells or ROS release [94]. For instance, cationic polymers, which contain both positive and hydrophobic functional groups [95], kill bacteria via electrostatic interactions between the cationic groups and negatively charged bacterial cells (Figure 3a). Because cationic polymers mimic the mechanism of action of antimicrobial peptides, the sequence through which they kill pathogens is the following: adsorption on the microbial cell surface, penetration into the cell wall, interaction with the cytoplasmic membrane, followed by irreversible damage of this structure, leakage of cytoplasmic components, such as electrolytes and nucleic acids, and, eventually, cell death. Unfortunately, antimicrobial surfaces that use cationic polymers are only suitable for short-term applications because their positive charges cause protein adhesion and accumulation of bacterial debris. They become inactive once covered by a (thick) layer of biomolecules and microbial fragments [96].

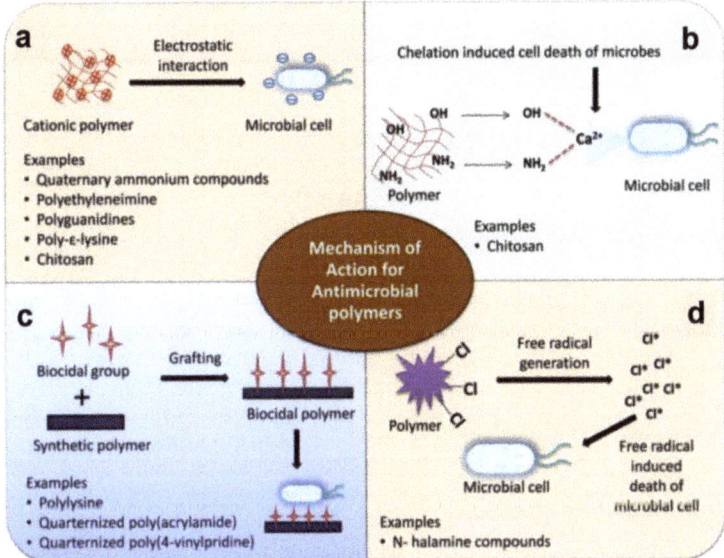

Figure 3. Schematic illustration of mechanisms of action of antimicrobial polymers based on electrostatic interactions (**a**), metal ion chelation (**b**), interactions with biocidal groups grafted on polymers (**c**), and release of halogen free radicals by N-halamine moieties (**d**). Adapted with permission from ref. [39].

In comparison to cationic polymers, polymers based on N-halamines contain nitrogen atoms that covalently bind halogens, such as chlorine, bromine or iodine. These halogens can be slowly released into the environment and lead to generating ROS (Figure 3d) that can kill faster a broad spectrum of bacteria. In general, the biocidal activity of N-halamines can be restored using halogen-donor compounds (sodium hypochlorite/hypobromite) [97]. The mechanism of action of N-halamines has been described in two different ways: either the halogen is directly transferred to the microbial cell wall, followed by oxidation, or the dissociation into water is followed by diffusion over the bacteria. Afterward, the oxidative action of the halogen is directed to a biological receptor of bacterial cells (e.g., thiol or amino protein groups), causing metabolic inhibition or cell death. Therefore, in contrast to cationic polymers, the biocidal properties do not arise from the polymer itself but from the N-halamine functional group [39,98].

Other mechanisms used to kill microbes include chelation (Figure 3b) and interactions with biocidal groups grafted on polymer chains (Figure 3c). While the former mechanism is rather related to biopolymers, such as chitosan (not a topic of discussion in this work), the latter mechanism can be used to kill bacteria by grafting various biocidal groups on synthetic polymers, leading to biocidal polymers, such as quaternized poly(acrylamides) or quaternized poly(4-vinylpiridines). Of course, instead of grafting, various biocide moieties and disinfectants (metallic ions, quaternary ammonium compounds, antibiotics, antimicrobial peptides, or triclosan) can be loaded within inactive polymer matrixes to obtain biocide-releasing polymers. These can be further employed in the manufacture of leaching polymer surfaces [39] characterized by the mechanism of action depending on the released component [96]. Often, the antimicrobial agents can be released when certain physical, chemical, or biological stimuli are present. For example, temperature-responsive poly(N-isopropylacrylamide) surfaces can reduce bacterial adhesion or induce toxicity at particular temperatures. Instead, bio-responsive surfaces change their microstructure when exposed to enzymes or other constituents of the biological fluids. Similarly, pH-responsive surfaces are based on pH-dependent antimicrobial compounds, which are activated at low pH values and can be used for targeting acid-producing bacteria [40]. Due to the tremendous amount of data available in the literature on antimicrobial polymers, we further discuss within this review only the most recent antimicrobial polymeric structures assembled in/on coatings/surfaces.

3. Recent Antimicrobial Structures Assembled from Polymers on Surfaces

Several well-established polymer-based antimicrobial mechanisms are challenging researchers to employ synthetic polymers to generate efficient antimicrobial structures in various environments. A prominent category of such structures is represented by drug-loaded and gel-like structures [42,53,99–102]. Another equally important category of antimicrobial structures includes those obtained from polymeric systems assembled on specific surfaces [103–116]. In this work, we further review only the most recent antimicrobial structures falling within the last category.

We start by emphasizing that structures generating antimicrobial surfaces can be obtained through the assembly of both synthetic polymers and biopolymers, such as cellulose [117], chitin [118], chitosan [119], furcellaran [120], pectin [121], pullulan [122] or starch [123], etc. However, in this work, we present only polymer-based antimicrobial surface structures. Based on various antimicrobial mechanisms [75,76], these surface protecting structures rely on three categories of polymers. The first category comprises polymers employed to prevent the adhesion of microbes on surfaces rather than killing them by minimizing the protein adsorption (i.e., the biopassive antiadhesive class of polymers; see the antimicrobial mechanisms in Section 2.3.1). The second category is based on polymers, eventually covalently decorated with antimicrobial moieties and conjugates, that kill microbes when in contact (i.e., the bioactive antimicrobial class of polymers; see the antimicrobial mechanisms in Section 2.3.2). The third category is represented by antiadhesive polymers combined with various antimicrobial agents/polymers to exploit the synergistic effects against microbes and provide surface protecting structures with both microbe preventing and killing attributes.

3.1. Polymeric Structures Preventing the Adhesion of Microbes on Surfaces

Nowadays, it is of paramount importance to prevent various microorganisms from contacting, adhering and flourishing on specific surfaces in order, for example, to keep various medical devices/facilities uncontaminated [124–127], to generate efficient implants [105], or to avoid biocorrosion of marine-related applications [128]. Polymers can be employed to engineer films and coatings with tuned surface properties that can prevent the adhesion of microorganisms. One such polymer is PEG, a polyether that can be synthesized in various molecular weights and that can display a diversity of hierarchically ordered nanostructures at multiple-length scales, especially when incorporated in block copolymers [129,130]. This ability to adopt various highly ordered intrachain conformations endows PEG with protein repellent or attractive properties at low or high compressive loads, respectively [131]. It makes it a versatile material that can be used to efficiently fight microbes when coated on surfaces [103–105].

Coating various surfaces with PEG can be mainly realized either by employing various direct deposition techniques [104,105] or through the grafting procedure, i.e., by covalently anchoring the polymer chains to surfaces as brushes [103]. The latter procedure has the advantage of facile tunning of the microbial adhesion through the control of polymer chain length and temperature at which experiments are performed [103] and can reduce the adhesion of microbes by several to tens of times (Figure 4a and Table 2) [104]. For example, PEG can be utilized to construct multilayered films on stainless steel, nylon, titanium oxide or silicon oxide substrates, previously covered with mussel-inspired polydopamine (PDA) [104]. Moreover, to target marine biofouling (i.e., problematic accumulation of diverse microorganisms, plants and/or algae on surfaces immersed in water that damages boats or various underwater constructions and devices) on the above-mentioned surfaces, PEG catechols, obtained by coupling the amine groups of 6-arm-PEG-amine and the carboxylic group of 3,4-dihydroxyhydrocinnamic acid [132], can be crosse-linked with PDA through catechol-catechol based interactions to become insoluble. The resulting multilayered films spin-cast from PEG catechol solution is not only stable under marine environments but also exhibits high resistance towards marine *A. coffeaeformis* (Figure 4a) [104]. Crosslinking process can further be employed to develop anticellular and bacterial repellent PEG-based coatings on bare and sandpapered titanium surfaces [105]. More exactly, nanofibers of PEG prepared via electrospinning can be coated on titanium surfaces and then rendered insoluble through using a photo-crosslinking agent. Results have shown that titanium surfaces covered with PEG nanofibers display enhanced antiadhesive properties against fibroblastic preosteoblasts and *S. epidermidis* compared to their counterpart blank surfaces [105].

Another class of polymers, known for their role in impeding the adhesion of microbes on surfaces, is represented by the zwitterionic polymers, such as polysulfobetaine (PSB) [133]. Moreover, methacrylate-based PSB (PSBMA) can be grafted on glass substrates by employing an atom transfer radical polymerization (ATRP) route and can lead to 10-15 nm thick brush-like structures of specific molecular conformations. These nontoxic structures can efficiently inhibit the adhesion of marine green algae, spores, sporelings and diatoms, such as *Navicula*, over various periods [106]. More recently, the microbe repellent properties of zwitterionic PSBMA were extended to *S. aureus* and *S. epidermidis* through the realization of thin brush-like films of PSBMA and PCBMA on polydimethylsiloxane (PDMS) surfaces [134]. PSBMA and PCBMA were covalently attached to PDMS substrates by employing a grafting method based on the exposure to UV light of the zwitterionic monomers in the presence of photoinitiators and crosslinking agents. Resulting PSBMA and PCBMA films decreased the adhesion of the above-mentioned bacteria under dry conditions by an order of magnitude [134]. A significant bacteria antiadhesive effect was also demonstrated for zwitterionic brushes grafted on brominated stainless steel and polymerized from 2-methacryloyloxyethyl phosphorylcholine and *N*-(3-sulfopropyl)-*N*-(methacryloxyethyl)-*N*,*N*-dimethylammonium betaine [135]. The thickness of these brushes was evaluated to be 53 and 132 nm, while the static water contact angle was measured to be only 14° and 11°, respectively.

The same ATRP reaction can be utilized along with photochemical grafting to grow protein-repelling zwitterionic brushes of poly [3-(methacryloylamino)propyl]dimethyl(3-sulfopropyl)ammonium hydroxide (PMPDSAH) onto indium thin oxide substrates [136]. Interestingly, the antifouling performance of PMPDSAH coated surfaces is outperforming that of the PEG surfaces [137]. Moreover, this polymer can be functionalized by a 1.9 nm thick metal–polyphenol coating to make it attractive to proteins and cells [138]. This is particularly important when targeting the spatio-selective functionalization of a specific surface. For instance, by patterning the PMPDSAH surface with a metal–polyphenol coating using a microcontact printing technique, surfaces displaying alternating hundred micrometers sized regions with fouling and antifouling properties can be fabricated [138].

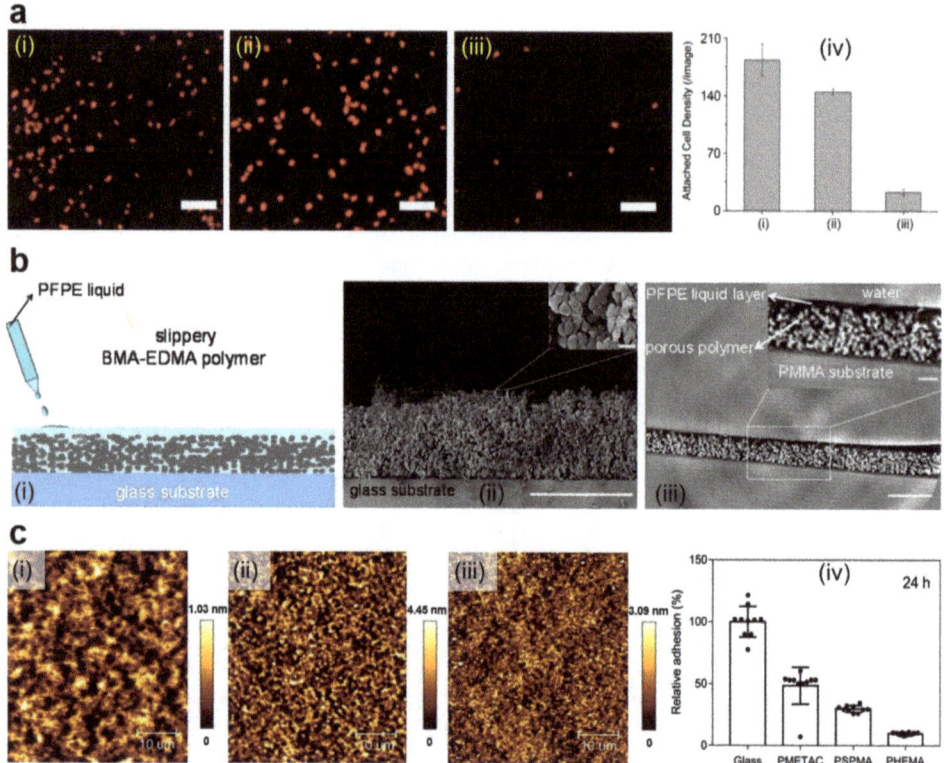

Figure 4. (a) Fluorescence images (i–iii) and quantification (iv) of *A. coffeaeformis* diatoms that succeeded to attach on the (i) untreated stainless steel, (ii) PDA-coated stainless steel, and (iii) PEG-based film. Each point is the mean of 60 counts on three replicate samples in (iv). Scale bars represent 50 μm. (b) Schematic representation of the fabrication of slippery P(BMA-co-EDMA) surface by infusion of the porous polymer with the PFPE fluid (i), scanning electron microscopy (SEM) images depict a cross-section and the porous morphology of the P(BMA-EDMA) surface (ii) and reconstructed X-ray propagation phase-contrast tomography image displaying the cross-section of the slippery P(BMA-EDMA) surface underwater (iii). Scale bars are 100 μm (ii,iii), 2 μm (ii) and 20 μm (iii) in the insets, respectively. (c) atomic force microscopy (AFM) height images depict PMETAC (i), PSPMA (ii), and PHEMA (iii) polymer brushes attached to the glass coverslips and their antiadhesive performance against *E. coli* compared to naked glass, after an incubation time of 24 h (iv). The relative adhesion of bacteria on the above substrates was normalized with respect to the adhesion on the glass substrate for 24 h. Adapted with permission from ref. [104] (a), ref. [107] (b) and ref. [139] (c).

Besides PEG and zwitterionic polymers, polyelectrolytes can also be used to create microbe repellent surfaces. Generally, the polyelectrolytes are structured within a multilayer configuration (also known as polyelectrolyte multilayer; PEM) through the layer-by-layer (LbL) deposition, assembly and/or functionalization procedures [140–142]. A relevant example of PEM structures was given by Schmolke and coworkers, who have combined poly(diallyldimethylammonium chloride) (PDADMAC) with poly(acrylic acid) (PAA) and also poly(allylamine hydrochloride) (PAH) with PAA. They have further adsorbed these systems on PDMS substrates by following an LbL assembling procedure. Resulting PDAD-MAC/PAA and PAH/PAA PEMs have demonstrated a decreased adhesion strength of *S. cerevisiae* on their surface [140]. Further adsorption of a pegylated PAA (PAA-g-PEG) on top of PAH/PAA or PDADMAC/PAA PEMs led to highly efficient coatings able to reduce the adhesion of microorganism cells up to two orders of magnitude.

Other microbe repellent polymers worth mentioning include poly(butyl methacrylate-*co*-ethylene dimethacrylate) (P(BMA-*co*-EDMA)) [107], poly(2-methyl-2-oxazoline) dimethylacrylate (PMOXDA) [108], negatively charged poly(3-sulfopropyl methacrylate potassium salt) (PSPMA) or neutral poly(2-hydroxyethyl methacrylate) (PHEMA) [139] (Table 2). While porous films of P(BMA-*co*-EDMA) can be infused with the liquid perfluoropolyether (PFPE) to fabricate slippery hydrophobic surfaces that efficiently prevent microbe adhesion and biofilm formation (Figure 4b) [107], PMOXDA can be copolymerized with different amounts of positively charged monomers of (2-(methacryloyloxy)ethyl)-trimethylammonium chloride (METAC) to lead to PMOXDA-*co*-PMETAC films displaying negative, neutral and positive surface zeta potential. Such films are then able to significantly reduce bacterial adhesion [108]. Moreover, PMETAC, PSPMA and PHEMA can be grafted onto glass surfaces by employing a surface-initiated ATRP reaction to develop polymer brushes adopting specific molecular conformations: surfaces are covered with a random few micrometer-sized PMETAC, micrometer-sized PSPMA and hundreds of nanometer-sized PHEMA features, respectively. Such brushes can control the surface-bacteria interactions and thus, can discourage the bacterial adhesion (Figure 4c) [139]. More details on polymers with antiadhesive properties and on strategies to prevent bacterial adhesion can be further consulted in the literature [37,38,78–80,86–89,141].

Table 2. Summary of polymeric systems and configurations that can be employed to prevent the adhesion of various microbes on surfaces. Following abbreviations were used: *Staphylococcus (S.), Pseudomonas (P.), Candida (C.), Amphora (A.), Saccharomyces (Sc.)* and *Escherichia (E.)*. Efficacy for biopassive systems refers to how many times fewer microbes were attached to the antiadhesive surfaces in a certain time as compared to their analogs.

Antimicrobial Polymer	Configuration/ Nanostructure	Dimension	Antimicrobial Mechanism	Efficacy	Microbe of Interest	Ref.
PEG	Nanofibers	167–184 nm diameter	Biopassive	~2–7 times	*S. epidermidis*	[105]
PEG	Brushes	2.8–23.7 nm length	Biopassive	~6 times ~6–8 times ~25 times ~4 times	*P. aeruginosa* *C. albicans* *S. epidermidis* *C. tropicalis*	[103]
PEG catechol	Multilayered films	5.2 nm thick	Biopassive	~8 times	*A. coffeaeformis*	[104]
PSBMA	Brushes	10–15 nm thick	Biopassive	~12 times ~7 times	Marine alga *Ulva* Diatom *Navicula*	[106]
PMPDSAH	Brushes	4.4 nm thick	Biopassive	-	Marine diatoms	[138]
PSBMA PCBMA	Thin-films	<25 μm thick	Biopassive	~10 times ~10 times	*S. aureus* *S. epidermidis*	[134]
PDADMAC/PAA, PAH/PAA	PEM structure	30–150 nm thick	Biopassive	~5 times	*Sc. cerevisiae*	[140]
PBMA-*co*-EDMA	Porous films	<100 μm thick	Biopassive	-	*P. aeruginosa*	[107]
PMOXDA-*co*-PMETAC	Thin-films	200 nm diameter, 230 nm thick	Biopassive	- -	*E. coli* *S. aureus*	[108]
PSPMA, PHEMA	Brushes	57 nm diameter, 43 nm thick	Biopassive	~9 times	*E. coli*	[139]

3.2. Polymeric Structures Employed to Kill Microbes on Surfaces

Bioactive polymers exhibit their antimicrobial attributes due to their intrinsic nature and/or their decorative biocidal moieties and conjugates incorporated into their backbone and/or side-chains [143–145]. An example of highly biocidal conjugates is given by the short antimicrobial peptides (AMPs), i.e., highly biocidal cationic units of gene-encoded peptide antibiotics possessing a low rate in driving antimicrobial resistance [146]. Generally, AMPs can be incorporated into polymeric device coatings and released to kill bacteria. Nonetheless, the time-limited antimicrobial effect is dictated by the elution of AMPs [147]. Instead, decorating polymers with AMPs might lead to augmented efficiency in targeting and killing the pathogens and can confer longer-term antibacterial properties while reducing the toxicity of the resulting systems against mammalian cells. In earlier studies, AMPs, such as Tet-213 (KRWWKWWRRC) [143] and other similar [144] were designed and tethered to poly-(N,N-dimethylacrylamide-co-N-(3-aminopropyl)-methacrylamide hydrochloride) P(DMA-co-APMA) copolymer chains that were grafted, in various molecular conformations and densities, from titanium surfaces (Figure 5a) [143,144]. The morphology of resulting surfaces consisted of different features exhibiting a roughness of about 6 nm. Moreover, higher copolymer brush densities rendered the resulting surfaces with more crowded AMPs and, thus, with better bactericidal properties [143].

More recently, Yu and coworkers have modulated the functionality of AMPs along with that of polymer brushes to generate high antimicrobial peptide potency that could be used in developing infection-resistant implant surfaces [148]. Their results revealed that the antimicrobial activity of brush coatings tethered with AMPs depended on the polymer brush chemistry (which has an impact on changes in the secondary structure of the AMPs) as well as on the AMP molecular conformations (which determine the density of AMPs on the polymer brushes and their microstructure) [148]. The microstructure of AMPs is further important because it regulates the rather unwanted interactions between peptides and biomolecules, such as blood proteins. An approach to eliminate such interactions, and thus, to allow peptides to attach effectively to the negatively charged bacterial surface, is to actually covalently connect few units of helical antimicrobial peptides with radial amphiphilicity into short polypeptides with a hydrophobic helical core and a charged shell [149]. Consequently, the resulting polypeptides can adopt peculiar conformations and form nanostructures exhibiting one to two orders of magnitude enhancement in their antimicrobial activity against both Gram-positive and Gram-negative bacteria [149].

Enhanced antimicrobial activity was further noticed for similar spherical nanostructures made of polylysine-b-poly(2-hydroxypropyl methacrylate) (PLys-b-PHPMA) block copolymer [150]. These nanostructures were created via polymerization-induced self-assembly of HPMA by employing PLys as the macrochain transfer agent and are comprised of a PHPMA core and a PLys shell. For instance, simple vesicles or vesicles with rather few and short branched worms formed when long or shorter PHPMA polymer chains were employed, respectively (Figure 5b). The obtained spherical nanostructures exhibited significant antimicrobial properties in thin-film membranes. This was possible due to the positively charged nature of PLys chains forming the shell of the spherical particles [150].

Another widely-used biocidal moiety that can be tethered to various polymer chains to render them antimicrobial is the quaternary ammonium (QA) salt, a disinfectant used already for many years to kill microbes, such as bacteria, yeasts or molds [151]. QA moiety can be attached to various copolymers made of polymethylhydrosiloxane (PMHS) and PDMS via a quaternization procedure based on 1-iodooctane. This procedure leads to crosslinked QA-functional polymers containing different concentrations of QA moieties that can generate highly homogeneous films. The latter can be then optimized with respect to the moisture curability and used to correlate the QA concentration with the biocidal activity toward the marine bacterium *Cellulophaga lytica* and algae *Navicula incerta* [151]. Optimized samples exhibit about 80% in biofilm retention and 90% reduction in biofilm growth for the two microbes, respectively. Similarly, QA-based antimicrobial coatings can be prepared by dip-coating the surface of interest directly into a solution made of amphiphilic poly((dopamine methacrylamide)-(methoxyethyl acrylate)-dodecyl QA) (P(DMA-MEA-DQA)) containing various amounts of antimicrobial dodecyl QA, hydrophobicity tuning methoxyethyl and immobilizing catechol groups [109]. The resulting antimicrobial polymer films display a smooth morphology comprised of dodecyl chains localized rather at the air-surface interface and with the phenyl groups of the catechols oriented with respect to the substrate surface. Instead, the most hydrophobic films made of polymers containing no methoxyethyl side-chains were comprised of polymeric domains of an average size of hundreds of nanometers that exhibited high surface roughness. All films proved themselves highly biocidal against various microbes, inclusively due to their adhesive functionality of catechol groups that have prevented the leaching of polymers [109].

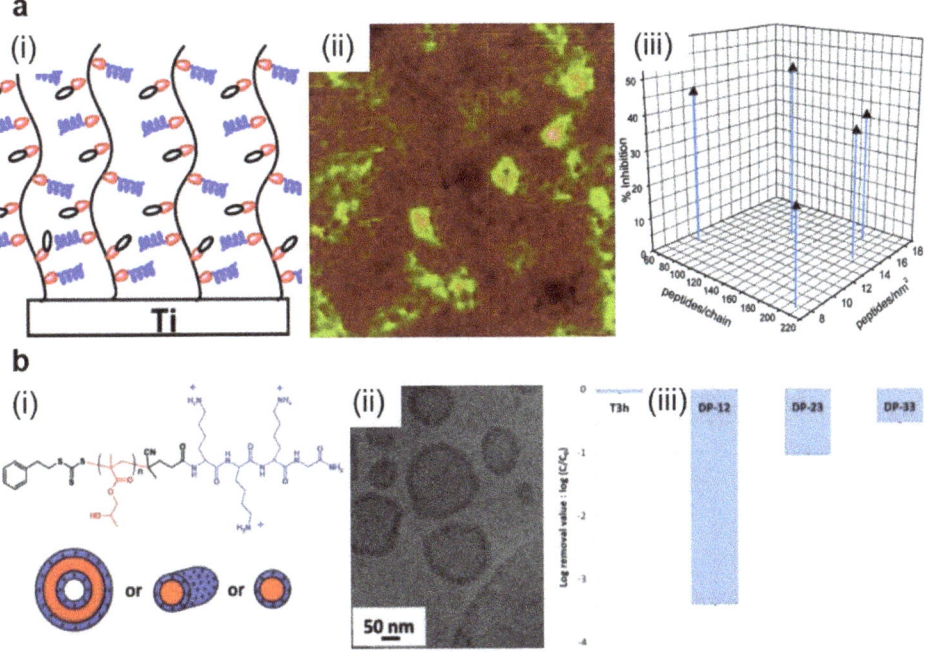

Figure 5. (a) Schematic representation (i) and AFM topography (ii) depict P(DMA-*co*-APMA) copolymer chains adopting brush-like conformations on titanium surfaces. Depending on the initial composition ratio between the DMA and APMA monomers and on the peptide density, P(DMA-*co*-APMA) copolymer exhibits more or less inhibitive effects against bacteria (iii). The size of the AFM image is 3×3 μm^2. (b) Schematic representation (i) and a cryo-transmission electron microscopy (cryo-TEM) image (ii) depict spherical core–shell vesicles that could self-assemble from PLys-*b*-PHPMA block copolymer. The chemical structure of the copolymer is presented on top of (i). Experimentally observed structures are then able to act against *S. epidermidis* with high efficiency (iii). Log removal data were obtained after 3 h for the control reactor without any material (T3h) and the reactor with various PLys-*b*-PHPMA block copolymers. Adapted with permission from ref. [149] (a) and ref. [150] (b).

QA moiety can be further attached to polymers, such as polyurethane (PU) [145] and poly(2-(dimethylamino)-ethyl methacrylate-*co*-methyl methacrylate) (P(DMEMA-*co*-MMA)) [152]. In the first case, QA salt moieties and hydroxyl groups are introduced to the backbone of the soybean oil-based polyols and then are reacted with diisocyanate monomers to obtain PUs. PU coatings containing more QA salt moieties exhibit the best antibacterial activity by killing about 95% of bacteria [145]. In the second case, a solution of partially quaternized P(DMEMA-*co*-MMA) mixed with ethylene glycol dimethacrylate (EGDMA) and photoinitiator 2-hydroxy-4-(2-hydroxyethoxy)-2-methylpropiophenone (HHMP) is spin-cast on glass slides and cured under UV light to generate a semi-interpenetrating network (SIPN) of P(DMEMA-co-MMA) and polymerized EGDMA [152]. This QA-garnished polymer-based coating not only displays strong bacteria-killing properties but can prevent the biocidal QA moieties from leaching.

The fact that ammonium-based polymers are capable to efficiently kill bacteria was recently further demonstrated by Sanches and coworkers, who generated core–shell NPs by decorating the poly(methyl methacrylate) (PMMA) NPs with antimicrobial poly(diallyldimethylammonium chloride) (PDDA) via emulsion polymerization reaction (Figure 6a) [153]. Resulting cationic NPs, self-assembled under low ionic strength conditions (Figure 6b), can access the inner layers of the cell and its membrane through the antimicrobial action of the PDDA shell and, thus, can kill microbes very efficiently (Figure 6c). The killing efficiency is nonetheless depending on the hydrophobic-hydrophilic balance of PDDA, as well as on the type of microbe, possibly due to the microstructural differences of the microbial cell walls [153]. More recently, PDDA/PMMA core–shell NPs were further optimized and utilized to fabricate antimicrobial coatings on substrates of interest by drop or spin casting [154]. Deposited hydrophilic coatings exhibited contact angles depending proportionally on the amount of PDDA in the NPs and reduced bacteria by about 7 logs.

A distinguished class of polymers that can be successfully used to generate antimicrobial structures on surfaces is represented by the cationic polymers. The main representants of this class of polymers are polyethylenimines (PEIs). The antimicrobial efficiency of PEIs, deriving from their cationic character, depends on optimizing their chemical structure [155]. For example, to enlarge their capability to perforate the hydrophobic membrane of bacteria, linear or branched PEIs with different molecular weights need to be synthesized [156]. While both types of PEIs possess enhanced antibacterial activity against *S. aureus*, only linear PEIs can induce depolarization of the bacterial membrane. Furthermore, PEIs can be N-alkylated with quaternary amino functional groups, such as hexyl, octadecyl or dodecyl, just to name a few, and rendered antimicrobial when deposited on surfaces or incorporated into NPs [155]. Besides N-alkyl, benzophenone can also be used to decorate PEIs. N-alkylated and benzophenone-based PEIs synthesized from poly(2-ethyl-2-oxazoline) (PEOX) can be then attached by photo-crosslinking to various surfaces, such as cotton, silicon oxide, or various other polymers, to fabricate leaching-free antimicrobial coatings for textile and plastic materials (Figure 6d) [110]. Exhibiting a morphology comprised of random, tens of nanometers sized features of the roughness of less than one nanometer (Figure 6e), resulting PEI-based coatings are capable to kill more than 98% of *S. aureus* or *E. coli*. (Figure 6f).

Besides PEIs, various other cationic polymers exhibiting important antimicrobial properties were reported. They were recently thoroughly reviewed by Alfei and Schito, and therefore, additional information on this topic can be found elsewhere [157]. We would only like to additionally emphasize the existence of a new water-soluble cationic copolymer synthesized from 4-ammoniumbuthylstyrene hydrochloride and reported only very recently [158]. This copolymer was shown to be able to display a rapid non-lytic bactericidal activity and thus, to be capable of acting against several bacteria, including *E. coli*, *Acinetobacter baumannii* and *Stenotrophomonas maltophilia*.

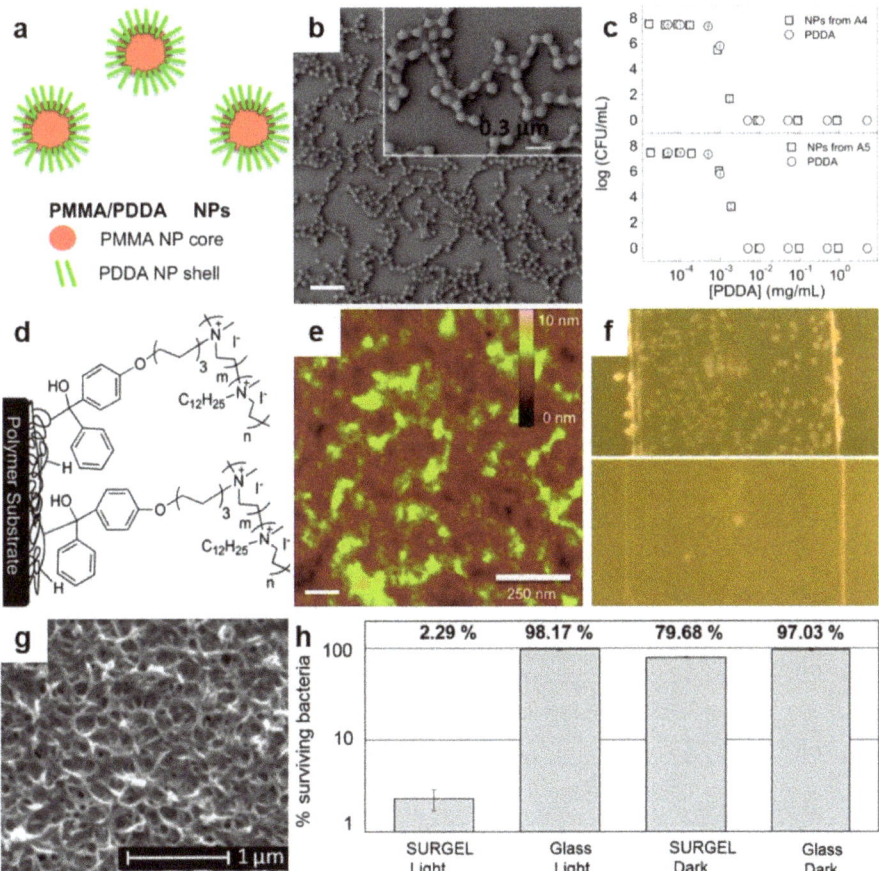

Figure 6. (**a,b**) Schematic representation (**a**) and SEM morphology (**b**) of the core–shell PMMA/PDDA NPs assembled at low ionic strength. The scale bar represents 1 μm. (**c**) Antimicrobial activity of NPs against *E. coli*. Representation of cell viability (log10 of colony-forming units CFU/mL) for *E. coli* with respect to the concentration of free PDDA or PDDA in the PMMA/PDDA NPs exhibiting sizes of 112 nm and 164 nm in diameter, respectively. (**d**) Schematics depict the covalent attachment of the benzophenone-based PEI copolymer to various surfaces and plastics. (**e**) Tapping-mode AFM height image depicts the surface of a thin-film of benzophenone-based PEI copolymer after sonication. (**f**) Digital images of a control glass substrate (top) and of a glass substrate modified with benzophenone-based PEI copolymer (bottom) sprayed with *S. aureus* and incubated for 24 h. (**g**) SEM micrograph depicts the surface of a porphyrin-based SURMOF after crosslinking and subsequent treatment with EDTA solution. (**h**) Antibacterial activity of the porphyrin SURGEL thin-films against *E. coli* using the LIVE/DEAD BacLight bacterial viability kit. Adapted with permission from ref. [153] (**a**–**c**), ref. [110] (**d,e**) and ref. [159] (**g,h**).

Antimicrobial polymers can be further synthesized by incorporating in their chemical structure the N-halamine, a biocidal moiety capable of almost instant and total sterilization over a broad spectrum of microorganisms [160,161]. Importantly, N-halamine polymers do not form toxic products and do not release halogen unless they are in contact with bacteria [162]. The synthesis of N-halamine-based polymers and grafted copolymers are cheap and rely on employing 4-(alkyl acryloxymethyl)-4-ethyl-2-oxazolidinones and commercial monomers or polymers (the latter are well-known for their efficacy to kill bacteria in both granular forms and as surface, coatings covering glass or plastics [163]). Alternatively, any inert polymer, including bilayers of PS functionalized with a top surface layer of poly(styrene-*b*-

tert-butyl acrylate) (PS-*b*-PtBA), can be employed to generate the desired density of chemical groups containing amine bonds grafted from PS surfaces (i.e., PS/PS-PAA) and can be further chlorinated to N-halamine [164]. Resulting N-halamine polymeric surfaces are highly biocidal against *S. aureus* and *E. coli*. Unfortunately, such antimicrobial systems are not very stable and need re-chlorination, as a significant part of the chlorine is lost upon UV irradiation [164].

Other N-halamine polymer precursors of cationic homopolymer poly((3-acrylamidopropyl) trimethylammonium chloride) (PCHP) and of anionic homopolymer poly(2-acrylamido-2-methylpropane sulfonic acid sodium salt) (PAHP) [111] or of poly[5,5-dimethyl-3-(3′-triethoxysilylpropyl)-hydantoin] (PSPH) [165] can be synthesized and coated onto cotton fabrics or mesoporous molecular sieves via LbL deposition or grafting techniques. The resulting N-halamine coatings can be further rendered biocidal upon their exposure to household bleach. Both concepts lead to chlorinated coatings of well-defined roughness that can inactivate 100% of *S. aureus* and *E. coli* [111,165], with significant log reductions within the first minute of contact [165]. Again, washing of the fabrics coated with N-halamine polymers is accompanied by a reduction of chlorine. This drawback is compensated by the fact that coated fabrics produce no irritations to rabbit skin, displaying thus important potential towards future biomedical applications [111]. More recently, the process of chlorine bleaching at different bleach concentrations on tape was further employed to transform polypyrrole (PPy) into N-halamines and to develop highly efficient antimicrobial coatings on stainless steel by taking advantage of the electrochemical deposition process [166].

At the end of this section, we note other peculiar polymers, such as porphyrins, which can be integrated as constituents into a surface anchored metal–organic framework (MOF). The resulting composition can be deposited on a substrate and crosslinked, leading to SURMOF (Figure 6g). The latter can be transformed by a treatment with ethylenediaminetetraacetic acid (EDTA) solution into a metal-free antimicrobial polymer-based coating abbreviated SURGEL that demonstrates significant antimicrobial activity by killing more than 97% of some bacteria (Figure 6h). This is possible due to ROS generation when thin polymer films based on porphyrin are exposed to visible light [159]. While a summary of results obtained on bioactive polymers exhibiting antimicrobial properties is presented in Table 3, additional details can be found in comprehensive scientific papers available in the literature [146,167,168].

Table 3. Summary of polymeric systems and configurations that can be employed to kill various microbes on surfaces. *Log* denotes log10 of colony-forming units CFU/mL and refers to the bacteria removal value defined as the logarithm ratio of the bacterial concentration measured at a specific time with respect to the initial bacterial concentration. Efficacy in % refers to the bacteria kill ratio.

Antimicrobial Polymer	Configuration/ Nanostructure	Dimension	Antimicrobial Mechanism	Efficacy	Microbe of Interest	Ref.
PLys-*b*-PHPMA	Spheres, worms, vesicles	50–200 nm diameter	Bioactive	3.4 *log*	*S. epidermidis*	[150]
P(DMA-*co*-APMA)	Brushes	~8–42 nm thick	Bioactive	-	*P. aeruginosa*	[143]
Polypeptides	α-Helical structure	~0.232 nm radius	Bioactive	~100%	*E. coli, S. aureus*	[149]
P(DMA-MEA-DQA)	Oriented thin-films with catechols	-	Bioactive	100% 85%	*E. coli* *S. aureus*	[109]
P(DMEMA-*co*-MMA)/PEGDMA	SIPN	800 nm thick	Bioactive	5 *log* 5 *log*	*S. epidermidis* *E. coli*	[152]
PDDA/PMMA	Core-shell NPs	A few hundred nm diameter (in aqueous medium)	Bioactive	8 *log* 7 *log* 2 *log*	*E. coli* *S. aureus* *C. albicans*	[153]
PDDA/PMMA	Core-shell NPs	94 nm thick	Bioactive	7 *log*	*E. coli, S. aureus*	[154]
PEI-PEOX	Thin-films	77 nm thick	Bioactive	>98%	*S. aureus, E. coli*	[110]
N-halamine PCHP, PAHP	Multilayered films	-	Bioactive	100% 99.73%	*S. aureus* *E. coli*	[111]

3.3. Polymeric Surface Structures Exhibiting Microbe Antiadhesive and Killing Properties

An interesting strategy against the accumulation of microbes on surfaces relies on combining the antiadhesive and antimicrobial properties of polymers to develop more complex bifunctional surface systems capable of both repelling and killing microbes. This strategy can be implemented by conjugating antiadhesive polymers with various antimicrobial (biocidal) moieties or even polymers. A relevant example was given a decade ago by Muszanska and coworkers, who have synthesized a triblock copolymer with a central polypropylene oxide (PPO) block and two terminal antiadhesive PEG segments under the name of Pluronic F-127 (PF-127) [169]. At the telechelic groups of the PEG chains, they have further covalently attached the antimicrobial enzyme lysozyme conjugate. This triblock copolymer led to structures comprised of one or two lysozyme molecules per each PF-127 polymer chain that could adsorb on a hydrophobic surface by adopting a brush-like molecular conformation. Surfaces coated with such brushes showed both antiadhesive and antimicrobial properties. Intriguingly, the structures with less lysozyme coverage obtained from a mixture of unconjugated PF-127 and PF-127-lysozyme conjugates were more bactericidal than brushes realized only from PF-127-lysozyme conjugates [169].

Bifunctional brushes adopting a bottle-like conformation were further designed and developed from a block copolymer obtained through the conjugation of antimicrobial polyhexanide (PHMB) with allyloxy PEG of both low and higher molecular weight. $APEG_{1200}$-PHMB and $APEG_{2400}$-PHMB copolymers assembled into 25 nm-thick bottlebrush nanostructures when grafted, via surface-initiated polymerization, from silicone rubber surfaces [170]. Nanostructures assembled from both copolymers showed excellent antimicrobial properties against Gram-negative and Gram-positive bacteria, with the emphasis that the $APEG_{2400}$-PHMB coating exhibited improved antiadhesive properties, most probably due to more abundant PEG units incorporated in its chemical structure [170]. A similar strategy was used to synthesize block copolymers with the same amount of PEG units but attached to an antimicrobial polyhexamethylene guanidine (PHMG) block [171]. As expected, the nanostructures obtained by grafting these block copolymers from silicone surfaces are bifunctional 20 nm thick bottlebrushes displaying a "crinkled" morphology. In this case, too, the $APEG_{2400}$-PHMG coating could inhibit the adsorption of proteins and kill bacteria more efficiently than its counterpart (Figure 7) [171].

Antiadhesive PEG can further be used, along with a cationic PC combined either with tethering or with an adhesive functional block, to synthesize V- and S-shaped triblock copolymers by placing the tethering block centrally or at the end, respectively [172]. While the surfaces coated with V-shaped polymer exhibited antibacterial properties but without being able to prevent microbial adhesion, the surfaces coated with S-shaped polymer exhibited strong antibacterial and antiadhesive attributes [172]. In comparison, linear PEG-*b*-PC diblock copolymers were also reported to exhibit both antiadhesive and antimicrobial properties when grafted onto PDA-covered silicone rubber surfaces [173]. Furthermore, if PC is replaced with cationic antimicrobial polypeptides, PEG-*b*-polypeptide diblock amphiphilic polymer chains with both antimicrobial and antiadhesive segments can be obtained via ring-opening polymerization (ROP) of N-carboxyanhydrides [174]. These polymers can be then grafted onto the PDMS surface via surface-induced polymerization to form bottlebrush nanostructures able to repel and kill microbes, such as *E. coli*, *P. aeruginosa* or *S. aureus*.

An efficient approach to combine antiadhesive and antimicrobial (polymeric) entities is based on the LbL deposition technique. For example, as synthesized hydantoinyl acrylamide-*co*-trimethyl-2-methacryloxyethylammonium chloride and hydantoinyl acrylamide-*co*-2-acrylamido-2-methyl-1-propanesulfonic acid polyelectrolytes can be deposited one layer at a time onto polypropylene (PP) fabrics either as single or as multilayers [175]. Resulting copolymer-based layered structures, when embedded into dilute sodium hypochlorite solution, can reduce microbes by about 6 logs within the first two minutes of contact [175]. To increase the stability of multilayered polyelectrolytes, it looks more appealing to "click" the antiadhesive polymer with another antimicrobial polymer.

For instance, Yang and coworkers combined antiadhesive azido-functionalized polyethylene glycol methyl ether methacrylate-based (PEGMA) polymer chains with antimicrobial alkynyl-functionalized 2-(methacryloyloxy)ethyl trimethyl ammonium chloride-based (PMETA) polymer system via a click-based LbL technique [112]. Practically, by repetitive deposition of a layer of the antiadhesive polymer on top of a layer of the antimicrobial polymer, multilayered polymeric coatings were obtained. These coatings were demonstrated to be not only resistant to bacterial adhesion but also bactericidal to marine microorganisms. One advantage of such multilayered structures is given by their tunable antimicrobial efficiency that depends on the number of polymer layers [112]. Interestingly, AFM studies conducted on the topography of such polymeric films revealed that the surface roughness decreased by almost 100% when increasing the number of polymer bilayers from 1 to 11, indicating a more compact coating structure for the thicker films [112].

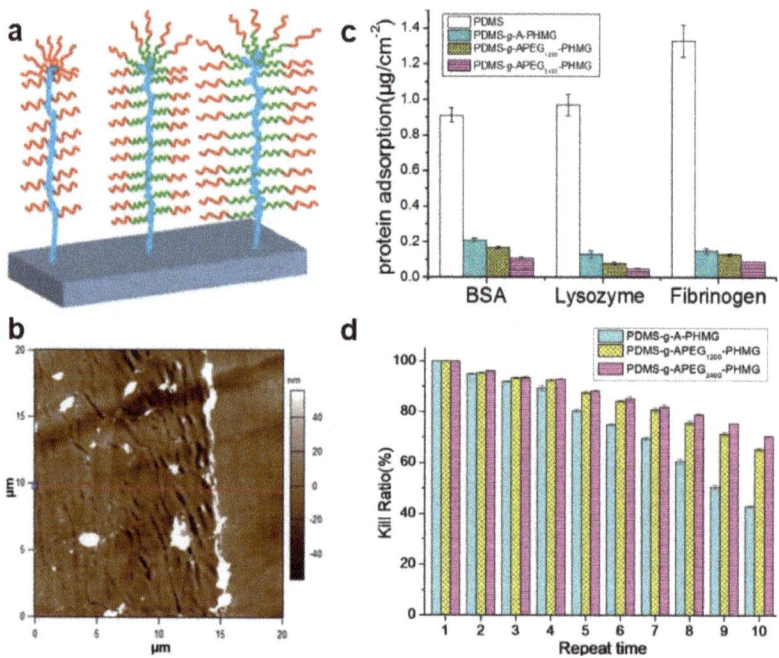

Figure 7. (a) Schematics of the APEG$_{2400}$–PHMG bottlebrushes grafted from a polymer-covered substrate. (b) AFM topography image depicts the surface morphology of the APEG$_{2400}$–PHMG coating (on the left side) and of pristine PDMS (on the right side). The red line corresponds to a height cross-section used to evaluate surface roughness. (c,d) Protein adsorption (c) and long-term reusable antibacterial properties (d) of pristine PDMS and of PDMS surfaces coated with allyl terminated PHMG or with APEG–PHMG. The antibacterial properties were tested against *P. aeruginosa*. Adapted with permission from ref. [171].

Coatings with compact structures simultaneously exhibiting antiadhesive and antimicrobial properties can also be obtained when spin casting, on top of PP/PP-graft-maleic anhydride hot-pressed coupons, branched PEI and styrene-maleic anhydride (SMA) copolymer in a PEI/SMA/PEI configuration [176]. The resulting structure exhibiting pores of about 100 nm in diameters (Figure 8a) was formed from hydrophobic styrene subunits, cationic primary amine groups with intrinsic antimicrobial properties and chlorinated N-halamine-based groups exhibiting enhanced antimicrobial attributes (Figure 8b). Experiments have revealed no evidence of *E. coli* adhesion on the PEI/SMA/PEI-coated surface [176].

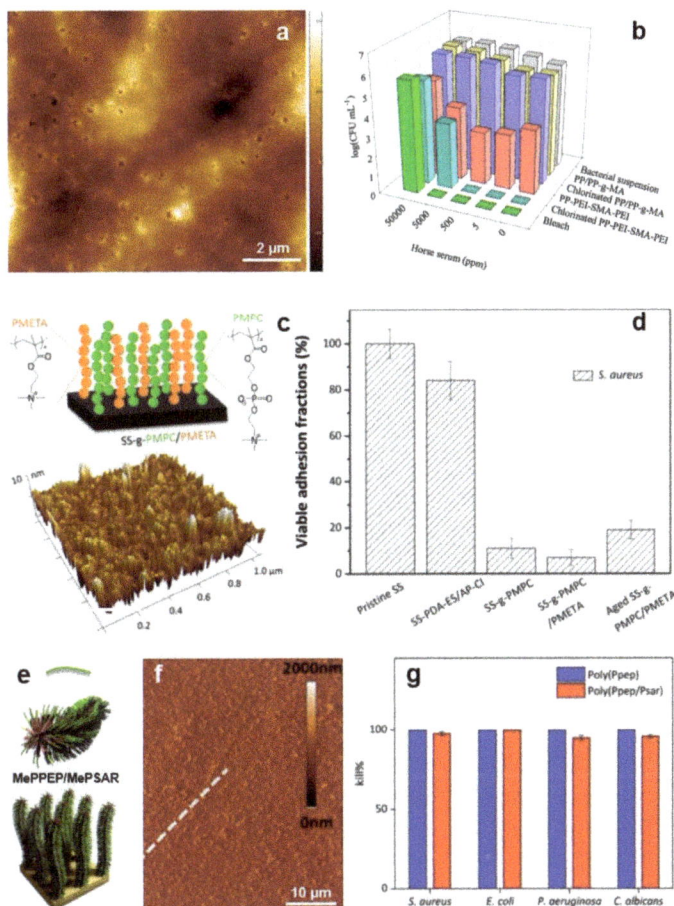

Figure 8. (**a**) AFM topography image depicts the morphology of the as-prepared PEI/SMA/PEI surface. (**b**) Antimicrobial properties of non-chlorinated and chlorinated PEI/SMA/PEI systems with respect to their analogs. (**c**) Schematic representation (top) and surface morphology, as revealed by AFM (bottom), of the PMPC/a-PMETA binary polymer brushes. (**d**) Percentage of the adhered *S. aureus* cells on the pristine and PMPC/a-PMETA surfaces compared to their analogs before and after aging, after exposure to the bacterial suspension in artificial seawater for 4 h. (**e**,**f**) Schematics (**e**) and AFM topography image (**f**) depict MePPEP/MePSAR binary brushes on a solid surface. White broken line in (**f**) was used to estimate the film roughness. (**g**) Contact bactericidal activity of the MePPEP and MePPEP/MePSAR coatings against various microbes. Adapted with permission from ref. [176] (**a**,**b**), ref. [177] (**c**,**d**) and ref. [178] (**e**–**g**).

Often, the bilayer/multilayer film configurations are unable to allow full percolation on the top surface of both the antiadhesive and antimicrobial polymer components. This inconvenience can be avoided by using peculiar grafting techniques. Here, polymer systems with antiadhesive properties can be grafted from a brush-like substrate displaying antimicrobial properties [179], or both antiadhesive and antimicrobial polymer components can be grafted from the same substrate, leading to surfaces comprised of the mixed brush "forests" [177]. The first approach can deliver bifunctional materials comprised of antiadhesive poly(oxonorbornene)-based zwitterions grafted onto the brush-like polymer network of antimicrobial cationic poly(oxonorbornene). The resulting structures are 30–40 nm thick while displaying a roughness of 3–4 nm. Their morphology consists of 5 nm deep pores randomly and homogeneously distributed over the whole surface [179] (more details on how polyzwitterions can be grafted onto a carpet of polycationic antimicrobial polymers can be found elsewhere [180]). The second approach, based on specific surface modifications, favors the assembly of zwitterionic poly(2-methacryloyloxyethyl phosphorylcholine) (PMPC) and alkynyl-modified cationic poly(2-(methacryloyloxy) ethyl trimethylammonium chloride) (a-PMETA) binary polymer brushes onto PDA-anchored stainless steel surfaces through thiol-ene and azide-alkyne graft polymerizations, respectively (Figure 8c). Resulting PMPC/a-PMETA binary polymer brushes not only display a smooth surface with a roughness of ~1.2 nm but are also highly stable in seawater. More important, they endow the stainless steel surfaces with antiadhesive and antimicrobial attributes, especially against Gram-positive *S. aureus* (Figure 8d) and Gram-negative *Pseudomonas* sp. bacteria [177].

Antimicrobial a-PMETA can also be coupled with antiadhesive alkyne-functionalized poly(N-hydroxyethyl acrylamide) (a-PHEAA) and then further grafted from stainless steel to obtain bifunctional brushes able to fight against Gram-negative *E. coli* and Gram-positive *S. epidermidis* [181]. Similarly, antiadhesive methacrylate-ended polysarcosine (MePSAR) and antimicrobial cationic methacrylate-ended polypeptides (MePPEP), synthesized via ROP of N-carboxyanhydrides, can be assembled on PDA coated substrates via grafting initiated under UV irradiation. The resulting MePPEP/MePSAR binary brushes can display, on flat surfaces, a roughness as high as 44 nm and can exhibit highly antiadhesive and antimicrobial properties against several microbes (Figure 8e–g) [178]. Relevant information of this subsection is summarized in Table 4.

Table 4. Summary of polymeric systems and configurations that exhibit both antiadhesive and antimicrobial properties on surfaces. Following abbreviations were used: *Fusarium (F.)*, *Bacillus (B.)* and *species* (sp.). Efficacy in % refers to the bacteria kill ratio or bacterial adhesion reduction.

Antimicrobial Polymer	Configuration/ Nanostructure	Dimension	Antimicrobial Mechanism	Efficacy	Microbe of Interest	Ref.
APEG$_{2400}$–PHMB	Bottlebrushes	25 nm thick	Biopassive + bioactive	5 log	*E. coli*	[170]
APEG$_{2400}$–PHMG	Bottlebrushes	20 nm thick	Biopassive + bioactive	>99.9%	*P. aeruginosa*, *S. aureus*, *F. solani*	[171]
PF-127	Binary brushes	~7–14 nm thick	Biopassive + bioactive	85%	*B. subtilis*	[169]
PEI/SMA/PEI	Pores in thin films	~100 nm diameter	Biopassive + bioactive	99.99%	*E. coli*	[176]
PEGMA/PMETA	Multilayered films	10–14 nm bilayer	Biopassive + bioactive	97%	*Pseudomonas* sp.	[112]
PEG-polypeptides	Bottlebrushes	1–2 µm thick	Biopassive + bioactive	>99%	*E. coli*, *S. aureus*, *P. aeruginosa*	[174]
PMPC/a-PMETA	Assembled binary brushes	59 nm thick	Biopassive + bioactive	93% 93%	*S. aureus* *A. coffeaeformis*	[177]
MePPEP/ MePSAR	Assembled binary brushes	310 nm thick	Biopassive + bioactive	99%/97% 99%/99% 99%/94% 99%/95%	*S. aureus* *E. coli* *P. aeruginosa* *C. albicans*	[178]
PEG-*b*-PC	Brushes	5–7 nm thick	Biopassive + bioactive	100%	*S. aureus*	[173]

4. Antimicrobial Surface Structures Developed from Polymer Blends

Recently, an alternative concept to develop antimicrobial structures on surfaces has emerged. Researchers have observed that blending polymers possessing antiadhesive or antimicrobial properties leads to peculiar nanostructures and chain conformations that enhance the above attributes when fighting microbes (Table 5). This concept can be implemented in both thin film [113,182] and brush configurations [183] by blending, for instance, an "inert" polymer with a copolymer deriving from it but decorated with antimicrobial moieties. Such a procedure facilitates an efficient control of the resulting film microstructure and thus, favors the optimization of the final antimicrobial properties. In this sense, it is possible to blend thermally stable polyacrylonitrile (PAN) with antimicrobial methacrylic copolymers bearing cationic moieties with 1,3-thiazolium and 1,2,3-thiazolium side-chain groups P(AN-co-MTA) (Figure 9a). While the morphology of the PAN/P(AN-co-MTA) films is smooth and homogeneous with rather few irregular pores, the surface wettability depends on the chemical compositions of the copolymers, i.e., on their flexibility and polarity, rendering PAN/P(AN-co-MTA) films with good antimicrobial properties (Figure 9b) [113]. Moreover, by increasing the density of the positive charges in such blend systems, the biocidal capacity can be increased to almost 100% cell-killing efficiency of bacteria and yeasts (Figure 9c).

Biocidal surfaces with high positive charge density were further fabricated from polymer blends containing PS and block copolymers comprised of PS and an antimicrobial block poly(4-(1-(2-(4-methylthiazol-5-yl)ethyl)-1H-1,2,3-triazol-4-yl)butyl methacrylate) decorated with either methyl or butyl groups (PS-b-PTTBM) [182]. By employing the breath figures approach (see details on surface relief patterns elsewhere [184]), porous yet ordered PS/PS-b-PTTBM films with narrow pore size distribution and low content of antimicrobial moieties were obtained. These structures showed increased antibacterial efficiencies against microbes, such as bacteria and fungi [182].

In analogy to biocidal polymers, polymers with protein repellent properties (e.g., PEG) can also be blended with other "inert" polymers (e.g., PAA) to control the adhesive properties of the former. This can be done by developing brush-like films of controlled PEG/PAA ratio and by utilizing PEGs of optimized molar mass [183]. Simply, by dip-coating a gold surface into a solution containing both polymers, mixed PEG/PAA polymer brushes are obtained. When PEG content is low, brushes exhibit higher affinity towards proteins, and the latter can irreversibly adsorb on the surface. Instead, when the PEG content is increased to over 25 PEG units per squared nanometer, the adsorption process becomes completely reversible, i.e., the adsorbed proteins can also be removed [183]. Moreover, PEG can be blended with antimicrobial biopolymers, such as chitosan, to obtain electrospun bifunctional nanofibers able not only to kill bacteria but also to promote osteoconductive activity [185].

Furthermore, one can blend microbe repellent polymers with microbe-killing polymers to obtain polymeric structures exhibiting both biopassive and bioactive attributes. This was demonstrated recently by Muszanska and coworkers, who have fabricated bifunctional brush-like coatings by dip-coating silicone rubber surfaces into solutions containing a mixture of antiadhesive PF-127, PF-127 conjugated with AMPs and PF-127 decorated with arginine-glycine-aspartate (RGDs) peptides [186]. The resulting brushes possess not only antiadhesive and antimicrobial attributes but are also capable of supporting mammalian cell growth. Because bacteria are significantly smaller than the mammalian cells and a limited number of RGDs may be enough to favor cell adhesion without notably affecting the adhesion of bacteria, a balanced optimization of the ratio of the three polymeric components can endow these brushes with antimicrobial attributes of either a more repelling or a predominant killing nature [186].

Finally, it is possible to blend one antimicrobial polymer with another antimicrobial polymer to create more complex structures with strengthened biocidal attributes able to kill microbes aggressively. This was recently demonstrated by Datta and coworkers, who blended a peptide-based antimicrobial poly(Boc-Phe-Ala-oxyethyl methacrylate) P(Boc-FA-HEMA) with antimicrobial polyoxometalate (POM) and obtained cationic/anionic P(Boc-FA-HEMA)/POM amphiphilic supra-assemblies (Figure 9d) [187]. These supra-assemblies are a result of the electrostatic attraction between the P(Boc-FA-HEMA) and POM, which transforms the cationic β-sheets (Figure 9e) to cationic multivalent nanorods (Figure 9f). Interestingly, while the peptide-based polymers kill microbes by disrupting their membrane, multivalent nanorods seem to also induce free radical-mediated cell damage, thus amplifying the antimicrobial efficacy [187].

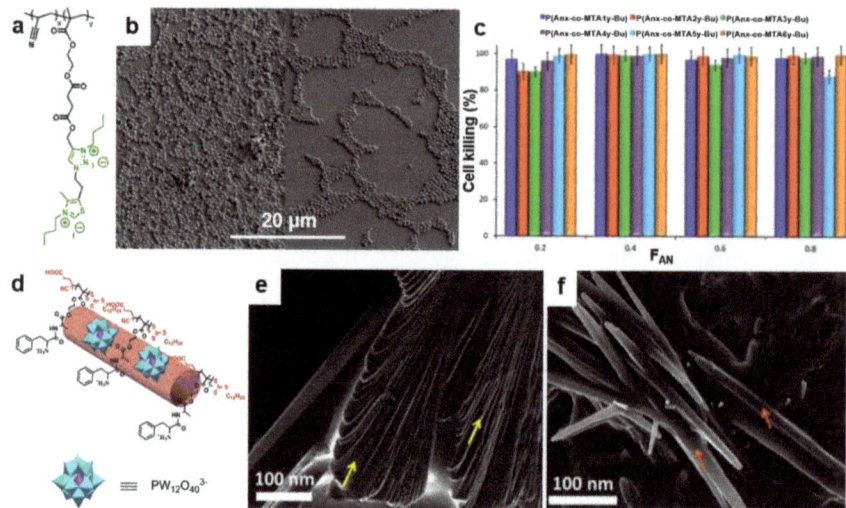

Figure 9. (a) Chemical structure of quaternized P(AN-*co*-MTA) acrylonitrile-based copolymers. (b) Field emission SEM micrographs depict *S. aureus* on PAN films in the absence (left) and in the presence (right) of a P(AN-*co*-MTA) antimicrobial copolymer after 2 h of contact. (c) The percentage of cell-killing for *C. parapsilosis* microorganism when in contact with antimicrobial films. (d) Schematic representation of the cationic/anionic P(Boc-FA-HEMA)/POM amphiphilic supra-assemblies. (e,f) Field emission SEM micrographs depict the morphology of P(Boc-FA-HEMA) before (e) and at 90 min after the addition of POM (f). The morphology of P(Boc-FA-HEMA) is comprised of individual sheets indicated by yellow arrows in (e). These sheets dissolve after the addition of POM and lead to forming nanorods, indicated by red arrows in (f). Adapted with permission from ref. [113] (a–c) and ref. [187] (d–f).

Table 5. Summary of the polymeric blends that can be used to fight microbes on surfaces.

Blend/Composite	Configuration/ Nanostructure	Dimension	Antimicrobial Mechanism	Efficacy	Microbe of Interest	Ref.
PS/PS-*b*-PTTBM	Porous films (breath figures)	5–11 μm diameter	Bioactive	99.99% 90%	*S. aureus* *C. parapsilosis*	[182]
PAN/P(AN-*co*-MTA)	Thin films	-	Bioactive	~100% ~100% ~100%	*S. aureus* *P. aeruginosa* *C. parapsilosis*	[113]
PEG/chitosan	Nanofibers	294 nm diameter	Bioactive	-	*S. epidermidis*	[185]
PF-127/PF-127-AMPs/PF-127-RGDs	Blend brushes	-	Biopassive + bioactive	- -	*S. epidermidis* *P. aeruginosa*	[186]
P(Boc-FA-HEMA)/POM	Multivalent nanorods	10–20 nm wide, a few μm long	Double bioactive	100% -	*E. coli* *B. subtilis*	[187]

5. Antimicrobial Surfaces Generated from Polymer-Based Nanocomposites

Antimicrobial properties of polymers can be significantly reinforced by exploiting various synergistic effects appearing when mixing these polymers with other materials displaying biocidal properties. For instance, antimicrobial polymers, such as PEG [114], PEI [188], zwitterionic PCBMA [189] or cationic poly(2-(tert-butylaminoethyl) methacrylate) (PTBAM) [190] can be utilized along with biocidal metallic NPs (Ag [74], copper/Cu [191], etc.) or carbon-based nanostructures [97] to develop highly biocidal nanocomposites (Table 6). Metallic NPs are increasingly used as an alternative to antibiotics for a myriad of purposes, including antibacterial coatings for implantable devices, to prevent infections [41]. Their main advantage consists in the possibility to overcome the ability of bacteria to develop resistance due to the fact that most of the antibacterial mechanisms (see more details in Section 2.2) are simultaneous and do not allow a bacterium to develop mutations on multiple genes [41]. Furthermore, to expand the biocompatibility and bioavailability of inorganic NPs, their further coating with various functional polymers can be performed (coating of NPs increases the oxidation stability of the core and diminishes particle aggregation in solution [192]). In earlier studies, the biologically inert silica-based core was coated with a layer of bactericidal polymers, but more recently, the attention was focused on the synergistic action of bactericidal metal-based NPs along with polymers with inherent antimicrobial activity [42] (the latter can easily integrate metal NPs of different shapes and sizes owing to their adjustable surface, morphology and porosity).

Table 6. Summary of the polymer-based nanocomposites that can be efficiently employed against various microorganisms on surfaces.

Blend/Composite	Configuration/ Nanostructure	Dimension	Antimicrobial Mechanism	Efficacy	Microbe of Interest	Ref.
PMMA/Ag, PTBAM/Ag	Nanofibers	40 nm diameter, 10 μm long	Bioactive reinforced	-	E. coli, S. aureus	[190]
PEI/Ag	NPs grafted on SAM	10–14 nm thick (total)	Bioactive reinforced	~6 log 0.86 log	E. coli S. aureus	[193]
PVDF-g-PCBMA/Ag	Pores/brushes	-	Bioactive reinforced	-	E. coli, S. aureus	[189]
PLA/PEG, PLA/PEG/Ag	Films; NPs	~40 μm thick 25 nm thick	Biopassive + bioactive	-	E. coli, S. aureus	[114]
P2VP-b-PEG	Smart micelles	60 nm (unloaded)	Bioactive	-	-	[194]
PEI/Cu	Positively charged NPs	34 nm radius	Bioactive reinforced	87% 96% 80%	E. coli P. aeruginosa S. aureus	[188]
Pectin–PEI–Cu	Films with Cu NPs	100 μm thick	Bioactive reinforced	-	S. aureus, E. coli	[115]
PDMEMA-MWCNTs	Nanotubes	26 nm diameter	Bioactive reinforced	42% -	E. coli S. aureus	[195]
MWCNTs-APPI/MWCNTs-APPI–Ag NPs	Nanotubes Ag NPs	15 nm diameter 15 nm diameter	Bioactive reinforced	96%/99% 96%/99% 87%/93%	B. subtilis S. aureus E. coli	[196]
PE/PEG/GO–NH2	Films	-	Bioactive reinforced	90%	E. coli	[116]

Ag NPs, typically displaying a diameter in the range of 1 to 100 nm, were demonstrated to be highly efficient antibacterial agents [197]. Thus, their further combination with antimicrobial polymers is expected to increase the bacteria-killing efficiency. Because Ag NPs may possess various shapes and surface properties, the corresponding bacteria-killing mechanism is rather complex and may occur following multiple pathways (see the antimicrobial mechanisms of Ag NPs described in Section 2.2) [74]. An example of the reinforcement of antimicrobial properties of polymers with biocidal Ag NPs was given by Song and coworkers, who successfully embedded Ag NPs in PMMA and in PTBAM nanofibers previously assembled via radical-mediated dispersion polymerization (Figure 10a,b) [190]. Both Ag/nanofiber-based composites exhibited great antimicrobial performance against *E. coli* and *S. aureus*, but, as compared to PMMA/Ag, the PTBAM/Ag nanocomposite exhibited better bactericidal attributes. Most probably, this was due to the antibacterial nature of the PTBAM substrate (in comparison, the PMMA substrate exhibits no antibacterial properties) [190]. The two bacteria mentioned above can be further targeted by a similar structure comprised of Ag NPs firmly grafted onto a self-assembled monolayer (SAM) of antimicrobial PEI on a glass substrate [193].

More recently, other antimicrobial nanocomposites containing Ag were developed. PCBMA/Ag and polylactide (PLA)/PEG/Ag nanocomposites are just two examples in which Ag was used to reinforce the antimicrobial properties of zwitterionic PCBMA [189] and the antiadhesive properties of PEG [114] polymers, respectively. In the first example,

Li and coworkers have coordinated the Ag$^+$ to the carbonyl groups located on PCBAM brushes, previously grafted on poly(vinylidene fluoride) (PVDF) membranes. Ag$^+$ was then reduced to Ag NPs, leading to PVDF-g-PCBMA/Ag nanocomposites. Due to synergistic effects, the latter displayed improved hydrophilicity and antimicrobial properties than membranes made of only PCBMA brushes or pure Ag NPs [189]. In the second example, Turalija and coworkers have prepared antimicrobial films made of PLA and containing 5% of PEG (PLA/PEG). Employing a procedure of surface modification based on plasma technology, they have further incorporated Ag into PLA/PEG films, finally obtaining PLA/PEG/Ag film nanocomposites. By comparing the antimicrobial properties of these two types of nanocomposites against *S. aureus* and *E. coli*, it was found that plasma-deposited thin-films containing Ag exhibited enhanced hydrophilicity and better antimicrobial properties, most probably due to the good antimicrobial attributes provided additionally by Ag [114].

Besides Ag, Cu is also well-known for its antimicrobial properties [191], and thus, it can be used to reinforce the antimicrobial properties of polymer-based composites (in this work, we refer to examples where Cu is only used along with antimicrobial polymers, for other Cu-based antimicrobial systems, the readers are advised to look elsewhere [127,198–200]). Cu can kill bacteria by several mechanisms, including the release of Cu ions [201] or Cu NPs [202], as well as biofilm inhibition [203]. More details on these mechanisms can be consulted in the excellent review of Tamayo and coworkers [191]. Typical examples of the reinforcement of antimicrobial properties of polymers with biocidal Cu are depicted by the PEI/Cu NPs and pectin–Cu^{2+}–PEI composites. The first example emphasizes an antimicrobial composite that was prepared by irreversibly binding positively charged Cu NPs, previously synthesized utilizing PEI as the capping agent, on the negatively charged surface of a polyamide-based membrane (Figure 10c) [188]. This composite was able, after 1 h of contact, to reduce about 80%, 87% and 96% of attached live *E. Coli*, *P. aeruginosa* and *S. aureus* bacteria, respectively (Figure 10d). The second example is depicting a Cu-containing polymer film composite prepared by thermal reduction of Cu^{2+} ions to metallic Cu in the pectin–Cu^{2+}–PEI interpolyelectrolyte–metal complexes, i.e., by transferring electrons from the nitrogen atoms of amino groups of PEI to Cu^{2+} ions accompanied by the destruction of the metal-based complexes [115]. Resulting pectin-PEI-Cu heterogeneously structured composite exhibited better antimicrobial attributes against *S. aureus* and *E. coli* than did its counterpart pectin-PEI.

Other inorganic NPs with antimicrobial properties that can be employed to prepare antimicrobial polymer-based composites include Cu oxide, titanium dioxide, zinc oxide, Au, etc. Most of these NPs need to be incorporated into polymers due to their tendency to form aggregates, which significantly diminishes the antimicrobial potential. Moreover, polymeric chains can boost the germicidal effects by enabling the inclusion of additional antimicrobial moieties [204]. For instance, TiO$_2$ compounds can disrupt bacterial cell membranes and kill bacteria by producing ROS in the presence of light. Coating TiO$_2$ compounds with polymers leads to composites that are biocidal irrespective of light conditions [205]. Au NPs are considered to be weak bactericidal agents. However, recyclable antibacterial Au NP-polymer composites with targeted efficacy against pathogenic *E. coli* were manufactured [206]. Unfortunately, these composites are less potent against *S. aureus*, on which they cannot specifically attach due to the absence of sugar-binding fimbriae on the Gram-positive bacterial cells. Further relevant information on antimicrobial composites based on inorganic NPs can be found in the literature [204].

Antimicrobial polymers can be further combined with carbon-based nanomaterials. The latter include nanostructures, such as carbon nanotubes (CNTs) [195], graphene [207], graphene oxide (GO) [208], reduced graphene oxide (RGO) [209] and others, and are well-known for their applications in biomedical engineering [210] and other biological applications [211], including drug delivery [212,213], sequential enrichment of peptides [214], osteoporotic bone regeneration [215], enzyme immobilization [216], biomaterials and bionics [217], generation of neurons [218], cellular migration [219], etc. Most of these nanomaterials can be routinely produced [220,221] or synthesized [222–224], even in common organic solvents used to also dissolve polymers [209] and possess important antimicrobial properties [225–230]. Such materials kill microbes through their physical interaction with the microbial surface leading to localized degradation of microbial cell walls [231] through wrapping, insertion or nano-knife-like processes [232].

An example depicting a polymer-CNTs antimicrobial composite was given by Joo and coworkers, who have grafted various contents of poly(2-dimethylaminoethyl methacrylate) (PDMAEMA), a polymer known for its antimicrobial properties [233–235], onto bromine-functionalized multiwalled CNTs (MWCNTs) via an ATRP method [195]. The resulting PDMAEMA-MWNT composite (Figure 10e), optimized with respect to its content of PDMAEMA, exhibited clear antibacterial properties when tested against *S. aureus* and *E. Coli* [195]. Similarly, MWCNTs can be functionalized with antimicrobial amphiphilic poly(propyleneimine) (APPI) dendrimer to obtain reinforced antimicrobial MWCNTs-APPI composites that can efficiently kill *B. subtilis*, *S. aureus*, and *E. coli*. Bacteria killing efficiency can be further reinforced by adding Ag NPs to MWCNTs-APPI composite [196].

Besides CNTs, a low percentage of amine-terminated GO (GO–NH_2) can also be used to reactively harmonize blends of low-density PE and PEG and to further develop porous antimicrobial membranes [116]. The resulting PE/PEG/GO–NH_2 uniformly dispersed composite proved itself more efficient in destroying *E. coli* than its analog PE/PEG blend (demonstrated antibacterial efficiencies were 90% and 20%, respectively). More detailed information on nanocomposites based on antimicrobial polymers, as well as on recent strategies to kill microbes, can be further consulted in excellent publications available in the literature [74,97,191,232].

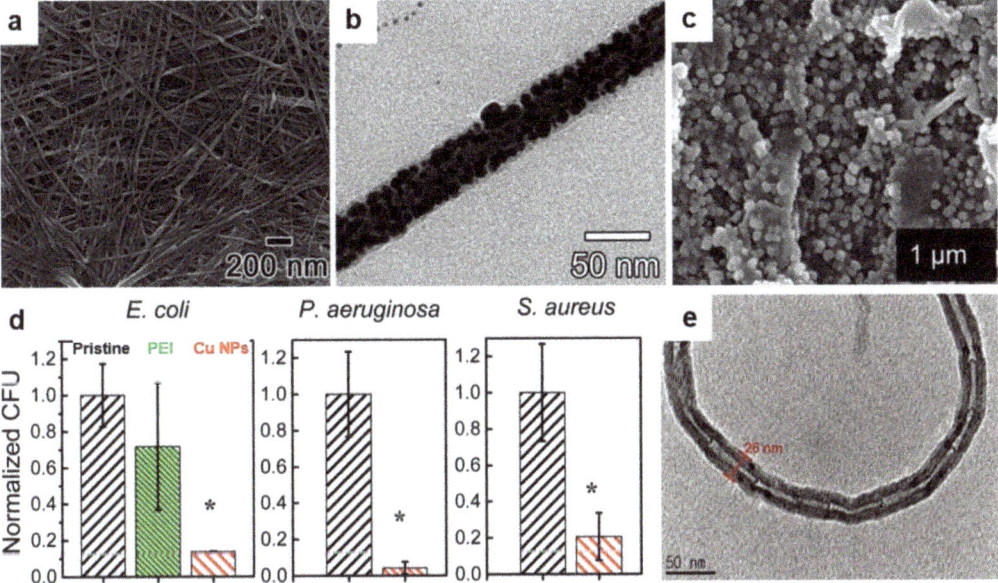

Figure 10. (a,b) Field emission SEM (a) and TEM (b) micrographs depict biocidal Ag NPs embedded in antimicrobial PTBAM nanofibers. (c) SEM micrograph depicts a membrane with bound Cu NPs after its sonication for 5 min in deionized water. (d) Number of attached live bacteria on pristine, PEI alone, and Cu NPs-based membrane for Gram-negative *E. coli* and *P. aeruginosa* and Gram-positive *S. aureus* bacteria. Asterisks (*) are emphasizing statistically significant differences observed between the functionalized and pristine membranes. (e) TEM micrograph depicts the morphology of PDMEMA-MWNT nanocomposite containing almost 25 wt % of PDMEMA. Adapted with permission from ref. [190] (a,b), ref. [188] (c,d) and ref. [195] (e).

6. Conclusions

State-of-the-art medical devices, implants, wound dressings and contact lenses are just a few of the biomedical applications with significant impacts on human health and comfort. These applications require microbe-free surfaces that can be obtained inclusively via coating with polymers possessing antiadhesive and/or antimicrobial properties. A first advantage of the polymer-coating approach is the possibility to assemble polymer chains into well-defined structures and, thus, to tune the antiadhesive and antimicrobial properties of resulting surfaces. The choices are to design and synthesize polymers either with only antiadhesive (biopassive) or antimicrobial (bioactive) properties or with both microbes repelling and killing attributes. Here, experiments have proved that various polymeric structures (brushes, nanofibers, NPs, worms, vesicles, etc.) can reduce the adhesion of microbes on different surfaces from several to few tens of times and/or can totally kill them after the contact. A second advantage of utilizing polymers to generate sterile surfaces is the possibility to either blend polymers exhibiting antiadhesive and/or antimicrobial properties to obtain new structures with enhanced antimicrobial attributes or to exploit the synergistic antimicrobial effects within nanocomposites made of microbe repelling/killing polymers and other materials displaying biocidal attributes (inorganic NPs, carbon-based materials, etc.). In these cases, blends and nanocomposites can exhibit either bifunctional (i.e., bioactive and biopassive) antimicrobial mechanisms or can simply reinforce the bioactive antimicrobial mechanisms. Both situations lead to polymer-based systems with enhanced capability to prevent and suppress the growth of various undesired microorganisms, including bacteria. Moreover, these systems exhibit antimicrobial mechanisms that cannot be outwitted by pathogens and thus, can combat, for example, the bacterial resistance to antibiotics. Consequently, polymer-based antimicrobial systems could become a viable alternative to antibiotics and disinfectants, especially if the research efforts in this direction further intensify. More exactly, significant improvements could be made by developing novel and more efficient biocidal moieties, by designing and optimizing antimicrobial polymer chains with the ability to form unique ordered structures and to attach to well-defined surfaces, or by synthesizing carbon-based materials in organic solvents with the aim to boost the preparation of the next-generation of antimicrobial polymer-based nanocomposites.

Author Contributions: I.B. (Iulia Babutan) and A.-D.L. contributed equally to this work by conducting the literature search and writing the original draft. I.B. (Ioan Botiz) edited the final manuscript. All authors have read and agreed to the published version of the manuscript.

Funding: This work was supported by a grant from the Ministry of Research, Innovation and Digitization, CNCS/CCCDI–UEFISCDI, project number PN-III-P2-2.1-PED-2019-3995, within PNCDI III.

Institutional Review Board Statement: Not applicable.

Informed Consent Statement: Not applicable.

Conflicts of Interest: The authors declare no conflict of interest.

References

1. Sugden, R.; Kelly, R.; Davies, S. Combatting antimicrobial resistance globally. *Nat. Microbiol.* **2016**, *1*, 16187. [CrossRef] [PubMed]
2. Morales, E.; Cots, F.; Sala, M.; Comas, M.; Belvis, F.; Riu, M.; Salvadó, M.; Grau, S.; Horcajada, J.P.; Montero, M.M.; et al. Hospital costs of nosocomial multi-drug resistant Pseudomonas aeruginosa acquisition. *BMC Health Serv. Res.* **2012**, *12*, 122. [CrossRef] [PubMed]
3. Nathwani, D.; Raman, G.; Sulham, K.; Gavaghan, M.; Menon, V. Clinical and economic consequences of hospital-acquired resistant and multidrug-resistant Pseudomonas aeruginosa infections: A systematic review and meta-analysis. *Antimicrob. Resist. Infect. Control* **2014**, *3*, 32. [CrossRef]
4. Levy, S.B.; Marshall, B. Antibacterial resistance worldwide: Causes, challenges and responses. *Nat. Med.* **2004**, *10*, S122–S129. [CrossRef] [PubMed]
5. Tan, S.Y.; Tatsumura, Y. Alexander Fleming (1881–1955): Discoverer of penicillin. *Singap. Med. J.* **2015**, *56*, 366–367. [CrossRef] [PubMed]
6. Stewart, P.S.; William Costerton, J. Antibiotic resistance of bacteria in biofilms. *Lancet* **2001**, *358*, 135–138. [CrossRef]

7. Shlaes, D.M. Research and Development of Antibiotics: The Next Battleground. *ACS Infect. Dis.* **2015**, *1*, 232–233. [CrossRef]
8. Darouiche, R.O. Treatment of Infections Associated with Surgical Implants. *N. Engl. J. Med.* **2004**, *350*, 1422–1429. [CrossRef]
9. Barriere, S.L. Clinical, economic and societal impact of antibiotic resistance. *Expert Opin. Pharmacother.* **2015**, *16*, 151–153. [CrossRef]
10. Donlan, R.M.; Costerton, J.W. Biofilms: Survival Mechanisms of Clinically Relevant Microorganisms. *Clin. Microbiol. Rev.* **2002**, *15*, 167–193. [CrossRef]
11. Wang, M.; Tang, T. Surface treatment strategies to combat implant-related infection from the beginning. *J. Orthop. Translat.* **2019**, *17*, 42–54. [CrossRef]
12. Costerton, J.W.; Stewart, P.S.; Greenberg, E.P. Bacterial Biofilms: A Common Cause of Persistent Infections. *Science* **1999**, *284*, 1318–1322. [CrossRef]
13. Muñoz-Bonilla, A.; Fernández-García, M. Polymeric materials with antimicrobial activity. *Prog. Polym. Sci.* **2012**, *37*, 281–339. [CrossRef]
14. Milner, S.T.; McLeish, T.C.B. Reptation and Contour-Length Fluctuations in Melts of Linear Polymers. *Phys. Rev. Lett.* **1998**, *81*, 725–728. [CrossRef]
15. Botiz, I.; Grozev, N.; Schlaad, H.; Reiter, G. The influence of protic non-solvents present in the environment on structure formation of poly(γ-benzyl-L-glutamate in organic solvents. *Soft Matter.* **2008**, *4*, 993–1002. [CrossRef]
16. Jahanshahi, K.; Botiz, I.; Reiter, R.; Scherer, H.; Reiter, G. Reversible Nucleation, Growth, and Dissolution of Poly(γ-benzyl l-glutamate) Hexagonal Columnar Liquid Crystals by Addition and Removal of a Nonsolvent. *Cryst. Growth Des.* **2013**, *13*, 4490–4494. [CrossRef]
17. Strobl, G.; Cho, T.Y. Growth kinetics of polymer crystals in bulk. *Eur. Phys. J. E* **2007**, *23*, 55–65. [CrossRef]
18. Botiz, I.; Codescu, M.-A.; Farcau, C.; Leordean, C.; Astilean, S.; Silva, C.; Stingelin, N. Convective self-assembly of π-conjugated oligomers and polymers. *J. Mater. Chem. C* **2017**, *5*, 2513–2518. [CrossRef]
19. Rahimi, K.; Botiz, I.; Stingelin, N.; Kayunkid, N.; Sommer, M.; Koch, F.P.V.; Nguyen, H.; Coulembier, O.; Dubois, P.; Brinkmann, M.; et al. Controllable Processes for Generating Large Single Crystals of Poly(3-hexylthiophene). *Angew. Chem. Int. Ed.* **2012**, *51*, 11131–11135. [CrossRef]
20. Jahanshahi, K.; Botiz, I.; Reiter, R.; Thomann, R.; Heck, B.; Shokri, R.; Stille, W.; Reiter, G. Crystallization of Poly(γ-benzyl L-glutamate) in Thin Film Solutions: Structure and Pattern Formation. *Macromolecules* **2013**, *46*, 1470–1476. [CrossRef]
21. Botiz, I.; Schlaad, H.; Reiter, G. Processes of Ordered Structure Formation in Polypeptide Thin Film Solutions. In *Advances in Polymer Science: Self-Organized Nanostructures of Amphiphilic Block Copolymers*; Springer: Berlin/Heidelberg, Germany, 2011; Volume 242, pp. 117–149.
22. Marsh, H.S.; Reid, O.G.; Barnes, G.; Heeney, M.; Stingelin, N.; Rumbles, G. Control of polythiophene film microstructure and charge carrier dynamics through crystallization temperature. *J. Polym. Sci. B Polym. Phys.* **2014**, *52*, 700–707. [CrossRef]
23. Botiz, I.; Freyberg, P.; Leordean, C.; Gabudean, A.-M.; Astilean, S.; Yang, A.C.-M.; Stingelin, N. Emission properties of MEH-PPV in thin films simultaneously illuminated and annealed at different temperatures. *Synth. Met.* **2015**, *199*, 33–36. [CrossRef]
24. Babel, A.; Zhu, Y.; Cheng, K.-F.; Chen, W.-C.; Jenekhe, S.A. High Electron Mobility and Ambipolar Charge Transport in Binary Blends of Donor and Acceptor Conjugated Polymers. *Adv. Funct. Mater.* **2007**, *17*, 2542–2549. [CrossRef]
25. Alam, M.M.; Tonzola, C.J.; Jenekhe, S.A. Nanophase-Separated Blends of Acceptor and Donor Conjugated Polymers. Efficient Electroluminescence from Binary Polyquinoline/Poly(2-methoxy-5-(2′-ethylhexyloxy)-1,4-phenylenevinylene) and Polyquinoline/Poly(3-octylthiophene) Blends. *Macromolecules* **2003**, *36*, 6577–6587. [CrossRef]
26. Todor-Boer, O.; Petrovai, I.; Tarcan, R.; David, L.; Astilean, S.; Botiz, I. Control of microstructure in polymer: Fullerene active films by convective self-assembly. *Thin Solid Films* **2020**, *697*, 137802. [CrossRef]
27. Todor-Boer, O.; Petrovai, I.; Tarcan, R.; Vulpoi, A.; David, L.; Astilean, S.; Botiz, I. Enhancing Photoluminescence Quenching in Donor–Acceptor PCE11:PPCBMB Films through the Optimization of Film Microstructure. *Nanomaterials* **2019**, *9*, 1757. [CrossRef]
28. Kukula, H.; Schlaad, H.; Tauer, K. Linear and Star-Shaped Polystyrene-block-poly(sodium glutamate)s as Emulsifiers in the Heterophase Polymerization of Styrene. *Macromolecules* **2002**, *35*, 2538–2544. [CrossRef]
29. Arora, A.; Mishra, A. Antibacterial Polymers—A Mini Review. *Mater. Today Proc.* **2018**, *5*, 17156–17161. [CrossRef]
30. Gao, H. Development of Star Polymers as Unimolecular Containers for Nanomaterials. *Macromol. Rapid Commun.* **2012**, *33*, 722–734. [CrossRef]
31. Fukushima, K.; Tan, J.P.K.; Korevaar, P.A.; Yang, Y.Y.; Pitera, J.; Nelson, A.; Maune, H.; Coady, D.J.; Frommer, J.E.; Engler, A.C.; et al. Broad-Spectrum Antimicrobial Supramolecular Assemblies with Distinctive Size and Shape. *ACS Nano* **2012**, *6*, 9191–9199. [CrossRef]
32. Fukushima, K.; Liu, S.; Wu, H.; Engler, A.C.; Coady, D.J.; Maune, H.; Pitera, J.; Nelson, A.; Wiradharma, N.; Venkataraman, S.; et al. Supramolecular high-aspect ratio assemblies with strong antifungal activity. *Nat. Commun.* **2013**, *4*, 2861. [CrossRef] [PubMed]
33. Zhang, C.; Zhu, Y.; Zhou, C.; Yuan, W.; Du, J. Antibacterial vesicles by direct dissolution of a block copolymer in water. *Polym. Chem.* **2013**, *4*, 255–259. [CrossRef]
34. Ganewatta, M.S.; Tang, C. Controlling macromolecular structures towards effective antimicrobial polymers. *Polymer* **2015**, *63*, A1–A29. [CrossRef]

35. Yang, K.; Han, Q.; Chen, B.; Zheng, Y.; Zhang, K.; Li, Q.; Wang, J. Antimicrobial hydrogels: Promising materials for medical application. *Int. J. Nanomed.* **2018**, *13*, 2217–2263. [CrossRef]
36. Salomé Veiga, A.; Schneider, J.P. Antimicrobial hydrogels for the treatment of infection. *Pept. Sci.* **2013**, *100*, 637–644. [CrossRef]
37. Singha, P.; Locklin, J.; Handa, H. A review of the recent advances in antimicrobial coatings for urinary catheters. *Acta Biomater.* **2017**, *50*, 20–40. [CrossRef]
38. Zheng, S.; Bawazir, M.; Dhall, A.; Kim, H.-E.; He, L.; Heo, J.; Hwang, G. Implication of Surface Properties, Bacterial Motility, and Hydrodynamic Conditions on Bacterial Surface Sensing and Their Initial Adhesion. *Front. Bioeng. Biotechnol.* **2021**, *9*, 643722. [CrossRef]
39. Kyzioł, A.; Khan, W.; Sebastian, V.; Kyzioł, K. Tackling microbial infections and increasing resistance involving formulations based on antimicrobial polymers. *Chem. Eng. J.* **2020**, *385*, 123888. [CrossRef]
40. Cavallaro, A.; Taheri, S.; Vasilev, K. Responsive and "smart" antibacterial surfaces: Common approaches and new developments (Review). *Biointerphases* **2014**, *9*, 029005. [CrossRef]
41. Wang, L.; Hu, C.; Shao, L. The antimicrobial activity of nanoparticles: Present situation and prospects for the future. *Int. J. Nanomed.* **2017**, *12*, 1227–1249. [CrossRef]
42. Lam, S.J.; Wong, E.H.H.; Boyer, C.; Qiao, G.G. Antimicrobial polymeric nanoparticles. *Prog. Polym. Sci.* **2018**, *76*, 40–64. [CrossRef]
43. Lam, S.J.; O'Brien-Simpson, N.M.; Pantarat, N.; Sulistio, A.; Wong, E.H.H.; Chen, Y.-Y.; Lenzo, J.C.; Holden, J.A.; Blencowe, A.; Reynolds, E.C.; et al. Combating multidrug-resistant Gram-negative bacteria with structurally nanoengineered antimicrobial peptide polymers. *Nat. Microbiol.* **2016**, *1*, 16162. [CrossRef]
44. Wiltshire, J.T.; Qiao, G.G. Recent Advances in Star Polymer Design: Degradability and the Potential for Drug Delivery. *Aust. J. Chem.* **2007**, *60*, 699–705. [CrossRef]
45. Yang, C.; Krishnamurthy, S.; Liu, J.; Liu, S.; Lu, X.; Coady, D.J.; Cheng, W.; De Libero, G.; Singhal, A.; Hedrick, J.L.; et al. Broad-Spectrum Antimicrobial Star Polycarbonates Functionalized with Mannose for Targeting Bacteria Residing inside Immune Cells. *Adv. Healthc. Mater.* **2016**, *5*, 1272–1281. [CrossRef]
46. Wong, E.H.H.; Khin, M.M.; Ravikumar, V.; Si, Z.; Rice, S.A.; Chan-Park, M.B. Modulating Antimicrobial Activity and Mammalian Cell Biocompatibility with Glucosamine-Functionalized Star Polymers. *Biomacromolecules* **2016**, *17*, 1170–1178. [CrossRef]
47. Kapoor, G.; Saigal, S.; Elongavan, A. Action and resistance mechanisms of antibiotics: A guide for clinicians. *J. Anaesthesiol. Clin. Pharmacol.* **2017**, *33*, 300–305. [CrossRef]
48. Guitor, A.K.; Wright, G.D. Antimicrobial Resistance and Respiratory Infections. *CHEST* **2018**, *154*, 1202–1212. [CrossRef]
49. Aguilar, Z.P. Chapter 5—Targeted Drug Delivery. In *Nanomaterials for Medical Applications*; Aguilar, Z.P., Ed.; Elsevier: Amsterdam, The Netherlands, 2013; pp. 181–234.
50. Nederberg, F.; Zhang, Y.; Tan, J.P.K.; Xu, K.; Wang, H.; Yang, C.; Gao, S.; Guo, X.D.; Fukushima, K.; Li, L.; et al. Biodegradable nanostructures with selective lysis of microbial membranes. *Nat. Chem.* **2011**, *3*, 409–414. [CrossRef]
51. Qiao, Y.; Yang, C.; Coady, D.J.; Ong, Z.Y.; Hedrick, J.L.; Yang, Y.-Y. Highly dynamic biodegradable micelles capable of lysing Gram-positive and Gram-negative bacterial membrane. *Biomaterials* **2012**, *33*, 1146–1153. [CrossRef]
52. Lin, C.; Liu, D.; Luo, W.; Liu, Y.; Zhu, M.; Li, X.; Liu, M. Functionalization of chitosan via single electron transfer living radical polymerization in an ionic liquid and its antimicrobial activity. *J. App. Polym. Sci.* **2015**, *132*, 42754. [CrossRef]
53. Hisey, B.; Ragogna, P.J.; Gillies, E.R. Phosphonium-Functionalized Polymer Micelles with Intrinsic Antibacterial Activity. *Biomacromolecules* **2017**, *18*, 914–923. [CrossRef] [PubMed]
54. Waschinski, C.J.; Tiller, J.C. Poly(oxazoline)s with Telechelic Antimicrobial Functions. *Biomacromolecules* **2005**, *6*, 235–243. [CrossRef] [PubMed]
55. Oda, Y.; Kanaoka, S.; Sato, T.; Aoshima, S.; Kuroda, K. Block versus Random Amphiphilic Copolymers as Antibacterial Agents. *Biomacromolecules* **2011**, *12*, 3581–3591. [CrossRef] [PubMed]
56. Engler, A.C.; Tan, J.P.K.; Ong, Z.Y.; Coady, D.J.; Ng, V.W.L.; Yang, Y.Y.; Hedrick, J.L. Antimicrobial Polycarbonates: Investigating the Impact of Balancing Charge and Hydrophobicity Using a Same-Centered Polymer Approach. *Biomacromolecules* **2013**, *14*, 4331–4339. [CrossRef]
57. Kiss, É.; Heine, E.T.; Hill, K.; He, Y.C.; Keusgen, N.; Pénzes, C.B.; Schnöller, D.; Gyulai, G.; Mendrek, A.; Keul, H.; et al. Membrane Affinity and Antibacterial Properties of Cationic Polyelectrolytes With Different Hydrophobicity. *Macromol. Biosci.* **2012**, *12*, 1181–1189. [CrossRef]
58. Antonietti, M.; Förster, S. Vesicles and Liposomes: A Self-Assembly Principle Beyond Lipids. *Adv. Mater.* **2003**, *15*, 1323–1333. [CrossRef]
59. Zhu, H.; Geng, Q.; Chen, W.; Zhu, Y.; Chen, J.; Du, J. Antibacterial high-genus polymer vesicle as an "armed" drug carrier. *J. Mater. Chem. B* **2013**, *1*, 5496–5504. [CrossRef]
60. Grayson, S.M.; Fréchet, J.M.J. Convergent Dendrons and Dendrimers: from Synthesis to Applications. *Chem. Rev.* **2001**, *101*, 3819–3868. [CrossRef]
61. Gupta, U.; Perumal, O. Chapter 15—Dendrimers and Its Biomedical Applications. In *Natural and Synthetic Biomedical Polymers*; Kumbar, S.G., Laurencin, C.T., Deng, M., Eds.; Elsevier: Oxford, UK, 2014; pp. 243–257.
62. Chen, C.Z.; Beck-Tan, N.C.; Dhurjati, P.; van Dyk, T.K.; LaRossa, R.A.; Cooper, S.L. Quaternary Ammonium Functionalized Poly(propylene imine) Dendrimers as Effective Antimicrobials: Structure−Activity Studies. *Biomacromolecules* **2000**, *1*, 473–480. [CrossRef]

63. Jain, A.; Duvvuri, L.S.; Farah, S.; Beyth, N.; Domb, A.J.; Khan, W. Antimicrobial Polymers. *Adv. Healthc. Mater.* **2014**, *3*, 1969–1985. [CrossRef]
64. Chen, C.Z.; Cooper, S.L. Recent Advances in Antimicrobial Dendrimers. *Adv. Mater.* **2000**, *12*, 843–846. [CrossRef]
65. Gillies, E.R.; Fréchet, J.M.J. Dendrimers and dendritic polymers in drug delivery. *Drug Discov. Today* **2005**, *10*, 35–43. [CrossRef]
66. Schito, A.M.; Schito, G.C.; Alfei, S. Synthesis and Antibacterial Activity of Cationic Amino Acid-Conjugated Dendrimers Loaded with a Mixture of Two Triterpenoid Acids. *Polymers* **2021**, *13*, 521. [CrossRef]
67. Schito, A.M.; Alfei, S. Antibacterial Activity of Non-Cytotoxic, Amino Acid-Modified Polycationic Dendrimers against Pseudomonas aeruginosa and Other Non-Fermenting Gram-Negative Bacteria. *Polymers* **2020**, *12*, 1818. [CrossRef]
68. Alfei, S.; Schito, A.M. From Nanobiotechnology, Positively Charged Biomimetic Dendrimers as Novel Antibacterial Agents: A Review. *Nanomaterials* **2020**, *10*, 2022. [CrossRef]
69. Jiang, Q.; Xu, J.; Li, T.; Qiao, C.; Li, Y. Synthesis and Antibacterial Activities of Quaternary Ammonium Salt of Gelatin. *J. Macromol. Sci. B* **2014**, *53*, 133–141. [CrossRef]
70. Li, Y.-Q.; Han, Q.; Feng, J.-L.; Tian, W.-L.; Mo, H.-Z. Antibacterial characteristics and mechanisms of ε-poly-lysine against Escherichia coli and Staphylococcus aureus. *Food Control* **2014**, *43*, 22–27. [CrossRef]
71. Reithofer, M.R.; Lakshmanan, A.; Ping, A.T.K.; Chin, J.M.; Hauser, C.A.E. In situ synthesis of size-controlled, stable silver nanoparticles within ultrashort peptide hydrogels and their anti-bacterial properties. *Biomaterials* **2014**, *35*, 7535–7542. [CrossRef]
72. Vasile, B.S.; Oprea, O.; Voicu, G.; Ficai, A.; Andronescu, E.; Teodorescu, A.; Holban, A. Synthesis and characterization of a novel controlled release zinc oxide/gentamicin–chitosan composite with potential applications in wounds care. *Int. J. Pharm.* **2014**, *463*, 161–169. [CrossRef]
73. Marchesan, S.; Qu, Y.; Waddington, L.J.; Easton, C.D.; Glattauer, V.; Lithgow, T.J.; McLean, K.M.; Forsythe, J.S.; Hartley, P.G. Self-assembly of ciprofloxacin and a tripeptide into an antimicrobial nanostructured hydrogel. *Biomaterials* **2013**, *34*, 3678–3687. [CrossRef] [PubMed]
74. Guo, L.; Yuan, W.; Lu, Z.; Li, C.M. Polymer/nanosilver composite coatings for antibacterial applications. *Colloids Surf. A Physicochem. Eng. Asp.* **2013**, *439*, 69–83. [CrossRef]
75. Kamaruzzaman, N.F.; Tan, L.P.; Hamdan, R.H.; Choong, S.S.; Wong, W.K.; Gibson, A.J.; Chivu, A.; Pina, M.D.F. Antimicrobial Polymers: The Potential Replacement of Existing Antibiotics? *Int. J. Mol. Sci.* **2019**, *20*, 2747. [CrossRef] [PubMed]
76. Charnley, M.; Textor, M.; Acikgoz, C. Designed polymer structures with antifouling–antimicrobial properties. *React. Funct. Polym.* **2011**, *71*, 329–334. [CrossRef]
77. Song, F.; Koo, H.; Ren, D. Effects of Material Properties on Bacterial Adhesion and Biofilm Formation. *J. Dent. Res.* **2015**, *94*, 1027–1034. [CrossRef]
78. Li, L.; Chen, S.; Zheng, J.; Ratner, B.D.; Jiang, S. Protein Adsorption on Oligo(ethylene glycol)-Terminated Alkanethiolate Self-Assembled Monolayers: The Molecular Basis for Nonfouling Behavior. *J. Phys. Chem. B* **2005**, *109*, 2934–2941. [CrossRef]
79. Hamilton-Brown, P.; Gengenbach, T.; Griesser, H.J.; Meagher, L. End Terminal, Poly(ethylene oxide) Graft Layers: Surface Forces and Protein Adsorption. *Langmuir* **2009**, *25*, 9149–9156. [CrossRef]
80. Barbey, R.; Lavanant, L.; Paripovic, D.; Schüwer, N.; Sugnaux, C.; Tugulu, S.; Klok, H.-A. Polymer Brushes via Surface-Initiated Controlled Radical Polymerization: Synthesis, Characterization, Properties, and Applications. *Chem. Rev.* **2009**, *109*, 5437–5527. [CrossRef]
81. Sofia, S.J.; Premnath, V.; Merrill, E.W. Poly(ethylene oxide) Grafted to Silicon Surfaces: Grafting Density and Protein Adsorption. *Macromolecules* **1998**, *31*, 5059–5070. [CrossRef]
82. Andruzzi, L.; Senaratne, W.; Hexemer, A.; Sheets, E.D.; Ilic, B.; Kramer, E.J.; Baird, B.; Ober, C.K. Oligo(ethylene glycol) Containing Polymer Brushes as Bioselective Surfaces. *Langmuir* **2005**, *21*, 2495–2504. [CrossRef]
83. Prime, K.L.; Whitesides, G.M. Adsorption of proteins onto surfaces containing end-attached oligo(ethylene oxide): A model system using self-assembled monolayers. *J. Am. Chem. Soc.* **1993**, *115*, 10714–10721. [CrossRef]
84. Kingshott, P.; Wei, J.; Bagge-Ravn, D.; Gadegaard, N.; Gram, L. Covalent Attachment of Poly(ethylene glycol) to Surfaces, Critical for Reducing Bacterial Adhesion. *Langmuir* **2003**, *19*, 6912–6921. [CrossRef]
85. Chen, S.; Zheng, J.; Li, L.; Jiang, S. Strong Resistance of Phosphorylcholine Self-Assembled Monolayers to Protein Adsorption: Insights into Nonfouling Properties of Zwitterionic Materials. *J. Am. Chem. Soc.* **2005**, *127*, 14473–14478. [CrossRef]
86. Feng, W.; Brash, J.L.; Zhu, S. Non-biofouling materials prepared by atom transfer radical polymerization grafting of 2-methacryloyloxyethyl phosphorylcholine: Separate effects of graft density and chain length on protein repulsion. *Biomaterials* **2006**, *27*, 847–855. [CrossRef]
87. Zhang, Z.; Zhang, M.; Chen, S.; Horbett, T.A.; Ratner, B.D.; Jiang, S. Blood compatibility of surfaces with superlow protein adsorption. *Biomaterials* **2008**, *29*, 4285–4291. [CrossRef]
88. Kuang, J.; Messersmith, P.B. Universal Surface-Initiated Polymerization of Antifouling Zwitterionic Brushes Using a Mussel-Mimetic Peptide Initiator. *Langmuir* **2012**, *28*, 7258–7266. [CrossRef]
89. Harding, J.L.; Reynolds, M.M. Combating medical device fouling. *Trends Biotechnol.* **2014**, *32*, 140–146. [CrossRef]
90. Zhang, Z.; Chao, T.; Chen, S.; Jiang, S. Superlow Fouling Sulfobetaine and Carboxybetaine Polymers on Glass Slides. *Langmuir* **2006**, *22*, 10072–10077. [CrossRef]
91. Zhang, X.; Wang, L.; Levänen, E. Superhydrophobic surfaces for the reduction of bacterial adhesion. *RSC Adv.* **2013**, *3*, 12003–12020. [CrossRef]

92. Rzhepishevska, O.; Hakobyan, S.; Ruhal, R.; Gautrot, J.; Barbero, D.; Ramstedt, M. The surface charge of anti-bacterial coatings alters motility and biofilm architecture. *Biomater. Sci.* **2013**, *1*, 589–602. [CrossRef]
93. Freschauf, L.R.; McLane, J.; Sharma, H.; Khine, M. Shrink-Induced Superhydrophobic and Antibacterial Surfaces in Consumer Plastics. *PLoS ONE* **2012**, *7*, e40987. [CrossRef] [PubMed]
94. Kaur, R.; Liu, S. Antibacterial surface design—Contact kill. *Prog. Surf. Sci.* **2016**, *91*, 136–153. [CrossRef]
95. Yang, Y.; Cai, Z.; Huang, Z.; Tang, X.; Zhang, X. Antimicrobial cationic polymers: From structural design to functional control. *Polym. J.* **2018**, *50*, 33–44. [CrossRef]
96. Riga, E.K.; Vöhringer, M.; Widyaya, V.T.; Lienkamp, K. Polymer-Based Surfaces Designed to Reduce Biofilm Formation: From Antimicrobial Polymers to Strategies for Long-Term Applications. *Macromol. Rapid Commun.* **2017**, *38*, 1700216. [CrossRef] [PubMed]
97. Yañez-Macías, R.; Muñoz-Bonilla, A.; De Jesús-Tellez, M.A.; Maldonado-Textle, H.; Guerrero-Sánchez, C.; Schubert, U.S.; Guerrero-Santos, R. Combinations of Antimicrobial Polymers with Nanomaterials and Bioactives to Improve Biocidal Therapies. *Polymers* **2019**, *11*, 1789. [CrossRef]
98. Gour, N.; Ngo, K.X.; Vebert-Nardin, C. Anti-Infectious Surfaces Achieved by Polymer Modification. *Macromol. Mater. Eng.* **2014**, *299*, 648–668. [CrossRef]
99. Ayres Cacciatore, F.; Dalmás, M.; Maders, C.; Ataíde Isaía, H.; Brandelli, A.; da Silva Malheiros, P. Carvacrol encapsulation into nanostructures: Characterization and antimicrobial activity against foodborne pathogens adhered to stainless steel. *Food Res. Int.* **2020**, *133*, 109143. [CrossRef]
100. Rahman, M.A.; Jui, M.S.; Bam, M.; Cha, Y.; Luat, E.; Alabresm, A.; Nagarkatti, M.; Decho, A.W.; Tang, C. Facial Amphiphilicity-Induced Polymer Nanostructures for Antimicrobial Applications. *ACS Appl. Mater. Interfaces* **2020**, *12*, 21221–21230. [CrossRef]
101. Su, Y.-R.; Yu, S.-H.; Chao, A.-C.; Wu, J.-Y.; Lin, Y.-F.; Lu, K.-Y.; Mi, F.-L. Preparation and properties of pH-responsive, self-assembled colloidal nanoparticles from guanidine-containing polypeptide and chitosan for antibiotic delivery. *Colloids Surf. A Physicochem. Eng. Asp.* **2016**, *494*, 9–20. [CrossRef]
102. Hou, Z.; Shankar, Y.V.; Liu, Y.; Ding, F.; Subramanion, J.L.; Ravikumar, V.; Zamudio-Vázquez, R.; Keogh, D.; Lim, H.; Tay, M.Y.F.; et al. Nanoparticles of Short Cationic Peptidopolysaccharide Self-Assembled by Hydrogen Bonding with Antibacterial Effect against Multidrug-Resistant Bacteria. *ACS Appl. Mater. Interfaces* **2017**, *9*, 38288–38303. [CrossRef]
103. Roosjen, A.; van der Mei, H.C.; Busscher, H.J.; Norde, W. Microbial Adhesion to Poly(ethylene oxide) Brushes: Influence of Polymer Chain Length and Temperature. *Langmuir* **2004**, *20*, 10949–10955. [CrossRef]
104. Kim, S.; Gim, T.; Jeong, Y.; Ryu, J.H.; Kang, S.M. Facile Construction of Robust Multilayered PEG Films on Polydopamine-Coated Solid Substrates for Marine Antifouling Applications. *ACS Appl. Mater. Interfaces* **2018**, *10*, 7626–7631. [CrossRef]
105. Şimşek, M.; Aldemir, S.D.; Gümüşderelioğlu, M. Anticellular PEO coatings on titanium surfaces by sequential electrospinning and crosslinking processes. *Emerg. Mater.* **2019**, *2*, 169–179. [CrossRef]
106. Zhang, Z.; Finlay, J.A.; Wang, L.; Gao, Y.; Callow, J.A.; Callow, M.E.; Jiang, S. Polysulfobetaine-Grafted Surfaces as Environmentally Benign Ultralow Fouling Marine Coatings. *Langmuir* **2009**, *25*, 13516–13521. [CrossRef]
107. Li, J.; Kleintschek, T.; Rieder, A.; Cheng, Y.; Baumbach, T.; Obst, U.; Schwartz, T.; Levkin, P.A. Hydrophobic Liquid-Infused Porous Polymer Surfaces for Antibacterial Applications. *ACS Appl. Mater. Interfaces* **2013**, *5*, 6704–6711. [CrossRef]
108. He, T.; Jańczewski, D.; Jana, S.; Parthiban, A.; Guo, S.; Zhu, X.; Lee, S.S.-C.; Parra-Velandia, F.J.; Teo, S.L.-M.; Vancso, G.J. Efficient and robust coatings using poly(2-methyl-2-oxazoline) and its copolymers for marine and bacterial fouling prevention. *J. Polym. Sci. A Polym. Chem.* **2016**, *54*, 275–283. [CrossRef]
109. Han, H.; Wu, J.; Avery, C.W.; Mizutani, M.; Jiang, X.; Kamigaito, M.; Chen, Z.; Xi, C.; Kuroda, K. Immobilization of Amphiphilic Polycations by Catechol Functionality for Antimicrobial Coatings. *Langmuir* **2011**, *27*, 4010–4019. [CrossRef]
110. Dhende, V.P.; Samanta, S.; Jones, D.M.; Hardin, I.R.; Locklin, J. One-Step Photochemical Synthesis of Permanent, Nonleaching, Ultrathin Antimicrobial Coatings for Textiles and Plastics. *ACS Appl. Mater. Interfaces* **2011**, *3*, 2830–2837. [CrossRef]
111. Liu, Y.; Li, J.; Cheng, X.; Ren, X.; Huang, T.S. Self-assembled antibacterial coating by N-halamine polyelectrolytes on a cellulose substrate. *J. Mater. Chem. B* **2015**, *3*, 1446–1454. [CrossRef]
112. Yang, W.J.; Pranantyo, D.; Neoh, K.-G.; Kang, E.-T.; Teo, S.L.-M.; Rittschof, D. Layer-by-Layer Click Deposition of Functional Polymer Coatings for Combating Marine Biofouling. *Biomacromolecules* **2012**, *13*, 2769–2780. [CrossRef]
113. Tejero, R.; Gutiérrez, B.; López, D.; López-Fabal, F.; Gómez-Garcés, J.L.; Muñoz-Bonilla, A.; Fernández-García, M. Tailoring Macromolecular Structure of Cationic Polymers towards Efficient Contact Active Antimicrobial Surfaces. *Polymers* **2018**, *10*, 241. [CrossRef]
114. Turalija, M.; Bischof, S.; Budimir, A.; Gaan, S. Antimicrobial PLA films from environment friendly additives. *Compos. Part B Eng.* **2016**, *102*, 94–99. [CrossRef]
115. Demchenko, V.; Riabov, S.; Rybalchenko, N.; Goncharenko, L.; Kobylinskyi, S.; Shtompel, V. X-ray study of structural formation, thermomechanical and antimicrobial properties of copper-containing polymer nanocomposites obtained by the thermal reduction method. *Eur. Polym. J.* **2017**, *96*, 326–336. [CrossRef]
116. Mural, P.K.S.; Banerjee, A.; Rana, M.S.; Shukla, A.; Padmanabhan, B.; Bhadra, S.; Madras, G.; Bose, S. Polyolefin based antibacterial membranes derived from PE/PEO blends compatibilized with amine terminated graphene oxide and maleated PE. *J. Mater. Chem. A* **2014**, *2*, 17635–17648. [CrossRef]

117. Shafipour Yordshahi, A.; Moradi, M.; Tajik, H.; Molaei, R. Design and preparation of antimicrobial meat wrapping nanopaper with bacterial cellulose and postbiotics of lactic acid bacteria. *Int. J. Food Microbiol.* **2020**, *321*, 108561. [CrossRef]
118. Shankar, S.; Reddy, J.P.; Rhim, J.-W.; Kim, H.-Y. Preparation, characterization, and antimicrobial activity of chitin nanofibrils reinforced carrageenan nanocomposite films. *Carbohydr. Polym.* **2015**, *117*, 468–475. [CrossRef]
119. Kumorek, M.; Minisy, I.M.; Krunclová, T.; Voršiláková, M.; Venclíková, K.; Chánová, E.M.; Janoušková, O.; Kubies, D. pH-responsive and antibacterial properties of self-assembled multilayer films based on chitosan and tannic acid. *Mater. Sc. Eng. C* **2020**, *109*, 110493. [CrossRef]
120. Jamróz, E.; Khachatryan, G.; Kopel, P.; Juszczak, L.; Kawecka, A.; Krzyściak, P.; Kucharek, M.; Bębenek, Z.; Zimowska, M. Furcellaran nanocomposite films: The effect of nanofillers on the structural, thermal, mechanical and antimicrobial properties of biopolymer films. *Carbohydr. Polym.* **2020**, *240*, 116244. [CrossRef]
121. Biddeci, G.; Cavallaro, G.; Di Blasi, F.; Lazzara, G.; Massaro, M.; Milioto, S.; Parisi, F.; Riela, S.; Spinelli, G. Halloysite nanotubes loaded with peppermint essential oil as filler for functional biopolymer film. *Carbohydr. Polym.* **2016**, *152*, 548–557. [CrossRef] [PubMed]
122. Trinetta, V.; Cutter, C.N. Chapter 30—Pullulan: A Suitable Biopolymer for Antimicrobial Food Packaging Applications. In *Antimicrobial Food Packaging*; Barros-Velázquez, J., Ed.; Academic Press: San Diego, CA, USA, 2016; pp. 385–397.
123. Abreu, A.S.; Oliveira, M.; de Sá, A.; Rodrigues, R.M.; Cerqueira, M.A.; Vicente, A.A.; Machado, A.V. Antimicrobial nanostructured starch based films for packaging. *Carbohydr. Polym.* **2015**, *129*, 127–134. [CrossRef]
124. Zander, Z.K.; Becker, M.L. Antimicrobial and Antifouling Strategies for Polymeric Medical Devices. *ACS Macro Lett.* **2018**, *7*, 16–25. [CrossRef]
125. Kazemzadeh-Narbat, M.; Cheng, H.; Chabok, R.; Alvarez, M.M.; de la Fuente-Nunez, C.; Phillips, K.S.; Khademhosseini, A. Strategies for antimicrobial peptide coatings on medical devices: A review and regulatory science perspective. *Crit. Rev. Biotechnol.* **2021**, *41*, 94–120. [CrossRef] [PubMed]
126. Chamsaz, E.A.; Mankoci, S.; Barton, H.A.; Joy, A. Nontoxic Cationic Coumarin Polyester Coatings Prevent Pseudomonas aeruginosa Biofilm Formation. *ACS Appl. Mater. Interfaces* **2017**, *9*, 6704–6711. [CrossRef] [PubMed]
127. Zuniga, J.M.; Cortes, A. The role of additive manufacturing and antimicrobial polymers in the COVID-19 pandemic. *Expert Rev. Med. Devices* **2020**, *17*, 477–481. [CrossRef] [PubMed]
128. Neoh, K.G.; Kang, E.T. Combating Bacterial Colonization on Metals via Polymer Coatings: Relevance to Marine and Medical Applications. *ACS Appl. Mater. Interfaces* **2011**, *3*, 2808–2819. [CrossRef]
129. Grozev, N.; Botiz, I.; Reiter, G. Morphological instabilities of polymer crystals. *Eur. Phys. J. E* **2008**, *27*, 63–71. [CrossRef]
130. Darko, C.; Botiz, I.; Reiter, G.; Breiby, D.W.; Andreasen, J.W.; Roth, S.V.; Smilgies, D.M.; Metwalli, E.; Papadakis, C.M. Crystallization in diblock copolymer thin films at different degrees of supercooling. *Phys. Rev. E* **2009**, *79*, 041802. [CrossRef]
131. Sheth, S.R.; Leckband, D. Measurements of attractive forces between proteins and end-grafted poly(ethylene glycol) chains. *Proc. Natl. Acad. Sci. USA* **1997**, *94*, 8399–8404. [CrossRef]
132. Jeong, J.-H.; Hong, S.W.; Hong, S.; Yook, S.; Jung, Y.; Park, J.-B.; Khue, C.D.; Im, B.-H.; Seo, J.; Lee, H.; et al. Surface camouflage of pancreatic islets using 6-arm-PEG-catechol in combined therapy with tacrolimus and anti-CD154 monoclonal antibody for xenotransplantation. *Biomaterials* **2011**, *32*, 7961–7970. [CrossRef]
133. Ladd, J.; Zhang, Z.; Chen, S.; Hower, J.C.; Jiang, S. Zwitterionic Polymers Exhibiting High Resistance to Nonspecific Protein Adsorption from Human Serum and Plasma. *Biomacromolecules* **2008**, *9*, 1357–1361. [CrossRef]
134. Shen, N.; Cheng, E.; Whitley, J.W.; Horne, R.R.; Leigh, B.; Xu, L.; Jones, B.D.; Guymon, C.A.; Hansen, M.R. Photograftable Zwitterionic Coatings Prevent Staphylococcus aureus and Staphylococcus epidermidis Adhesion to PDMS Surfaces. *ACS Appl. Bio Mater.* **2021**. [CrossRef]
135. Pranantyo, D.; Xu, L.Q.; Neoh, K.-G.; Kang, E.-T.; Ng, Y.X.; Teo, S.L.-M. Tea Stains-Inspired Initiator Primer for Surface Grafting of Antifouling and Antimicrobial Polymer Brush Coatings. *Biomacromolecules* **2015**, *16*, 723–732. [CrossRef]
136. Li, Y.; Giesbers, M.; Gerth, M.; Zuilhof, H. Generic Top-Functionalization of Patterned Antifouling Zwitterionic Polymers on Indium Tin Oxide. *Langmuir* **2012**, *28*, 12509–12517. [CrossRef]
137. Cho, W.K.; Kong, B.; Park, H.J.; Kim, J.; Chegal, W.; Choi, J.S.; Choi, I.S. Long-term stability of cell micropatterns on poly((3-(methacryloylamino)propyl)-dimethyl(3-sulfopropyl)ammonium hydroxide)-patterned silicon oxide surfaces. *Biomaterials* **2010**, *31*, 9565–9574. [CrossRef]
138. Kim, S.; Kwak, S.; Lee, S.; Cho, W.K.; Lee, J.K.; Kang, S.M. One-step functionalization of zwitterionic poly[(3-(methacryloylamino)propyl) dimethyl(3-sulfopropyl)ammonium hydroxide] surfaces by metal–polyphenol coating. *Chem. Commun.* **2015**, *51*, 5340–5342. [CrossRef]
139. Oh, Y.J.; Khan, E.S.; Campo, A.D.; Hinterdorfer, P.; Li, B. Nanoscale Characteristics and Antimicrobial Properties of (SI-ATRP)-Seeded Polymer Brush Surfaces. *ACS Appl. Mater. Interfaces* **2019**, *11*, 29312–29319. [CrossRef]
140. Schmolke, H.; Demming, S.; Edlich, A.; Magdanz, V.; Büttgenbach, S.; Franco-Lara, E.; Krull, R.; Klages, C.-P. Polyelectrolyte multilayer surface functionalization of poly(dimethylsiloxane) (PDMS) for reduction of yeast cell adhesion in microfluidic devices. *Biomicrofluidics* **2010**, *4*, 044113. [CrossRef]
141. Séon, L.; Lavalle, P.; Schaaf, P.; Boulmedais, F. Polyelectrolyte Multilayers: A Versatile Tool for Preparing Antimicrobial Coatings. *Langmuir* **2015**, *31*, 12856–12872. [CrossRef]

142. Guzmán, E.; Rubio, R.G.; Ortega, F. A closer physico-chemical look to the Layer-by-Layer electrostatic self-assembly of polyelectrolyte multilayers. *Adv. Colloid Interface Sci.* **2020**, *282*, 102197. [CrossRef]
143. Gao, G.; Yu, K.; Kindrachuk, J.; Brooks, D.E.; Hancock, R.E.W.; Kizhakkedathu, J.N. Antibacterial Surfaces Based on Polymer Brushes: Investigation on the Influence of Brush Properties on Antimicrobial Peptide Immobilization and Antimicrobial Activity. *Biomacromolecules* **2011**, *12*, 3715–3727. [CrossRef]
144. Gao, G.; Lange, D.; Hilpert, K.; Kindrachuk, J.; Zou, Y.; Cheng, J.T.J.; Kazemzadeh-Narbat, M.; Yu, K.; Wang, R.; Straus, S.K.; et al. The biocompatibility and biofilm resistance of implant coatings based on hydrophilic polymer brushes conjugated with antimicrobial peptides. *Biomaterials* **2011**, *32*, 3899–3909. [CrossRef]
145. Bakhshi, H.; Yeganeh, H.; Mehdipour-Ataei, S.; Shokrgozar, M.A.; Yari, A.; Saeedi-Eslami, S.N. Synthesis and characterization of antibacterial polyurethane coatings from quaternary ammonium salts functionalized soybean oil based polyols. *Mater. Sci. Eng. C* **2013**, *33*, 153–164. [CrossRef] [PubMed]
146. Sun, H.; Hong, Y.; Xi, Y.; Zou, Y.; Gao, J.; Du, J. Synthesis, Self-Assembly, and Biomedical Applications of Antimicrobial Peptide–Polymer Conjugates. *Biomacromolecules* **2018**, *19*, 1701–1720. [CrossRef] [PubMed]
147. Sinclair, K.D.; Pham, T.X.; Farnsworth, R.W.; Williams, D.L.; Loc-Carrillo, C.; Horne, L.A.; Ingebretsen, S.H.; Bloebaum, R.D. Development of a broad spectrum polymer-released antimicrobial coating for the prevention of resistant strain bacterial infections. *J. Biomed. Mater. Res. A* **2012**, *100A*, 2732–2738. [CrossRef] [PubMed]
148. Yu, K.; Lo, J.C.Y.; Mei, Y.; Haney, E.F.; Siren, E.; Kalathottukaren, M.T.; Hancock, R.E.W.; Lange, D.; Kizhakkedathu, J.N. Toward Infection-Resistant Surfaces: Achieving High Antimicrobial Peptide Potency by Modulating the Functionality of Polymer Brush and Peptide. *ACS Appl. Mater. Interfaces* **2015**, *7*, 28591–28605. [CrossRef] [PubMed]
149. Xiong, M.; Lee, M.W.; Mansbach, R.A.; Song, Z.; Bao, Y.; Peek, R.M.; Yao, C.; Chen, L.-F.; Ferguson, A.L.; Wong, G.C.L.; et al. Helical antimicrobial polypeptides with radial amphiphilicity. *Proc. Natl. Acad. Sci. USA* **2015**, *112*, 13155–13160. [CrossRef] [PubMed]
150. Luppi, L.; Babut, T.; Petit, E.; Rolland, M.; Quemener, D.; Soussan, L.; Moradi, M.A.; Semsarilar, M. Antimicrobial polylysine decorated nano-structures prepared through polymerization induced self-assembly (PISA). *Polym. Chem.* **2019**, *10*, 336–344. [CrossRef]
151. Majumdar, P.; Lee, E.; Patel, N.; Stafslien, S.J.; Daniels, J.; Chisholm, B.J. Development of environmentally friendly, antifouling coatings based on tethered quaternary ammonium salts in a crosslinked polydimethylsiloxane matrix. *J. Coat. Technol. Res.* **2008**, *5*, 405–417. [CrossRef]
152. Zhao, J.; Ma, L.; Millians, W.; Wu, T.; Ming, W. Dual-Functional Antifogging/Antimicrobial Polymer Coating. *ACS Appl. Mater. Interfaces* **2016**, *8*, 8737–8742. [CrossRef]
153. Sanches, L.M.; Petri, D.F.S.; de Melo Carrasco, L.D.; Carmona-Ribeiro, A.M. The antimicrobial activity of free and immobilized poly (diallyldimethylammonium) chloride in nanoparticles of poly (methylmethacrylate). *J. Nanobiotechnol.* **2015**, *13*, 58. [CrossRef]
154. Galvão, C.N.; Sanches, L.M.; Mathiazzi, B.I.; Ribeiro, R.T.; Petri, D.F.S.; Carmona-Ribeiro, A.M. Antimicrobial Coatings from Hybrid Nanoparticles of Biocompatible and Antimicrobial Polymers. *Int. J. Mol. Sci.* **2018**, *19*, 2965. [CrossRef]
155. Lin, J.; Qiu, S.; Lewis, K.; Klibanov, A.M. Bactericidal Properties of Flat Surfaces and Nanoparticles Derivatized with Alkylated Polyethylenimines. *Biotechnol. Prog.* **2002**, *18*, 1082–1086. [CrossRef]
156. Gibney, K.A.; Sovadinova, I.; Lopez, A.I.; Urban, M.; Ridgway, Z.; Caputo, G.A.; Kuroda, K. Poly(ethylene imine)s as Antimicrobial Agents with Selective Activity. *Macromol. Biosci.* **2012**, *12*, 1279–1289. [CrossRef]
157. Alfei, S.; Schito, A.M. Positively Charged Polymers as Promising Devices against Multidrug Resistant Gram-Negative Bacteria: A Review. *Polymers* **2020**, *12*, 1195. [CrossRef]
158. Alfei, S.; Piatti, G.; Caviglia, D.; Schito, A.M. Synthesis, Characterization, and Bactericidal Activity of a 4-Ammoniumbuthylstyrene-Based Random Copolymer. *Polymers* **2021**, *13*, 1140. [CrossRef]
159. Zhou, W.; Begum, S.; Wang, Z.; Krolla, P.; Wagner, D.; Bräse, S.; Wöll, C.; Tsotsalas, M. High Antimicrobial Activity of Metal–Organic Framework-Templated Porphyrin Polymer Thin Films. *ACS Appl. Mater. Interfaces* **2018**, *10*, 1528–1533. [CrossRef]
160. Hui, F.; Debiemme-Chouvy, C. Antimicrobial N-Halamine Polymers and Coatings: A Review of Their Synthesis, Characterization, and Applications. *Biomacromolecules* **2013**, *14*, 585–601. [CrossRef]
161. Jang, J.; Kim, Y. Fabrication of monodisperse silica–polymer core–shell nanoparticles with excellent antimicrobial efficacy. *Chem. Commun.* **2008**, 4016–4018. [CrossRef]
162. Dong, A.; Huang, J.; Lan, S.; Wang, T.; Xiao, L.; Wang, W.; Zhao, T.; Zheng, X.; Liu, F.; Gao, G.; et al. Synthesis of N-halamine-functionalized silica–polymer core–shell nanoparticles and their enhanced antibacterial activity. *Nanotechnology* **2011**, *22*, 295602. [CrossRef]
163. Eknoian, M.W.; Worley, S.D. New N-Halamine Biocidal Polymers. *J. Bioact. Compat. Polym.* **1998**, *13*, 303–314. [CrossRef]
164. Chen, Y.; Han, Q. Designing N-halamine based antibacterial surface on polymers: Fabrication, characterization, and biocidal functions. *Appl. Surf. Sci.* **2011**, *257*, 6034–6039. [CrossRef]
165. Wang, Y.; Li, L.; Liu, Y.; Ren, X.; Liang, J. Antibacterial mesoporous molecular sieves modified with polymeric N-halamine. *Mater. Sci. Eng. C* **2016**, *69*, 1075–1080. [CrossRef] [PubMed]
166. Nautiyal, A.; Ren, M.Q.T.; Huang, T.-S.; Zhang, X.; Cook, J.; Bozack, M.J.; Farag, R. High-performance Engineered Conducting Polymer Film towards Antimicrobial/Anticorrosion Applications. *Eng. Sci.* **2018**, *4*, 70–78. [CrossRef]

167. Kugel, A.; Stafslien, S.; Chisholm, B.J. Antimicrobial coatings produced by "tethering" biocides to the coating matrix: A comprehensive review. *Prog. Org. Coat.* **2011**, *72*, 222–252. [CrossRef]
168. Bonduelle, C. Secondary structures of synthetic polypeptide polymers. *Polym. Chem.* **2018**, *9*, 1517–1529. [CrossRef]
169. Muszanska, A.K.; Busscher, H.J.; Herrmann, A.; van der Mei, H.C.; Norde, W. Pluronic–lysozyme conjugates as anti-adhesive and antibacterial bifunctional polymers for surface coating. *Biomaterials* **2011**, *32*, 6333–6341. [CrossRef]
170. Zhi, Z.; Su, Y.; Xi, Y.; Tian, L.; Xu, M.; Wang, Q.; Pandidan, S.; Li, P.; Huang, W. Dual-Functional Polyethylene Glycol-b-polyhexanide Surface Coating with in Vitro and in Vivo Antimicrobial and Antifouling Activities. *ACS Appl. Mater. Interfaces* **2017**, *9*, 10383–10397. [CrossRef]
171. Su, Y.; Zhi, Z.; Gao, Q.; Xie, M.; Yu, M.; Lei, B.; Li, P.; Ma, P.X. Autoclaving-Derived Surface Coating with In Vitro and In Vivo Antimicrobial and Antibiofilm Efficacies. *Adv. Healthc. Mater.* **2017**, *6*, 1601173. [CrossRef]
172. Voo, Z.X.; Khan, M.; Narayanan, K.; Seah, D.; Hedrick, J.L.; Yang, Y.Y. Antimicrobial/Antifouling Polycarbonate Coatings: Role of Block Copolymer Architecture. *Macromolecules* **2015**, *48*, 1055–1064. [CrossRef]
173. Ding, X.; Yang, C.; Lim, T.P.; Hsu, L.Y.; Engler, A.C.; Hedrick, J.L.; Yang, Y.-Y. Antibacterial and antifouling catheter coatings using surface grafted PEG-b-cationic polycarbonate diblock copolymers. *Biomaterials* **2012**, *33*, 6593–6603. [CrossRef]
174. Gao, Q.; Yu, M.; Su, Y.; Xie, M.; Zhao, X.; Li, P.; Ma, P.X. Rationally designed dual functional block copolymers for bottlebrush-like coatings: In vitro and in vivo antimicrobial, antibiofilm, and antifouling properties. *Acta Biomater.* **2017**, *51*, 112–124. [CrossRef]
175. Cerkez, I.; Worley, S.D.; Broughton, R.M.; Huang, T.S. Antimicrobial surface coatings for polypropylene nonwoven fabrics. *React. Funct. Polym.* **2013**, *73*, 1412–1419. [CrossRef]
176. Bastarrachea, L.J.; Goddard, J.M. Self-healing antimicrobial polymer coating with efficacy in the presence of organic matter. *Appl. Surf. Sci.* **2016**, *378*, 479–488. [CrossRef]
177. Xu, G.; Liu, P.; Pranantyo, D.; Xu, L.; Neoh, K.-G.; Kang, E.-T. Antifouling and Antimicrobial Coatings from Zwitterionic and Cationic Binary Polymer Brushes Assembled via "Click" Reactions. *Ind. Eng. Chem. Res.* **2017**, *56*, 14479–14488. [CrossRef]
178. Gao, Q.; Li, P.; Zhao, H.; Chen, Y.; Jiang, L.; Ma, P.X. Methacrylate-ended polypeptides and polypeptoids for antimicrobial and antifouling coatings. *Polym. Chem.* **2017**, *8*, 6386–6397. [CrossRef]
179. Zou, P.; Hartleb, W.; Lienkamp, K. It takes walls and knights to defend a castle—Synthesis of surface coatings from antimicrobial and antibiofouling polymers. *J. Mater. Chem.* **2012**, *22*, 19579–19589. [CrossRef]
180. Hartleb, W.; Saar, J.S.; Zou, P.; Lienkamp, K. Just Antimicrobial is not Enough: Toward Bifunctional Polymer Surfaces with Dual Antimicrobial and Protein-Repellent Functionality. *Macromol. Chem. Phys.* **2016**, *217*, 225–231. [CrossRef]
181. Yang, W.J.; Cai, T.; Neoh, K.-G.; Kang, E.-T.; Teo, S.L.-M.; Rittschof, D. Barnacle Cement as Surface Anchor for "Clicking" of Antifouling and Antimicrobial Polymer Brushes on Stainless Steel. *Biomacromolecules* **2013**, *14*, 2041–2051. [CrossRef]
182. Muñoz-Bonilla, A.; Cuervo-Rodríguez, R.; López-Fabal, F.; Gómez-Garcés, J.L.; Fernández-García, M. Antimicrobial Porous Surfaces Prepared by Breath Figures Approach. *Materials* **2018**, *11*, 1266. [CrossRef]
183. Bratek-Skicki, A.; Eloy, P.; Morga, M.; Dupont-Gillain, C. Reversible Protein Adsorption on Mixed PEO/PAA Polymer Brushes: Role of Ionic Strength and PEO Content. *Langmuir* **2018**, *34*, 3037–3048. [CrossRef]
184. Handrea-Dragan, M.; Botiz, I. Multifunctional Structured Platforms: From Patterning of Polymer-Based Films to Their Subsequent Filling with Various Nanomaterials. *Polymers* **2021**, *13*, 445. [CrossRef]
185. Boschetto, F.; Ngoc Doan, H.; Phong Vo, P.; Zanocco, M.; Zhu, W.; Sakai, W.; Adachi, T.; Ohgitani, E.; Tsutsumi, N.; Mazda, O.; et al. Antibacterial and Osteoconductive Effects of Chitosan/Polyethylene Oxide (PEO)/Bioactive Glass Nanofibers for Orthopedic Applications. *Appl. Sci.* **2020**, *10*, 2360. [CrossRef]
186. Muszanska, A.K.; Rochford, E.T.J.; Gruszka, A.; Bastian, A.A.; Busscher, H.J.; Norde, W.; van der Mei, H.C.; Herrmann, A. Antiadhesive Polymer Brush Coating Functionalized with Antimicrobial and RGD Peptides to Reduce Biofilm Formation and Enhance Tissue Integration. *Biomacromolecules* **2014**, *15*, 2019–2026. [CrossRef]
187. Datta, L.P.; Mukherjee, R.; Biswas, S.; Das, T.K. Peptide-Based Polymer–Polyoxometalate Supramolecular Structure with a Differed Antimicrobial Mechanism. *Langmuir* **2017**, *33*, 14195–14208. [CrossRef]
188. Ben-Sasson, M.; Zodrow, K.R.; Genggeng, Q.; Kang, Y.; Giannelis, E.P.; Elimelech, M. Surface Functionalization of Thin-Film Composite Membranes with Copper Nanoparticles for Antimicrobial Surface Properties. *Environ. Sci. Technol.* **2014**, *48*, 384–393. [CrossRef]
189. Li, J.-h.; Zhang, D.-b.; Ni, X.-x.; Zheng, H.; Zhang, Q.-Q. Excellent Hydrophilic and Anti-bacterial Fouling PVDF Membrane Based on Ag Nanoparticle Self-assembled PCDMA Polymer Brush. *Chin. J. Polym. Sci.* **2017**, *35*, 809–822. [CrossRef]
190. Song, J.; Kang, H.; Lee, C.; Hwang, S.H.; Jang, J. Aqueous Synthesis of Silver Nanoparticle Embedded Cationic Polymer Nanofibers and Their Antibacterial Activity. *ACS Appl. Mater. Interfaces* **2012**, *4*, 460–465. [CrossRef]
191. Tamayo, L.; Azócar, M.; Kogan, M.; Riveros, A.; Páez, M. Copper-polymer nanocomposites: An excellent and cost-effective biocide for use on antibacterial surfaces. *Mater. Sci. Eng. C* **2016**, *69*, 1391–1409. [CrossRef] [PubMed]
192. Ghosh Chaudhuri, R.; Paria, S. Core/Shell Nanoparticles: Classes, Properties, Synthesis Mechanisms, Characterization, and Applications. *Chem. Rev.* **2012**, *112*, 2373–2433. [CrossRef] [PubMed]
193. Dacarro, G.; Cucca, L.; Grisoli, P.; Pallavicini, P.; Patrini, M.; Taglietti, A. Monolayers of polyethylenimine on flat glass: A versatile platform for cations coordination and nanoparticles grafting in the preparation of antibacterial surfaces. *Dalton Trans.* **2012**, *41*, 2456–2463. [CrossRef] [PubMed]

194. Iurciuc-Tincu, C.-E.; Cretan, M.S.; Purcar, V.; Popa, M.; Daraba, O.M.; Atanase, L.I.; Ochiuz, L. Drug Delivery System Based on pH-Sensitive Biocompatible Poly(2-vinyl pyridine)-b-poly(ethylene oxide) Nanomicelles Loaded with Curcumin and 5-Fluorouracil. *Polymers* **2020**, *12*, 1450. [CrossRef]
195. Joo, Y.T.; Jung, K.H.; Kim, M.J.; Kim, Y. Preparation of antibacterial PDMAEMA-functionalized multiwalled carbon nanotube via atom transfer radical polymerization. *J. Appl. Polym. Sci.* **2013**, *127*, 1508–1518. [CrossRef]
196. Murugan, E.; Vimala, G. Effective functionalization of multiwalled carbon nanotube with amphiphilic poly(propyleneimine) dendrimer carrying silver nanoparticles for better dispersability and antimicrobial activity. *J. Colloid Interface Sci.* **2011**, *357*, 354–365. [CrossRef]
197. Graf, C.; Vossen, D.L.J.; Imhof, A.; van Blaaderen, A. A General Method To Coat Colloidal Particles with Silica. *Langmuir* **2003**, *19*, 6693–6700. [CrossRef]
198. Scully, J.R. The COVID-19 Pandemic, Part 1: Can Antimicrobial Copper-Based Alloys Help Suppress Infectious Transmission of Viruses Originating from Human Contact with High-Touch Surfaces? *Corrosion* **2020**, *76*, 523–527. [CrossRef]
199. Selvamani, V.; Zareei, A.; Elkashif, A.; Maruthamuthu, M.K.; Chittiboyina, S.; Delisi, D.; Li, Z.; Cai, L.; Pol, V.G.; Seleem, M.N.; et al. Hierarchical Micro/Mesoporous Copper Structure with Enhanced Antimicrobial Property via Laser Surface Texturing. *Adv. Mater. Interfaces* **2020**, *7*, 1901890. [CrossRef]
200. El Nahrawy, A.M.; Hemdan, B.A.; Abou Hammad, A.B.; Abia, A.L.K.; Bakr, A.M. Microstructure and Antimicrobial Properties of Bioactive Cobalt Co-Doped Copper Aluminosilicate Nanocrystallines. *Silicon* **2020**, *12*, 2317–2327. [CrossRef]
201. Thampi, V.V.A.; Thanka Rajan, S.; Anupriya, K.; Subramanian, B. Functionalization of fabrics with PANI/CuO nanoparticles by precipitation route for anti-bacterial applications. *J. Nanopart. Res.* **2015**, *17*, 57. [CrossRef]
202. Longano, D.; Ditaranto, N.; Cioffi, N.; Di Niso, F.; Sibillano, T.; Ancona, A.; Conte, A.; Del Nobile, M.A.; Sabbatini, L.; Torsi, L. Analytical characterization of laser-generated copper nanoparticles for antibacterial composite food packaging. *Anal. Bioanal. Chem.* **2012**, *403*, 1179–1186. [CrossRef]
203. Chapman, J.; Le Nor, L.; Brown, R.; Kitteringham, E.; Russell, S.; Sullivan, T.; Regan, F. Antifouling performances of macro- to micro- to nano-copper materials for the inhibition of biofouling in its early stages. *J. Mater. Chem. B* **2013**, *1*, 6194–6200. [CrossRef]
204. Alfredo, N.V.; Rodríguez-Hernández, J. Chapter 4—Antimicrobial Polymeric Nanostructures. In *Nanostructures for Antimicrobial Therapy*; Ficai, A., Grumezescu, A.M., Eds.; Elsevier: Amsterdam, The Netherlands, 2017; pp. 85–115.
205. Kong, H.; Song, J.; Jang, J. Photocatalytic Antibacterial Capabilities of TiO2–Biocidal Polymer Nanocomposites Synthesized by a Surface-Initiated Photopolymerization. *Environ. Sci. Technol.* **2010**, *44*, 5672–5676. [CrossRef]
206. Yuan, Y.; Liu, F.; Xue, L.; Wang, H.; Pan, J.; Cui, Y.; Chen, H.; Yuan, L. Recyclable Escherichia coli-Specific-Killing AuNP–Polymer (ESKAP) Nanocomposites. *ACS Appl. Mater. Interfaces* **2016**, *8*, 11309–11317. [CrossRef] [PubMed]
207. Lee, C.; Wei, X.; Kysar, J.W.; Hone, J. Measurement of the elastic properties and intrinsic strength of monolayer graphene. *Science* **2008**, *321*, 385–388. [CrossRef] [PubMed]
208. Sajjad, S.; Khan Leghari, S.A.; Iqbal, A. Study of Graphene Oxide Structural Features for Catalytic, Antibacterial, Gas Sensing, and Metals Decontamination Environmental Applications. *ACS Appl. Mater. Interfaces* **2017**, *9*, 43393–43414. [CrossRef] [PubMed]
209. Tarcan, R.; Handrea-Dragan, M.; Todor-Boer, O.; Petrovai, I.; Farcau, C.; Rusu, M.; Vulpoi, A.; Todea, M.; Astilean, S.; Botiz, I. A new, fast and facile synthesis method for reduced graphene oxide in N,N-dimethylformamide. *Synth. Met.* **2020**, *269*, 116576. [CrossRef]
210. Cha, C.; Shin, S.R.; Annabi, N.; Dokmeci, M.R.; Khademhosseini, A. Carbon-Based Nanomaterials: Multifunctional Materials for Biomedical Engineering. *ACS Nano* **2013**, *7*, 2891–2897. [CrossRef]
211. Tarcan, R.; Todor-Boer, O.; Petrovai, I.; Leordean, C.; Astilean, S.; Botiz, I. Reduced graphene oxide today. *J. Mater. Chem. C* **2020**, *8*, 1198–1224. [CrossRef]
212. Oz, Y.; Barras, A.; Sanyal, R.; Boukherroub, R.; Szunerits, S.; Sanyal, A. Functionalization of Reduced Graphene Oxide via Thiol–Maleimide "Click" Chemistry: Facile Fabrication of Targeted Drug Delivery Vehicles. *ACS Appl. Mater. Interfaces* **2017**, *9*, 34194–34203. [CrossRef]
213. Shao, L.; Zhang, R.; Lu, J.; Zhao, C.; Deng, X.; Wu, Y. Mesoporous Silica Coated Polydopamine Functionalized Reduced Graphene Oxide for Synergistic Targeted Chemo-Photothermal Therapy. *ACS Appl. Mater. Interfaces* **2017**, *9*, 1226–1236. [CrossRef]
214. Ma, W.; Zhang, F.; Li, L.; Chen, S.; Qi, L.; Liu, H.; Bai, Y. Facile Synthesis of Mesocrystalline SnO2 Nanorods on Reduced Graphene Oxide Sheets: An Appealing Multifunctional Affinity Probe for Sequential Enrichment of Endogenous Peptides and Phosphopeptides. *ACS Appl. Mater. Interfaces* **2016**, *8*, 35099–35105. [CrossRef]
215. Xiong, K.; Wu, T.; Fan, Q.; Chen, L.; Yan, M. Novel Reduced Graphene Oxide/Zinc Silicate/Calcium Silicate Electroconductive Biocomposite for Stimulating Osteoporotic Bone Regeneration. *ACS Appl. Mater. Interfaces* **2017**, *9*, 44356–44368. [CrossRef]
216. Patel, S.K.S.; Choi, S.H.; Kang, Y.C.; Lee, J.-K. Eco-Friendly Composite of Fe3O4-Reduced Graphene Oxide Particles for Efficient Enzyme Immobilization. *ACS Appl. Mater. Interfaces* **2017**, *9*, 2213–2222. [CrossRef]
217. Thompson, B.C.; Murray, E.; Wallace, G.G. Graphite Oxide to Graphene. Biomaterials to Bionics. *Adv. Mater.* **2015**, *27*, 7563–7582. [CrossRef]
218. Baek, S.; Oh, J.; Song, J.; Choi, H.; Yoo, J.; Park, G.-Y.; Han, J.; Chang, Y.; Park, H.; Kim, H.; et al. Generation of Integration-Free Induced Neurons Using Graphene Oxide-Polyethylenimine. *Small* **2017**, *13*, 1601993. [CrossRef]
219. Tian, X.; Yang, Z.; Duan, G.; Wu, A.; Gu, Z.; Zhang, L.; Chen, C.; Chai, Z.; Ge, C.; Zhou, R. Graphene Oxide Nanosheets Retard Cellular Migration via Disruption of Actin Cytoskeleton. *Small* **2017**, *13*, 1602133. [CrossRef]

220. Novoselov, K.S.; Geim, A.K.; Morozov, S.V.; Jiang, D.; Zhang, Y.; Dubonos, S.V.; Grigorieva, I.V.; Firsov, A.A. Electric Field Effect in Atomically Thin Carbon Films. *Science* **2004**, *306*, 666–669. [CrossRef]
221. Jiao, L.; Zhang, L.; Wang, X.; Diankov, G.; Dai, H. Narrow graphene nanoribbons from carbon nanotubes. *Nature* **2009**, *458*, 877–880. [CrossRef]
222. Marta, B.; Leordean, C.; Istvan, T.; Botiz, I.; Astilean, S. Efficient etching-free transfer of high quality, large-area CVD grown graphene onto polyvinyl alcohol films. *Appl. Surf. Sci.* **2016**, *363*, 613–618. [CrossRef]
223. Hummers, W.S.; Offeman, R.E. Preparation of Graphitic Oxide. *J. Am. Chem. Soc.* **1958**, *80*, 1339. [CrossRef]
224. Staudenmaier, L. Verfahren zur Darstellung der Graphitsäure. *Ber. Dtsch. Chem. Ges.* **1898**, *31*, 1481–1487. [CrossRef]
225. Maleki Dizaj, S.; Mennati, A.; Jafari, S.; Khezri, K.; Adibkia, K. Antimicrobial activity of carbon-based nanoparticles. *Adv. Pharm. Bull.* **2015**, *5*, 19–23.
226. Kang, S.; Herzberg, M.; Rodrigues, D.F.; Elimelech, M. Antibacterial Effects of Carbon Nanotubes: Size Does Matter! *Langmuir* **2008**, *24*, 6409–6413. [CrossRef] [PubMed]
227. Pangule, R.C.; Brooks, S.J.; Dinu, C.Z.; Bale, S.S.; Salmon, S.L.; Zhu, G.; Metzger, D.W.; Kane, R.S.; Dordick, J.S. Antistaphylococcal Nanocomposite Films Based on Enzyme−Nanotube Conjugates. *ACS Nano* **2010**, *4*, 3993–4000. [CrossRef] [PubMed]
228. Al-Jumaili, A.; Alancherry, S.; Bazaka, K.; Jacob, M.V. Review on the Antimicrobial Properties of Carbon Nanostructures. *Materials* **2017**, *10*, 1066. [CrossRef] [PubMed]
229. Nguyen, H.N.; Nadres, E.T.; Alamani, B.G.; Rodrigues, D.F. Designing polymeric adhesives for antimicrobial materials: Poly(ethylene imine) polymer, graphene, graphene oxide and molybdenum trioxide—A biomimetic approach. *J. Mater. Chem. B* **2017**, *5*, 6616–6628. [CrossRef]
230. Zhu, J.; Wang, J.; Hou, J.; Zhang, Y.; Liu, J.; Van der Bruggen, B. Graphene-based antimicrobial polymeric membranes: A review. *J. Mater. Chem. A* **2017**, *5*, 6776–6793. [CrossRef]
231. Kang, S.; Pinault, M.; Pfefferle, L.D.; Elimelech, M. Single-Walled Carbon Nanotubes Exhibit Strong Antimicrobial Activity. *Langmuir* **2007**, *23*, 8670–8673. [CrossRef]
232. Han, W.; Wu, Z.; Li, Y.; Wang, Y. Graphene family nanomaterials (GFNs)—promising materials for antimicrobial coating and film: A review. *Chem. Eng. J.* **2019**, *358*, 1022–1037. [CrossRef]
233. Green, J.-B.D.; Bickner, S.; Carter, P.W.; Fulghum, T.; Luebke, M.; Nordhaus, M.A.; Strathmann, S. Antimicrobial testing for surface-immobilized agents with a surface-separated live–dead staining method. *Biotechnol. Bioeng.* **2011**, *108*, 231–236. [CrossRef]
234. Wang, H.; Wang, L.; Zhang, P.; Yuan, L.; Yu, Q.; Chen, H. High antibacterial efficiency of pDMAEMA modified silicon nanowire arrays. *Colloids Surf. B* **2011**, *83*, 355–359. [CrossRef]
235. Rawlinson, L.-A.B.; Ryan, S.M.; Mantovani, G.; Syrett, J.A.; Haddleton, D.M.; Brayden, D.J. Antibacterial Effects of Poly(2-(dimethylamino ethyl)methacrylate) against Selected Gram-Positive and Gram-Negative Bacteria. *Biomacromolecules* **2010**, *11*, 443–453. [CrossRef]

Article

Paclitaxel-Loaded Folate-Targeted Albumin-Alginate Nanoparticles Crosslinked with Ethylenediamine. Synthesis and In Vitro Characterization

Ana María Martínez-Relimpio [1], Marta Benito [2], Elena Pérez-Izquierdo [3,*], César Teijón [4], Rosa María Olmo [5] and María Dolores Blanco [5]

[1] Facultad de Ciencias Experimentales, Universidad Francisco de Vitoria, Pozuelo de Alarcón, 28223 Madrid, Spain; am.martinez.prof@ufv.es
[2] Fundación San Juan de Dios, Centro de Ciencias de la Salud San Rafael, Universidad de Nebrija, Paseo de La Habana, 70, 28036 Madrid, Spain; mbenito@nebrija.es
[3] Department of Health Sciences, Faculty of Biomedical and Health Sciences, Universidad Europea de Madrid, Urbanización El Bosque, Calle Tajo, s/n, 28670 Villaviciosa de Odón, Spain
[4] Nursing Department, Faculty of Nursing, Physiotherapy and Podiatry, Universidad Complutense de Madrid, 28040 Madrid, Spain; cteijon@ucm.es
[5] Department of Biochemistry and Molecular Biology, Faculty of Medicine, Universidad Complutense de Madrid, 28040 Madrid, Spain; rmolmo@med.ucm.es (R.M.O.); mdblanco@med.ucm.es (M.D.B.)
* Correspondence: elena.perez2@universidadeuropea.es

Abstract: Among the different ways to reduce the secondary effects of antineoplastic drugs in cancer treatment, the use of nanoparticles has demonstrated good results due to the protection of the drug and the possibility of releasing compounds to a specific therapeutic target. The α-isoform of the folate receptor (FR) is overexpressed on a significant number of human cancers; therefore, folate-targeted crosslinked nanoparticles based on BSA and alginate mixtures and loaded with paclitaxel (PTX) have been prepared to maximize the proven antineoplastic activity of the drug against solid tumors. Nanometric-range-sized particles (169 ± 28 nm–296 ± 57 nm), with negative Z-potential values (between −0.12 ± 0.04 and −94.1± 0.4), were synthesized, and the loaded PTX (2.63 ± 0.19–3.56 ±0.13 µg PTX/mg Np) was sustainably released for 23 and 27 h. Three cell lines (MCF-7, MDA-MB-231 and HeLa) were selected to test the efficacy of the folate-targeted PTX-loaded BSA/ALG nanocarriers. The presence of FR on the cell membrane led to a significantly larger uptake of BSA/ALG–Fol nanoparticles compared with the equivalent nanoparticles without folic acid on their surface. The cell viability results demonstrated a cytocompatibility of unloaded nanoparticle–Fol and a gradual decrease in cell viability after treatment with PTX-loaded nanoparticle–Fol due to the sustainable PTX release.

Keywords: folate-targeted nanoparticles; BSA/alginate nanocarriers; paclitaxel; cellular uptake; cell viability

1. Introduction

Targeted treatments and personalized medicine are two of the axes of current oncology. They refer to the design of drugs that respond specifically to some genetic characteristics of each patient's cancer. More efficiency is pursued with fewer adverse effects, adapting the procedures to the characteristics of each patient. Numerous studies have focused on the development of targeted drug-delivery systems [1] for cancer treatment in order to reduce side effects due to the unspecific effect of anticancer drugs on healthy cells. Nanocarriers based on natural polymers have the advantage of being a priori biocompatible and also biodegradable. Among biological polymers, polysaccharides have been used for therapeutic delivery systems [2] due to their non-toxic and non-reactogenic properties, as well as their physicochemical properties that allow chemical modifications and, therefore, an easy preparation of nanocarriers. Their reactive groups, such as hydroxyl, carboxyl or amine, can attach different molecules and so introduce new physicochemical characteristics [3]. Anionic polysaccharides have an important role in the preparation of drug delivery systems [4]; they have extensively contributed to the development of several types of nanocarriers for cancer treatment and diagnosis. Among anionic polysaccharides, alginate, whose structure is based on a backbone of [1–4] linked β-D-mannuronic acid (M units) and α-L-guluronic acid (G units), have been used for preparing different nanocarriers for biomedical applications in the last years [5]. Alginate has been declared safe by the FDA [6] for application in humans as a dental impression material or wound dressing, and their hydrogels are the most assayed material for bone tissue engineering and bioprinting since they provide an appropriate niche for cell loading [7]. On the other hand, there is significant interest in developing anticancer drug carriers based on serum albumin [8] due to the therapeutic efficacy of Abraxane®, an albumin-bound form of paclitaxel, that rapidly dissociates in serum, losing some benefits of nanoformulations. Some studies have indicated that Abraxane® induced the overexpression of P-glycoprotein and did not allow it to overcome the common small molecule drug resistance problem mediated by P-gp [9]; in addition, when crosslinked and non-crosslinked PTX-loaded albumin nanoparticles were compared, differences in pharmacokinetics were observed, due to their different physiological ways of delivering the drug to the tumor [10]. In an attempt to improve the characteristics of nanocarriers, numerous studies have focused on the synthesis of nanosystems formed by proteins and polysaccharides; in this way, core–shell microcapsules based on BSA gel with a polyelectrolyte complex multilayer shell of hyaluronic acid and chitosan, encapsulating Sorafenib, have been designed for hepatocellular cancer therapy [11]. Ionic crosslinked nanoparticles based on chitosan and BSA were also developed as carriers for doxorubicin, showing their biocompatibility after intravenous administration in a *Wistar* rat model [12]. Nanocarriers can reach the internal part of solid tumors, taking advantage of their angiogenesis, but the uptake of these nanosystems improves significantly if they are targeted to overexpressed receptors on the surface of cancer cells. Among those receptors [13], the axis folic acid/folate receptor is one of the most important ligand–receptor interactions used to target cancer. Folic acid reaches normal tissues through two carriers, the reduced folate carrier (RFC) and the proton-coupled folate transporter, but either of them are able to bind folate conjugates (such as folate drugs or folate nanocarriers). Folate-bounded nanocarriers are internalized by the membrane-bound folate receptor (FR), which has a high affinity for folate (Kd ~ 0.1–1 nmol/L) although is expressed on few cell types. The FR-α isoform is overexpressed on 40% of human cancers (such as ovary, kidney, breast and liver tumors) where it is completely accessible to FA-bounded systems, unlike what happens in healthy cells, where the FR-α isoform is detected in the apical membrane of some epithelial cells, where folate conjugates do not have access. This fact makes folate-targeted nanosystems [14,15] very selective for cancer cells.

In the present study different compositions of folate-targeted nanoparticles based on BSA and alginate have been prepared for use as an active targeting strategy for the delivery of paclitaxel (PTX). Thus, the challenge of this study was the preparation of stable nanoparticles capable of specifically delivering that hydrophobic drug in a prolonged time in the environment of many solid tumors, avoiding the hypersensitivity, nephrotoxicity and cardiotoxicity reactions induced by the conventional use of Cremophor®EL [16] and drug resistance problem mediated by the aforementioned P-gp [9]. These nanoparticles were in vitro characterized to evaluate their application in targeted tumor therapy.

2. Materials and Methods

2.1. Materials

Folic acid (FA), alginic acid sodium salt (ALG; viscosity 15.00–20.00 cps, 1% H_2O), trypsin from bovine pancreas (13,000 units/mg solid), gentamicin (50 µg/mL), dimethyl sulfoxide (DMSO), methylthiazoletetrazolium (MTT), 6-coumarin, ethylenediamine (ED;

98%), N-hydroxysuccinimide (NHS), N,N'-dicyclohexylcarbodiimide (DCC) were purchased from Sigma-Aldrich (Barcelona, Spain). Sodium hydroxide (NaOH), hydrochloric acid (HCl; 35%), trichloroacetic acid, ethanol absolute, ethylenediaminetetraacetic acid (EDTA), Tween®80, triethylamine, anhydrous di-sodium hydrogen phosphate (Na_2HPO4), di-hydrogen potassium phosphate (KH_2PO4), diethylether, dichloromethane (DCM) and acetonitrile (HPLC-analysis grade) were supplied by Panreac (Madrid, Spain). Bovine serum albumin (BSA, Fraction V), sodium chloride (NaCl) and disodium hydrogen phosphate dehydrated were purchased from Merck (Barcelona, Spain). Fetal bovine serum (FBS), penicillin (50 U/mL), streptomycin (50 µg/mL) and 0.05% trypsin/0.53 mM EDTA were purchased from Invitrogen Life Technologies, Grand Island, NY. Dubecco's Modified Eagle Medium was purchased from Lonza Bioscience (Bornem, Belgium). Paclitaxel (Taxol) was supplied by Tocris Bioscience (Mw 889.95 > 99%). All water used in the assays was Millipore Milli Q grade.

2.2. Preparation of BSA/ALG Nanoparticles

Folate-targeted nanoparticles (Nps–Fol) based on bovine serum albumin (BSA) and alginate (ALG) mixtures were prepared and stabilized by amide bonds using ethylenediamine. BSA 5% (w/v) and ALG 1% (w/v) aqueous solutions were prepared in the presence of EDAC 50 mM (Sigma-Aldrich). Subsequently, several blends of BSA/ALG were obtained by mixing different volumes of the solutions prepared previously (Supplementary data Table S1). Ethylenediamine (ED) was added to each blend according to 2:1 ratio (polymer: ED, w/w) under intense stirring conditions (magnetic stirrer), and pH was then adjusted to 3 with HCl to attain amide bond formation. Stirring was maintained for 2 h and blends were finally centrifuged (10,000 rpm, 5 min; Sigma SM202 centrifuge). Pellets were freeze dried (Heto PowerDry LL1500 Freeze Dryer, Thermo Electro Corporation, Waltham, MA, USA) for 24 h at −110 °C.

The conjugation of folate to nanoparticles was carried out by a carbodimiide reaction with NHS–folate. The N-Hydroxysuccinimide ester of folic acid (NHS–folate) was prepared according to the method developed by Lee and co-workers [17]. Folic acid (2.5 g) was dissolved in 50 mL of dry dimethyl sulfoxide (DMSO) plus 1.25 mL of triethylamine and reacted with NHS (1.3 g) in the presence of dicyclohexylcarbodiimide (2.35 g) overnight at room temperature. The byproduct, dicyclohexylurea, was removed by filtration. The DMSO solution was then freeze-dried, and NHS–folate was precipitated in diethylether. The final product, NHS–folate, was washed several times with anhydrous ether, dried under vacuum, and yielded a yellow powder.

The conjugation of folate to nanoparticles was carried out following the method developed by Zhang and co-workers [18]: NHS–folate (50 mg/mL) was dissolved in 3 mL DMSO and added slowly to 6 mL of the stirred nanoparticle suspension (10 mg NP/mL, pH adjusted to 10 using 1 M carbonate/bicarbonate buffer). After stirring for 3 h at room temperature, the reaction mixture was centrifuged (10,000 rpm, 10 min) in order to separate the folate conjugated nanoparticles from unreacted folic acid and other byproducts and then washed with DMSO (10,000 rpm, 10 min). Finally, folate-conjugated nanoparticles (30BSA/70ALG–Fol, 40BSA/60ALG–Fol; 50BSA/50ALG–Fol) were freeze dried for 24 h at −110 °C.

2.3. Characterization of Nanoparticles

2.3.1. Composition of BSA/AlG Nanoparticles

Composition of nanoparticles was determined according to the method developed by Martinez and co-workers [19]. The composition of BSA/ALG nanoparticles was studied measuring the concentration of BSA and ALG in the supernatant obtained after the centrifugation of the samples in the synthesis process. The difference between the initial concentration of each polymer and their concentration in the supernatant was determined, to reveal the proportion of each polymer in nanoparticle formation. Concentration of alginate in the supernatant was determined by measuring the absorbance of the colored products obtained after a hydrolysis reaction of alginate. The hydrolysis of alginate was carried out in extreme conditions (10 N HCl, 5 h, 100 °C) and, after this reaction, samples were neutralized with 10 N NaOH and cooled. Absorbance of the samples was measured at 277.6 nm. A standard curve of alginate (0.5–4 mg/mL) was prepared. Concentration of BSA in the supernatant was determined using Bradford's method [20]. The absorbance of blue-colored products of this reaction could be measured at 595 nm. To avoid interactions between alginate and Bradford reagent, the BSA in the supernatant was precipitated with 2M trichloroacetic acid (TCA). After the precipitation process, samples were centrifuged and the supernatant was removed. The pellet was suspended in 1 mL 6M urea (urea solution in 2 N HCl). A standard curve of BSA (0.125–0.5 mg/mL) was prepared.

2.3.2. Thermogravimetric Analysis (TGA)

TGA curves of BSA/ALG nanoparticles, as well as raw materials, were obtained using a Mettler Toledo thermal analyzer (TGA-SDTA 851®, Switzerland). Samples (3 mg) were placed onto the balance, the temperature increased from 25 to 600 °C at 10 °C min^{-1} in a nitrogen atmosphere (nitrogen flow rate of 60 cm^3 min^{-1}) and the mass continuously recorded as a function of temperature.

2.3.3. Determination of Folate-Conjugated to BSA/ALG Nanoparticles

The amount of folate conjugated to the amine groups of albumin was determined by spectrophotometric analysis [18]. A quantity of 2 mg folate-conjugated nanoparticles (Nps–Fol) were hydrolysed by trypsin (0.05 mg/mg NP) with stirring at 37 °C for 2 h. After the digestion process, the quantification of folate-conjugation was performed by spectrophotometric measurement of its absorbance at 358 nm (folic acid ε = 8643.5 M^{-1} cm^{-1}).

2.3.4. Morphology, Size and Z-Potential of Nanoparticles

The morphology and size of Nps and Nps–Fol were studied by Transmission Electron Microscopy (TEM) (JOEL JEM 1010 microscope from ICTS Centro Nacional de Microscopía Electrónica, UCM). Size and zeta potential measurements of Nps and Nps–Fol were performed in deionized distilled water using a backscattered quasi-elastic light scattering device (Zetatrac NPA). Data were analyzed with Microtrac Flex Software.

2.3.5. Preparation of PTX–Folate-Conjugated Nanoparticles

The paclitaxel (PTX) load into Nps–Fol was carried out as follows: a suspension of 30 mg of NPs–Fol in 1 mL of the solution 1 mg PTX/mL ethanol was prepared and incubated for 22 h using a circular rotary stirrer in darkness at room temperature. Finally, the suspension was centrifuged (12,000 rpm, 15 min), the pellet was washed with water and nanoparticles were freeze-dried.

2.3.6. Estimation of PTX Content

Two different methods were carried out in order to determine the amount of PTX loaded into Nps–Fol. One of the methods consisted of a direct extraction of the drug from PTX-loaded Nps–Fol using ethanol: 3 mg of PTX-loaded Nps–Fol were suspended in 1 mL of ethanol under stirring conditions for 5 h. After this time, the suspension was centrifuged (13,000 rpm, 5 min) and the supernatant was collected. The samples were analyzed by HPLC. The other method consisted of an enzymatic digestion of PTX-loaded Nps–Fol with trypsin and a subsequent extraction of the drug with an organic solvent: 2 mg of PTX-loaded nanoparticles was suspended in 2 mL of phosphate-buffered saline (PBS, pH 7.4), and 50 µg of trypsin was added to the solution. The digestion process was performed under orbital stirring (100 rpm) at 37 °C for 24 h. After this time, PTX was extracted from PBS solution by the addition of 4 mL of dichloromethane. The tubes were mixed in a vortex mixer for 10 min and then centrifuged for 10 min at 10,000 rpm. The organic phase was transferred to flask vials and evaporated to dryness. The resulting residue was reconstituted in ethanol and then quantified by the HPLC technique.

The concentration of PTX was determined by high-performance liquid chromatography (HPLC) using a Spectra Physics HPLC system, Spectra 100 UV–Vis detector and a SP8800 pump. The mobile phase consisted of acetonitrile/water 60:40 (v/v) and was pumped through 25 mm × 4.6 mm RP-Spherisorb ODS2 C18 column (5 mm particle size, Waters) at a flow rate of 1 mL/min. The output at 227 nm was monitored and the peak area of each sample generated by ChromQuestTM software. The calibration curve was generated using PTX solutions between 0.1 and 25 µg/mL in acetonitrile/water (60:40, v/v), prepared from a PTX solution in ethanol, and a good linear correlation ($r^2 = 0.99$) was obtained. All quantifications were performed in triplicate.

2.3.7. In Vitro Drug Release Studies

The release of PTX from Nps–Fol was carried out in centrifuge tubes. Drug-loaded Nps–Fol (3 mg) was suspended in PBS (4 mL pH 7.4, 37 °C) containing 0.1% (w/v) Tween 80® [21,22]. Release studies were carried out in the dark with orbital shaking (100 rpm; Ecotron INFORS HT, Switzerland) for 96 h. Sink conditions were maintained during drug release experiments and all experimental conditions were performed in triplicate. At predetermined time intervals, the solutions were centrifuged at 10,000 rpm for 1 min and the supernatants collected for PTX concentration analysis. The Nps–Fol samples were dispersed in fresh PBS and placed back on the shaker. The PTX was extracted from the supernatants by DCM (1 mL) by stirring using a vortex mixer for 10 min at room temperature. The DCM layer collected was evaporated at room temperature for 24 h, and the dried PTX was dissolved in acetonitrile/water (1 mL; 60:40 (v/v)). The percentage of PTX released into the supernatant was determined by HPLC.

In order to analyze the drug delivery mechanism used by the NPs–Fol, the in vitro drug release data were fitted to various kinetic models. Higuchi, Korsemeyer–Peppas, zero-order and first-order models were selected for this study, applying the following set of equations [23]:

Higuchi model

$$M_t = K_H \sqrt{t}$$

Korsemeyer–Peppas

$$M_t/M_\infty = K t^n$$

Zero-order kinetic model

$$M_o - M_t = K_o t$$

First-order kinetic model

$$M_t/M_\infty = 1 - e^{-Kt}$$

where M_o, M_t and M_∞ corresponded to the drug amount at time zero, at a particular time and at an infinite time, respectively. The K terms referred to the release kinetic constants obtained from the linear curves of simple regression analysis.

2.3.8. Cell Culture Studies

Human breast adenocarcinoma (MCF7, MDA-MB-231) and human cervical cancer HeLa cell lines were obtained from ECACC (Sigma-Aldrich).These cells were chosen due to their expression of folate receptors (FRs) [24–26]. Cells were maintained in DMEM + GlutaMax-I supplemented with 10% heat inactivated fetal bovine serum, penicillin (50 U/mL), streptomycin (50 µg/mL) (Invitrogen Life Technologies, Grand Island, NY, USA) and gentamicin (50 µg/mL) (Sigma–Aldrich Company, Gillingham, UK) in a humidified incubator at 37 °C and 5% CO2 atmosphere (HERA cell, Sorval Heraeus, Kendro Laboratory Products GmbH, Hanau, Germany). Cells were plated in a 75 cm^2 flask (Sarstedt Ag and Co., Barcelona, Spain) and were passaged when they reached 90% confluence by gentle trypsinization (0.05% trypsin/0.53 mM EDTA; Invitrogen Life Technologies).

Cellular Uptake of Nanoparticles

Nps and Nps–Fol (10 mg) were loaded with 6-coumarin (98%, Sigma-Aldrich, Madrid, Spain) by immersion in 1 mL of a 1 mg/mL 6-coumarin solution (dissolved in ethanol) for 24 h at room temperature in the dark and under orbital shaking. Then, the suspension was centrifuged (7000 rpm, 5 min), and the 6-coumarin-loaded NPs were washed twice with ethanol and freeze-dried for 24 h at $-110\ °C$

Cellular uptake of Nps and Nps–Fol was investigated in a monolayer of MCF7, MDA-MB-231 and HeLa cells. Cells were seeded in 96 well flat-bottom plates at 10,000 cells/well for MCF7 and MDA-MB-231, and 5000 cells/well for HeLa. After 24 h, the medium was replaced with 100 µL of 1% FBS medium containing 6-coumarin-loaded Nps or Np–Fol and the plates were incubated in the 5% CO2 incubator at 37 °C for 2, 4 and 24 h. The concentration of nanoparticles was the equivalent to obtain a concentration of 1 µM of drug, considering the PTX load of each nanoparticle composition. Thus, the concentration of 6-coumarin-loaded Nps and Nps–Fol was 0.24 mg/mL (composition 50BSA/50ALG) and 0.325 mg/mL (composition 30BSA/70ALG) respectively. As a positive control, cells were grown with 0.2 µg/mL of 6-coumarin in FBS medium, in the absence of nanoparticles. At each time interval, cells were washed twice with PBS to remove any uninternalized nanoparticles and lysed with 100 µL lysis reagent (PBS with 2% SDS v/v and 50 mM EDTA). Internalization of nanoparticles was evaluated by fluorimetry at 488 nm, using a spectrophotometer (Varioskan, Thermo FisherScientific, Barcelona, Spain) [27]. Calibration curves (fluorescence versus concentration) were prepared with 6-coumarin (1 ng/mL^{-1} µg/mL) dissolved in the lysis medium in order to calculate the amount of 6-coumarin incorporated into nanoparticles. All experimental conditions were performed in quintuplicate.

Fluorescence Microscopy

Cellular uptake of Nps and Nps–Fol with 6-coumarin was observed by fluorescence microscopy. Cells were seeded in 24 well flat-bottom plates at 40,000 cells/well for MCF-7 and MDA-MB-231, and 20,000 cells/well for HeLa. After 24 h, the medium was replaced with 500 µL of 1% FBS medium containing 6-coumarin-loaded Nps or Np–Fol and the plates were incubated in the 5% CO$_2$ incubator at 37 °C for 4 h. Then, cells were washed twice with PBS to remove any uninternalized nanoparticles. Localization of Nps–Fol with 6-coumarin was examined by light fluorescence microscopy (Leica DMIL microscope, Leica Microsystems, Balgach, Switzerland). Cells were photographed with a Leica DFC 300FX digital camera and Leica Application Suite software was used for processing the pictures (Leica Microsystems Switzerland).

Cell Viability

Cell viability was evaluated by using the MTT method. All experimental conditions were performed in quintuplicate. Each experiment was carried out in triplicate.

In preliminary experiments, increasing concentrations of PTX from 0.1 to 500 nM were tested in culture. The IC50 of PTX was determined in the three cell lines tested. According to that, the selected concentrations of PTX in cell viability studies were 2.5 and 7.5 nM for HeLa and MCF7 cell lines, and 7.5 and 30 nM for MDA-MB-231 cells.

Cells were seeded in 96-well flat-bottom plates at 5000 cells/well in the case of MCF7 and MDA-MB-231, and at 2500 cells/well in the case of HeLa. After 24 h, the medium was replaced with 100 µL medium with 1% FBS containing Nps–Fol, PTX-loaded Nps–Fol or the drug in solution. The amount of Nps–Fol was the needed to obtain a final concentration of 30, 7.5 and 2.5 nM of PTX, considering the PTX load of each nanoparticle composition; for this purpose, a nanoparticle suspension of 0.24 mg/mL of 50BSA/50ALG and 0.325 mg/mL of 30BSA/70ALG Nps–Fol, (equivalent to 1000 nM PTX) was used. After 1, 3 and 6 days, 10 µL of MTT solution (5 mg/mL) were added to each well. After 2 h of incubation at 37 °C, 5% CO2, each well was replaced with 100 µL DMSO [24]. The cell viability was determined by measuring the absorbance at 570 nm using a spectrophotometer (Varioskan, Thermo

Fisher Scientific, Barcelona, Spain). Results are expressed as the percentage survival in relation to untreated control cells.

2.4. Statistical Analysis

Statistical comparisons were performed with the unpaired Student's t-test. A value of $p < 0.05$ was considered significant.

3. Results

3.1. Preparation of BSA/ALG Nanoparticles

3.1.1. Composition of Nanoparticles

Final nanoparticle composition was determined by the concentration of BSA and ALG in the supernatant collected after centrifugation, in the synthesis process. Thus, the contribution of each polymer within the final formulation of nanoparticles was calculated by the difference between the initial concentration and the final concentration after the synthesis protocol (Table 1). The BSA/ALG ratio was calculated considering the concentration of each polymer incorporated into nanoparticles. These results demonstrated that not all the BSA nor ALG added in the initial solution of synthesis was incorporated in the final nanoparticles. Surprisingly, an increased proportion of BSA (w/w) in the initial composition was not translated into a higher incorporation of BSA into Nps.

Table 1. Nanoparticle composition. Percentage of each polymer incorporated into nanoparticles. BSA/ALG ratio in nanoparticles.

Nanoparticles	Composition of Blend (w/w Ratio)	BSA Incorporated into Nanoparticles (%)	ALG Incorporated into Nanoparticles (%)	BSA/ALG Ratio in Nanoparticles
30BSA/70ALG	2BSA:1ALG	52 ± 4	65 ± 5	1/1
40BSA/60ALG	3BSA:1ALG	13 ± 4	50 ± 2	0.25/1
50BSA/50ALG	5BSA:1ALG	25 ± 3	54 ± 4	0.5/1

3.1.2. Thermogravimetric Analysis (TGA)

The thermal stability of raw materials and BSA/ALG nanoparticles was studied. The TGA first derivative of the materials is plotted in Figure S1 (Supplementary Materials). The Nps showed a thermal profile different from raw materials: Nps that included more BSA in their composition (30BSA/70ALG, ratio 1:1) showed a two-step degradation process: at 217 °C (close to the degradation peak of ALG) and 317 °C (close to the degradation peak of BSA). However, Nps with more ALG in their composition (40BSA/60ALG, ratio 0.25:1; 50BSA/50ALG, ratio 0.5:1) showed only one degradation peak at 212 °C (close to the degradation peak of ALG). These differences in thermal behavior between raw materials and the Nps would indicate the chemical interaction between both polymers in the Nps and would confirm the different ratio of polymers incorporated in each Np composition. Additionally, all Np compositions showed a mass loss above 200 °C which indicated that Nps were stable to use at routine work temperatures.

3.1.3. Determination of Folate Conjugated to Nanoparticles

The amount of folate conjugated to Nps–Fol was spectrophotometrically evaluated after the hydrolysis of nanoparticles by trypsin. Values of 30 ± 2, 9 ± 2 and 26 ± 2 µmol Fol/g Nps were respectively obtained from 30BSA/70ALG–Fol, 40BSA/60ALG–Fol and 50BSA/50ALG–Fol. Statistically significant differences did not exist between 30BSA/70ALG–Fol and 50BSA/50ALG–Fol nanoparticles ($p = 0.07$) (data not shown); however, significant differences were observed between 30BSA/70ALG–Fol and 40BSA/60ALG–Fol ($p = 0.0002$) and also between 50BSA/50ALG–Fol and 40BSA/60ALG–Fol ($p = 0.0004$). Therefore, no correlation between the amount of BSA in the nanoparticles and the extent of folate on their surface was established.

According to these results, the compositions 30BSA/70ALG–Fol and 50BSA/50ALG–Fol were finally chosen for further studies due to their greater content of incorporated folate.

3.2. Characterization of Folate-Conjugated BSA/ALG Nanoparticles

3.2.1. Morphology, Size and Z-Potential of Nanoparticles

The morphology of 30BSA/70ALG and 50BSA/50ALG Nps was analyzed by TEM. Micrographs showed nanometric-range sized particles (<100 nm), with a spherical appearance. This morphology was maintained after folate conjugation (Figure 1).

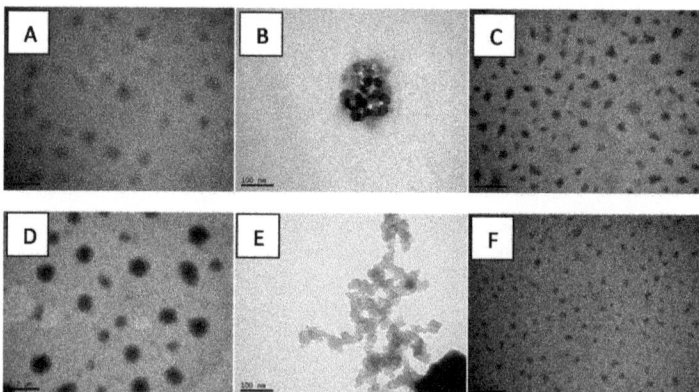

Figure 1. TEM micrographs of 30BSA/70ALG (**A**), 30BSA/70ALG–Fol (**B**), PTX-loaded 30BSA/70ALG–Fol (**C**), 50BSA/50ALG (**D**), 50BSA/50ALG–Fol (**E**) and PTX-loaded 50BSA/50ALG–Fol (**F**) nanoparticles.

These nanoparticles were also analyzed using quasi-elastic light scattering to characterize them in terms of mean size and surface charge (Table 2). The results showed that all synthesized systems had a nanometric size in the range of 169 ± 28 nm and 296 ± 57 nm and a polydispersity index between 1.2 ± 0.4 and 1.8 ± 0.4. Moreover, an increase in size was observed after the incorporation of folate and the loading of paclitaxel in both nanoparticle compositions. However, these differences in size were not statistically significant. Differences in nanoparticle size among TEM photographs and DLS data can be attributed to the different characteristics of these techniques. In this way, DLS measures are realized in a water solution without stirring, and nanoparticles tend to establish interactions and aggregate, increasing the observed size.

Table 2. Mean size, polydispersity index and zeta potential of nanoparticles and folate-conjugate nanoparticles determined by quasi-elastic light scattering. Data were shown as mean ± S.D.

Nanoparticles	Mean Size (nm)	Polydispersity Index	Zeta Potential (mV)
30BSA/70ALG	182 ± 82	1.4 ± 0.7	−0.12 ± 0.04 [a]
30BSA/70ALG–Fol	189 ± 81	1.7 ± 0.3	−69.3 ± 0.8 [a]
PTX-loaded 30BSA/70ALG–Fol	290 ± 126	1.8 ± 0.4	−67.3 ± 0.8 [a]
50BSA/50ALG	169 ± 28	1.5 ± 0.3	−0.43 ± 0.06 [b]
50BSA/50ALG–Fol	268 ± 102	1.4 ± 0.3	−66.2 ± 0.6 [bc]
PTX-loaded 50BSA/50ALG–Fol	296 ± 57	1.2 ± 0.4	−94.1 ± 0.4 [bc]

[a]: significant statistical differences ($p < 0.001$) between 30BSA/70ALG y 30BSA/70ALG–Fol, and between 30BSA/70ALG and PTX-loaded 30BSA/70ALG–Fol; [b]: significant statistical differences ($p < 0.001$) between 50BSA/50ALG y 50BSA/50ALG–Fol, and between 50BSA/50ALG and PTX-loaded 50BSA/50ALG–Fol; [c]: significant statistical differences ($p < 0.001$) between 50BSA/50ALG–Fol and PTX-loaded 50BSA/50ALG–Fol.

Results of Z-potential analysis (Table 2) showed a negative charge at the surface of all the tested compositions, like the results obtained in similar studies [28]. The negative value of this parameter was increased after the conjugation of folic acid and the incorporation of PTX in both formulations, those differences were statistically significant.

3.2.2. Estimation of PTX Content in Folate-Conjugate Nanoparticles

The estimation of PTX content in each formulation was determined by two methods: extraction with ethanol and enzymatic degradation using trypsin (Table 3). Independently of the method, results showed statistically significant differences ($p < 0.05$) in drug content comparing both compositions, showing a higher PTX load in 50BSA/50ALG–Fol nanoparticles.

Table 3. Estimation of PTX content in folate-conjugated nanoparticles (Nps–Fol).

Nanoparticles	Drug Content in Nps–Fol	
	By Extraction with Ethanol (µg PTX/mg Np)	By Tryptic Hydrolysis of the Nanoparticles (µg PTX/mg Np)
50BSA/50ALG–Fol	3.56 ± 0.13 [a]	3.28 ± 0.24 [b]
30BSA/70ALG–Fol	2.63 ± 0.19 [a]	2.42 ± 0.36 [b]

Data: mean ± S.D. (n = 3). [a]: significant statistical differences $p < 0.05$ ($p = 0.002$); [b]: significant statistical differences $p < 0.05$ ($p = 0.03$).

3.2.3. PTX Release from Folate-Conjugated Nanoparticles (Nps–Fol)

The cumulative release curve of PTX is represented in Figure 2. The maximum PTX release (82% and 88% of drug load) was achieved at 23 and 27 h from 30BSA/70ALG–Fol and 50BSA/50ALG–Fol, respectively. Drug release in the presence of Tween 80® occurred quickly in the first 4 h (Table 4), followed by a slower release rate in both types of nanosystem. In both stages of drug delivery, the release of PTX was faster from the 50BSA/50ALG–Fol composition.

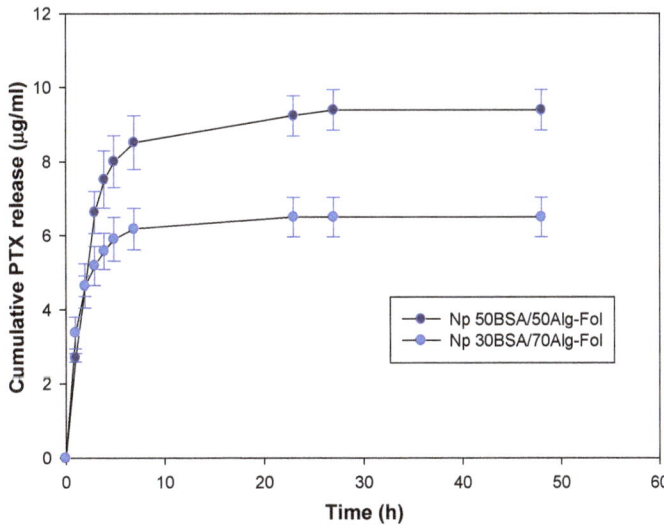

Figure 2. Cumulative release of PTX from BSA/ALG–Fol nanoparticles.

Table 4. Release rates of PTX from 50BSA/50ALG–Fol and 30BSA/70ALG–Fol nanoparticles.

Nanoparticles	Release Rate of PTX from Nanoparticle–Fol	
	First Stage: 0–4 h	Second Stage: 5–27 h
50BSA/50ALG–Fol	0.63 µg PTX/h per mg of Nps–Fol (r^2 0.973)	0.019 µg PTX/h per mg of Nps–Fol (r^2 0.935)
30BSA/70ALG–Fol	0.43 µg PTX/h per mg of Nps–Fol (r^2 0.823)	0.008 µg PTX/h per mg of Nps–Fol (r^2 0.877)

The PTX release was also evaluated in a Tween 80® free medium at 74 h. In this experimental condition, only 16% and 23% of the PTX load was released from 50BSA/50ALG–Fol and 30BSA/70ALG–Fol, respectively.

3.3. In Vitro Evaluation of Folate-Conjugated Nanoparticles in Tumour Cell Lines

3.3.1. Cellular Uptake of Nanoparticles

To evaluate the uptake of the synthesized nanosystems by cells, three different human tumor cell lines, MCF-7, MDA-MB-231 and HeLa, were selected. BSA/ALG–Fol nanoparticles and BSA/ALG nanoparticles were loaded with a fluorescent marker (6-coumarin) and added to cells in order to study their cellular uptake. The fluorescent signal into cells was quantified after 2, 4 and 24 h of incubation. Results are shown in Figure 3. As results showed, either targeted or non-targeted systems were internalized in cells. However, folate-conjugated nanoparticles enhanced their uptake in the three cell lines. The 50BSA/50ALG nanoparticles were less internalized than 30BSA/70ALG.

Besides the quantitative assay, the internalization of these systems was easily observed due to the green fluorescence of coumarin attached to the particles. Figure 3 shows the uptake when they were functionalized with folate (BSA50/50ALG–Fol) for 4 h, as an example of the images obtained.

3.3.2. Cell Viability

Tumor cell cytotoxicity induced by different concentrations of unloaded and PTX-loaded BSA/ALG–Fol nanoparticles was determined in MCF7, MDA-MB-231 and HeLa cells. IC50 values of PTX in solution were 7.32 nM, 7.71 nM and 38 nM for HeLa, MCF7 and MDA-MB-231 cells, respectively, after 24 h of incubation. According to these results, two concentrations of PTX (2.5 and 7.5 nM for the HeLa and MCF7 cell lines, and 7.5 and 30 nM for the MDA-MB-231 cell line) were selected to perform cell viability studies. Cell viability studies in the presence of unloaded and PTX-loaded 50BSA/50ALG–Fol and 30BSA/70ALG–Fol nanoparticles were carried out over one, three and six days using the equivalent concentration to 2.5 and 7.5 nM PTX for HeLa and MCF7 cell lines, and 7.5 and 30 nM PTX for the MDA-MB-231 cell line when the drug was administered by nanoparticles.

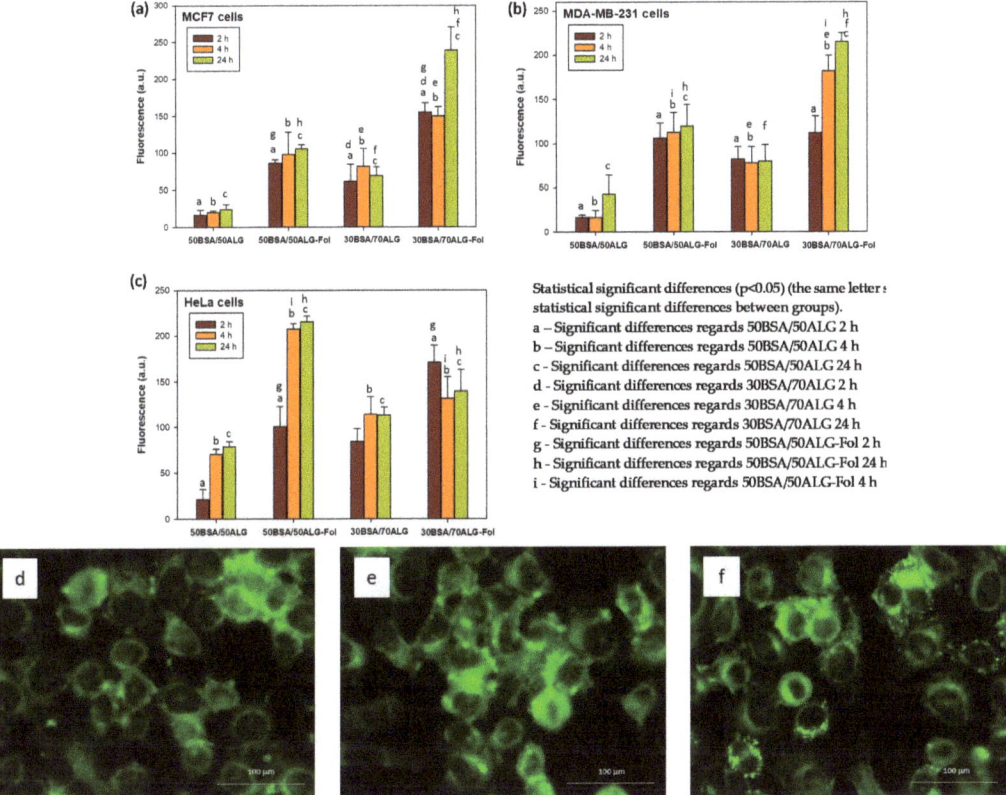

Figure 3. Quantitative comparison of coumarin-loaded internalized nanoparticles in (**a**) MCF-7, (**b**) MDA-MB-231 and (**c**) HeLa cells. Fluorescence microscopy images of coumarin-loaded internalized nanoparticles 50BSA/50ALG–Fol into MCF-7 (**d**), MDA-MB-231 (**e**) and HeLa (**f**) after 4 h of incubation.

Both unloaded nanoparticles (50BSA/50ALG–Fol and 30BSA/70ALG–Fol) were cytocompatible and no significant decrease in viability (cell survival between 91% ± 6% and 89% ± 3% for MCF7; 83% ± 5% and 84% ± 13% for MDA-MB-231; 100% ± 3% and 100% ± 12% for HeLa) was observed for the three tested cell lines at longer times of exposure.

PTX administered into 50BSA/50ALG–Fol and 30BSA/70ALG–Fol nanoparticles (Figure 4) caused a more gradual decrease in cell viability than PTX in solution from the first day, at the highest drug concentration assayed for MCF7, MDA-MB-231 and HeLa cells.

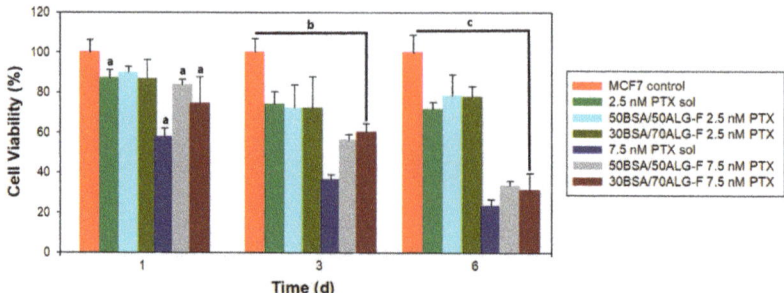

a – Significant differences (p<0.05) regards MCF7 control at day 1
b - Significant differences (p<0.05) regards MCF7 control at day 3 (all treatments show significant differences)
c - Significant differences (p<0.05) regards MCF7 control at day 6 (all treatments show significant differences)

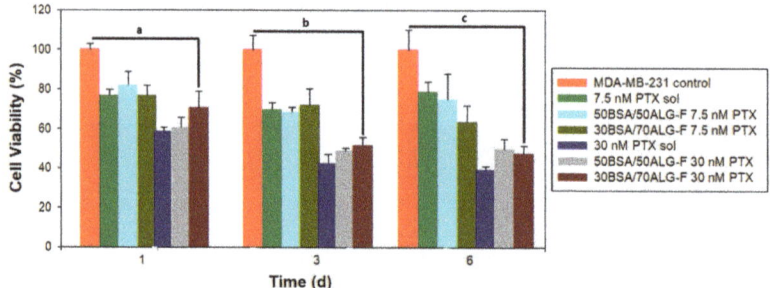

a – Significant differences (p<0.05) regards MDA-MB-231 control at day 1 (all treatments show significant differences)
b - Significant differences (p<0.05) regards MDA-MB-231 control at day 3 (all treatments show significant differences)
c - Significant differences (p<0.05) regards MDA-MB-231 control at day 6 (all treatments show significant differences)

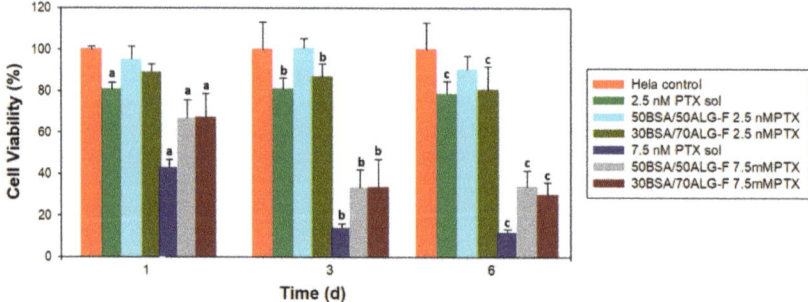

a – Significant differences (p<0.05) regards HeLa control at day 1
b - Significant differences (p<0.05) regards HeLa control at day 3
c - Significant differences (p<0.05) regards Hela control at day 6

Figure 4. Cell viability of MCF-7, MDA-MB-231 and HeLa cells in the presence of paclitaxel (PTX) in solution and PTX-loaded nanoparticles.

4. Discussion

Paclitaxel (PTX) is one of the most common drugs used for the treatment of different types of solid tumors [16]. PTX acts by promoting microtubule polymerization by binding to the β-subunit of tubulins, which causes mitotic arrest at the G2/M phase and results in cell death by the apoptosis pathway. Besides this, PTX exhibits cytotoxic properties that lead to the inhibition of cellular growth. Both characteristics contribute to the antitumor efficacy of PTX [29]. Since it is a hydrophobic drug, a vehicle is required to dissolve it, which causes numerous side effects, limiting its optimal clinical utility as an anticancer agent. For this reason, different nanocarriers have been prepared to improve the safety, efficacy and pharmacokinetic profile of PTX. Thus, in recent studies PTX has been encapsulated in nanoparticles [30,31], liposomes [32] and polymeric micelles [33]. On the other hand, the folate receptor is overexpressed on many tumors [34]. The FRα isoform is the most widely expressed of all the FR isoforms [35] and is overexpressed (100–300 times more than in healthy cells) in a large number of cancers of epithelial origin. Besides this, the high binding affinity (K_D = 0.1–1 nM) [36] of FRα for folic acid and other oxidized folates has led to the development of folate-conjugated nanosystems for targeting cancer cells [37,38]. For these reasons, folic acid was selected to target tumor cells with folate-bounded drug-loaded BSA/ALG nanoparticles.

4.1. Composition, Characterization and PTX Release from Folate-Conjugated Nanoparticles

Nanoparticles of different BSA/ALG composition, cross-linked with ethylenediamine (ED), were synthesized. BSA was better incorporated in 30BSA/70ALG (BSA:ALG 1:1) nanoparticles; meanwhile, alginate was similarly incorporated in all compositions, which can be attributed, at least in part, to the formation of amide bonds between ED and carboxylic groups of both ALG and BSA (each BSA molecule has 58 residues of Glu and 41 residues of Asp). TGA results confirmed the different ratio of polymers incorporated in each Np composition and could also indicate the chemical interaction between both polymers in Np.

Covalent attachment of folic acid to the three systems was achieved. There was no correlation between the amount of BSA in the nanoparticles and the extent of folate on their surface. Whereas 30BSA/70ALG–Fol and 50BSA/50ALG–Fol nanoparticles bound a similar amount of folic acid (FA), 40BSA/60ALG–Fol nanoparticles showed a significantly lower amount of FA attached. This fact could be related to the disposition of the amine groups of the side chain of amino acids in the nanoparticles; one BSA molecule has 60 residues of Lys and 26 residues of Arg, which can participate in FA attaching. Thus, the low amount of FA attached was the main reason to reject 40BSA/60ALG–Fol nanoparticles for later experiments.

TEM micrographs showed spherical, nano-sized particles without significant differences between nanoparticles of different BSA/ALG composition or those that were FA-bounded and not. The size of the unloaded nanosystems, determined by DLS (Table 2), was larger, but not statistically significantly so, for 50BSA/50ALG–Fol (268 ± 102 nm) than for 30BSA/70ALG–Fol (189 ± 81 nm), which could be due to the larger proportion of alginate in those nanoparticles, which would induce a higher swelling of the systems. However, the value of the polydispersity index was the same for both types of nanoparticles, and it was like the values obtained for other polymeric systems [12,39].

The Z-potential value of all studied nanosystems was negative (Table 2). The presence of alginate in 30BSA/70AGL and 50BSA/50ALG nanoparticles showed lightly negative Z-potential values. The FA-bounded nanosystems significantly increased Z-potential values (−69.3 mV and −66.2 mV for 30BSA/70ALG–Fol and 50BSA/50ALG–Fol, respectively), which were very much larger than −30 mV, the Z-potential value (positive or negative value) that has been established as the minimum for obtaining a physically stable nanosuspension and avoiding nanoparticle aggregation [40]. Besides this, when these nanosystems were loaded with PTX, Z-potential was of the same magnitude for PTX-loaded 30BSA/70ALG–Fol, and significantly larger (−94.1 mV) for PTX-loaded 50BSA/50ALG–Fol. This increase could be explained by the possible superficial distribution of part of the drug on the nanosystem. Thus, the low tendency to establish interactions among them, indicated by the Z-potential value of these unloaded and PTX-loaded FA-attached nanosystems, make them very suitable for later in vivo intravenous administration studies.

The PTX load of Fol-nanosystems was determined by extraction with ethanol and afterward by the enzymatic digestion of the nanoparticles with trypsin (Table 3). No differences were observed among the results obtained by both methods. Regardless of the method used, the PTX load of 50BSA/50ALG–Fol was larger (3.56–3.28 µg PTX/mg Np) than that of 30BSA/70ALG–Fol (2.63–2.42 µg PTX/mg Np). Interactions between PTX and nanosystems seem to be very similar in both types of nanoparticles since the value of the PTX load obtained by the two methods is almost the same. These observed results were in contrast with the expected results, where the drug load determined by ethanol extraction should be lower than that obtained by enzymatic digestion in the case of high PTX–nanoparticle interaction [24]. The PTX release studies showed a prolonged and faster drug release from 50BSA/50ALG–Fol nanoparticles; PTX release took place in two stages from both types of nanosystems (Figure 3). The first stage of release was during the first four hours (Table 4), and the release was faster from 50BSA/50ALG–Fol, which could be related to a part of the drug being located more superficially as indicated by the Z-potential values. A second stage of drug release was determined up to 27 h, and PTX release was faster from 50BSA/50ALG–Fol. In no case was the total amount of the PTX load released: 88% of the PTX load was released from 50BSA/50ALG–Fol and 82% of it from 30BSA/70ALG–Fol. In all these experiments, the release conditions were favored by the presence of Tween 80® in the medium; however, in the absence of the surfactant, an experimental condition closer to biological systems, a slower release occurred, which would make possible a significant amount of PTX loaded into the nanoparticle–Fol to be delivered to cancer cells in vivo. Similar in vitro PTX release behavior was observed when that drug was released from nanohydrogels [27]. In that case, a maximum release took place at 50 h, and it caused a significant decrease in cell viability of human cancer cell lines. These PTX-loaded nanoparticulated systems were subcutaneously injected in female athymic nude mice bearing HeLa human tumor xenografts, and they showed higher antitumor activity than PTX in solution [41]. Equivalently, tamoxifen-loaded folate-targeted protein/polysaccharide-based nanoparticles, which released the drug in the first 8 h in in vitro studies, were demonstrated to be highly effective in tumor remission assays in MCF7 cell xenograft mice after intravenous administration [24]. In that case, although in vitro drug release from nanoparticulated and folate-targeted nanoparticulated systems took place faster, the results obtained in xenograft models demonstrated the effectiveness of these types of nanosystems as an antitumoral therapy.

In order to determine the predominant drug release mechanism from BSA/ALG–Fol nanoparticles, the obtained data were fitted to different physical models with a significant correlation coefficient (r^2 = 0.791–0.979) (Supplementary Materials; Table S2). According to the results, the PTX release kinetics from 50BSA/50ALG–Fol nanoparticles fitted in the order of Korsemeyer–Peppas = Higuchi = First Order > Zero Order; PTX release from 30BSA/70ALG–Fol nanoparticles fitted in the order of Korsemeyer–Peppas > Higuchi > First Order > Zero Order. The Higuchi model indicates drug diffusion from the matrix, with no matrix dissolution and constant drug diffusivity [42]. The release of PTX from 30BSA/70ALG–Fol nanoparticles showed the lowest value of K_H, which implies a slower release in comparison with 50BSA/50ALG–Fol nanoparticles. An equivalent behavior has been observed for the release of naproxen from cyclodextrin-based hydrogel matrices [43]. The release of PTX was favored by the swelling capability of the nanoparticles, which made the diffusion of the entrapped drug easier. In the case of the Korsemeyer–Peppas model, the nanosystems showed values of the release exponent (n) different from the standard value for declaring Fickian release behavior (n = 0.43; considering spherical shape) [44,45]. This was 0.36 and 0.70 for 30BSA/70ALG–Fol and 50BSA/50ALG–Fol nanoparticles, respectively. The values of the n parameter lower than 0.43 can be related to a dispersion of nanoparticle sizes. For spherical particles, when the n value is between 0.43 and 0.85, an anomalous transport (non-Fickian; intermediate between Fickian and Case II transport, zero order) is considered, and then a mixed diffusion and chain relaxation mechanism takes place, which happened for PTX release from 50BSA/50ALG–Fol. This nanoparticle composition presents an acceptable fitting to the first order kinetic model.

4.2. In Vitro Evaluation of Folate-Conjugated Nanoparticles in Tumor Cell Lines

The presence of FRs in several human solid tumors, such as breast cancer, has been clearly demonstrated. Thus, these receptors provide an interesting therapeutic target for solid tumor treatment. PTX is widely used as a first-line treatment for patients with breast cancer [25]. The cancer cell lines of human breast adenocarcinoma MCF-7 and MDA-MB-231 express a low and high level of folate receptor, respectively [46]. However, whereas MCF-7 cells express estrogen receptors (ER), MDA-MB-231 cells are triple negative breast cancer (TNBC) cells; they do not express ER, progesterone receptor (PR) or the amplification of HER-2/Neu [47]. Besides this, among the drugs used to treat stage IVB cervical cancer is PTX [48]. HeLa cells are human cervical carcinoma cells, positive for FR, adequate to study the efficacy of these folate-targeted PTX-loaded BSA/ALG nanocarriers.

Therefore, these three different cellular lines expressing folate receptors (FRs) were chosen for evaluating the effectiveness of these BSA/ALG–Fol nanoparticles. Different studies carried out by our group [24] and other research [49] have shown that MCF7, MDA-MB-231 and HeLa cells express different levels of FRs on their surface, meaning they can be used as an in vitro test of folate-targeted nanosystem efficacy. In general, good cytocompatibility values were obtained with the two compositions of BSA/ALG–Fol nanoparticles for the three cell lines, although a slightly lower value was observed for MDA-MB-231 cells, which can be attributed to the larger amount of nanoparticles in the experiments with this cell line. Therefore, the uptake of unloaded nanoparticle–Fol observed in these cell lines seemed not to be harmful even after six days of incubation. The presence of FRs on the cell surface led to a larger uptake in these cell lines for both nanoparticle–Fol compositions in comparison with the corresponding nanoparticle composition without folic acid bounded on their surface. A larger amount of folate receptors on the cell surface made nanoparticle–Fol uptake easier as fluorescence images of HeLa cells showed. The uptake of Nps without FA on their surface revealed there was a nonspecific uptake mechanism that would act in addition to FR-mediated internalization. In this way, different studies indicate that negative-charged nanoparticles showed high cellular uptake due to electrostatic interactions established with some parts of the cellular membrane [24,50], which would be in accordance with the negative values of the Z-potential of these BSA/ALG nanoparticles. This nonspecific uptake was lower in the case of 50BSA/50ALG nanoparticles in the three cell lines studied, and so they would be less internalized into cells without FR on their surface, like non-cancer cells.

PTX-loaded nanoparticle–Fol caused a gradual decrease in cell viability values when PTX was released from both nanosystems, mainly at a higher PTX load (7.5 nM PTX for MCF7 and HeLa; 30 nM for MDA-MB-231) at a longer time of incubation. Significant differences in decreasing viability were not observed between PTX-loaded 50BSA/50ALG–Fol and 30BSA/70ALG–Fol nanoparticles in all cell lines studied. Although viability values obtained in the presence of PTX-loaded nanoparticle–Fol were similar to those of PTX in solution, it must be taken into account that the PTX amount within the cells treated with PTX-loaded nanosystems–Fol will be lower than the amount of drug available in cells treated with PTX solutions, in accordance with in vitro drug release studies. Once PTX-loaded nanoparticle–Fol was taken up into cells via the FR-mediated endocytosis pathway, the unligate FR may then recycle to the cell surface at a rate in the range of 0.5–5 h [51]. It is believed that more nanoparticles were taken up into the cells; PTX release would have take place mainly after the enzymatic degradation of nanoparticles, which would have caused a delay in the cytotoxic action of PTX-loaded nanoparticle–Fol compared to PTX in solution.

5. Conclusions

The prepared folate-targeted PTX–BSA/ALG nanoparticles seem to be a promising alternative to the conventional use of Cremophor® EL and would reduce the drug resistance problem mediated by P-gp. Preliminary cellular assays have demonstrated the effectiveness of these nanosystems against cancer cells overexpressing surface folate receptors. In this respect, a reduction in cell survival was observed in a cervical carcinoma cell line, a breast adenocarcinoma ER+ cell line and a TNBC cell line, which causes a very aggressive cancer type with a difficult treatment. These results allow us to think that the synthesized folate-targeted PTX–BSA/ALG nanoparticles could be considered for further in vivo biocompatibility and tumor growth suppression studies.

Supplementary Materials: The following are available online at https://www.mdpi.com/article/10.3390/polym13132083/s1, Figure S1: TGA first derivative curves of BSA, sodium alginate, 30BSA/70ALG Nps, 40BSA/60ALG Nps and 50BSA/50ALG Nps; Table S1: Percentages (v/v) of BSA (5% w/v) and ALG (1% w/v) solutions added to each blend; Table S2. Statistical parameters of nanoparticle formulations obtained after fitting the drug release data to the release kinetic models. Model fitting was attempted until 91% and 85% of drug released for PTX-30BSA/70ALG-Fol and PTX-50BSA/50ALG-Fol, respectively.

Author Contributions: Conceptualization, A.M.M.-R. and M.D.B.; methodology, A.M.M.-R., M.B. and R.M.O.; validation, R.M.O. and M.D.B.; investigation, A.M.M.-R., M.B., E.P.-I., C.T. and R.M.O.; resources, E.P.-I., C.T. and R.M.O.; writing—original draft preparation, E.P.-I. and M.D.B.; writing—review and editing, E.P.-I. and M.D.B.; visualization, A.M.M.-R., M.B., E.P.-I., C.T. and R.M.O.; supervision, A.M.M.-R. and M.D.B.; project administration, M.D.B. and A.M.M.-R.; funding acquisition, M.D.B. and A.M.M.-R. All authors have read and agreed to the published version of the manuscript.

Funding: This research was funded by Santander-Complutense University of Madrid (PR26/16-20273 and PR75/18-21575) and Francisco de Vitoria University (UFV).

Informed Consent Statement: Not applicable.

Data Availability Statement: The data presented in this study are available on request from the corresponding author.

Acknowledgments: The financial support of Santander-Complutense University of Madrid (PR26/16-20273 and PR75/18-21575) and Francisco de Vitoria University (UFV) are gratefully acknowledged.

Conflicts of Interest: The authors declare no conflict of interest.

References

1. Shah, A.; Aftab, S.; Nisar, J.; Ashiq, M.N.; Iftikhar, F.J. Nanocarriers for targeted drug delivery. *J. Drug Deliv. Sci. Technol.* **2021**, *62*, 102426. [CrossRef]
2. Barclay, T.G.; Day, C.M.; Petrovsky, N.; Garg, S. Review of polysaccharide particle-based functional drug delivery. *Carbohydr. Polym.* **2019**, *221*, 94–112. [CrossRef] [PubMed]
3. Zhang, E.; Xing, R.; Liu, S.; Qin, Y.; Li, K.; Li, P. Advances in chitosan-based nanoparticles for oncotherapy. *Carbohydr. Polym.* **2019**, *222*, 115004. [CrossRef]
4. Martínez, A.M.; Benito, M.; Pérez, E.; Teijón, J.M.; Blanco, M.D. The role of anionic polysaccharides in the preparation of nanomedicines with anticancer applications. *Curr. Pharm. Des.* **2016**, *22*, 3364–3379. [CrossRef]
5. Martínez, A.; Pérez, E.; Montero, N.; Teijón, C.; Olmo, R.; Blanco, M.D. Alginate-based drug carriers: Recent advances. In *Alginic Acid: Chemical Structure, Uses and Health Benefits*; Moore, A., Ed.; Nova Science Publishers, Inc.: Hauppauge, NY, USA, 2015; pp. 15–43, ISBN 978-1-63463-224-9.
6. Xu, Z.; Lam, M.T. Alginate application for heart and cardiovascular diseases. In *Alginates and Their Biomedical Applications*; Rehm, B., Moradali, M., Eds.; Springer Series in Biomaterials Science and Engineering; Springer: Singapore, 2018; Volume 11, pp. 185–212, ISBN 978-981-10-6910-9.
7. Hernández-González, A.C.; Téllez-Jurado, L.; Rodríguez-Lorenzo, L.M. Alginate hydrogels for bone tissue engineering, from injectables to bioprinting: A review. *Carbohydr. Polym.* **2020**, *229*, 115514. [CrossRef]
8. Chung, H.J.; Kim, H.J.; Hong, S.T. Tumor-specific delivery of a paclitaxel-loading HSA-haemin nanoparticle for cancer treatment. *Nanomed. Nanotechnol.* **2020**, *23*, 102089. [CrossRef] [PubMed]
9. Zhao, M.; Lei, C.; Yang, Y.; Bu, X.; Ma, H.; Gong, H.; Liu, J.; Fang, X.; Hu, Z.; Fang, Q. Abraxane, the nanoparticle formulation of paclitaxel can induce drug resistance by up-regulation of P-gp. *PLoS ONE* **2015**, *10*, e0131429. [CrossRef] [PubMed]
10. Li, C.; Li, Y.; Gao, Y.; Wei, N.; Zhao, X.; Wang, C.; Li, Y.; Xiu, X.; Cui, J. Direct comparison of two albumin-based paclitaxel-loaded nanoparticle formulations: Is the crosslinked version more advantageous? *Int. J. Pharm.* **2014**, *468*, 15–25. [CrossRef]

11. Paşcalău, V.; Tertis, M.; Pall, E.; Suciu, M.; Marinca, T.; Pustan, M.; Merie, V.; Rus, I.; Moldovan, C.; Topala, T.; et al. Bovine serum albumin gel/polyelectrolyte complex of hyaluronic acid and chitosan based microcarriers for Sorafenib targeted delivery. *J. Appl. Polym. Sci.* **2020**. [CrossRef]
12. Montero, N.; Pérez, E.; Benito, M.; Teijón, C.; Teijón, J.M.; Olmo, R.; Blanco, M.D. Biocompatibility studies of intravenously administered ionic-crosslinked chitosan-BSA nanoparticles as vehicles for antitumour drugs. *Int. J. Pharm.* **2019**, *554*, 337–351. [CrossRef]
13. Martínez, A.M.; Benito, M.; Pérez, E.; Blanco, M.D. Recent advances of folate-targeted anticancer therapies and diagnostics: Current status and future prospectives. In *Nanostructures for Cancer Therapy*; Nanostructures in Therapeutic Medicine Series; Ficai, A., Grumezescu, A.M., Eds.; Elsevier Inc.: Amsterdam, The Netherlands, 2017; pp. 329–350, ISBN 9780323461443/9780323461504.
14. Gong, Y.C.; Xiong, X.Y.; Ge, X.J.; Li, Z.L.; Li, Y.P. Effect of the folate ligand density on the targeting property of folated-conjugated polymeric nanoparticles. *Macromol. Biosci.* **2019**, *19*, 1800348. [CrossRef] [PubMed]
15. Vinothini, K.; Rajendran, N.K.; Ramu, A.; Elumalai, N.; Rajan, M. Folate receptor targeted delivery of paclitaxel to breast cancer cells via folic acid conjugated graphene oxide grafted methyl acrylate nanocarrier. *Biomed. Pharmacother.* **2019**, *110*, 906–917. [CrossRef] [PubMed]
16. Bernabeu, E.; Cagel, M.; Lagomarsino, E.; Moretton, M.; Chiappetta, D.A. Paclitaxel: What has been done and the challenges remain ahead. *Int. J. Pharm.* **2017**, *526*, 474–495. [CrossRef] [PubMed]
17. Lee, R.J.; Low, P.S. Delivery of liposomes into cultured KB cells via folate receptor-mediated endocytosis. *J. Biol. Chem.* **1994**, *269*, 3198–3204. [CrossRef]
18. Zhang, L.; Hou, S.; Mao, S.; Wei, D.; Song, X.; Lu, Y. Uptake of folate conjugated albumin nanoparticles to the SKOV3 cells. *Int. J. Pharm.* **2004**, *287*, 155–162. [CrossRef] [PubMed]
19. Martínez, A.; Iglesias, I.; Lozano, R.; Teijón, J.M.; Blanco, M.D. Synthesis and characterization of thiolated alginate-albumin nanoparticles stabilized by disulfide bonds. Evaluation as drug delivery systems. *Carbohydr. Polym.* **2011**, *83*, 1311–1321. [CrossRef]
20. Bradford, M.M. A rapid and sensitive method for the quantitation of microgram quantities of protein utilizing the principle of protein-dye binding. *Anal. Biochem.* **1976**, *72*, 248–254. [CrossRef]
21. Yang, T.; Cui, F.D.; Choi, M.K.; Cho, J.W.; Chung, S.J.; Shim, C.K.; Kim, D.D. Enhanced solubility and stability of PEGylated liposomal paclitaxel: In vitro and in vivo evaluation. *Int. J. Pharm.* **2007**, *338*, 317–326. [CrossRef] [PubMed]
22. López-Gasco, P.; Iglesias, I.; Benedí, J.; Lozano, R.; Teijón, J.M.; Blanco, M.D. Paclitaxel-loaded polyester nanoparticles prepared by spray-drying technology: In vitro bioactivity evaluation. *J. Microencapsul.* **2011**, *28*, 417–429. [CrossRef]
23. Ahuja, N.; Katare, O.P.; Singh, B. Studies on dissolution enhancement and mathematical modeling of drug release of a poorly water-soluble drug using water-soluble carriers. *Eur. J. Pharm. Biopharm.* **2007**, *65*, 26–38. [CrossRef]
24. Martínez, A.; Olmo, R.; Iglesias, I.; Teijón, J.M.; Blanco, M.D. Folate-targeted nanoparticles based on albumin and albumin/alginate mixtures as controlled release systems of tamoxifen: Synthesis and in vitro characterization. *Pharm. Res.* **2014**, *31*, 182–193. [CrossRef]
25. Barbosa, M.V.; Monteiro, L.O.F.; Carneiro, G.; Malagutti, A.R.; Vilela, J.M.C.; Andrade, M.S.; Oliveira, M.C.; Carvalho-Junior, A.D.; Leitea, E.A. Experimental design of a liposomal lipid system: A potential strategy for paclitaxel-based breast cancer treatment. *Colloids Surf. B* **2015**, *136*, 553–561. [CrossRef]
26. Rizk, N.; Christoforou, N.; Lee, S. Optimization of anti-cancer drugs and a targeting molecule on multifunctional gold nanoparticles. *Nanotechnology* **2016**, *27*, 185704. [CrossRef]
27. Pérez, E.; Fernández, A.; Olmo, R.; Teijón, J.M.; Blanco, M.D. pH and glutathion-responsive hydrogel for localized delivery of paclitaxel. *Colloids Surf. B* **2014**, *116*, 247–256. [CrossRef] [PubMed]
28. Martínez, A.; Benito-Miguel, M.; Iglesias, I.; Teijón, J.M.; Blanco, M.D. Tamoxifen-loaded thiolated alginate-albumin nanoparticles as antitumoral drug delivery systems. *J. Biomed. Mater. Res. Part A* **2012**, *100A*, 1467–1476. [CrossRef] [PubMed]
29. Fauzee, N.J.S.; Dong, Z.; Wang, Y. Taxanes: Promising anti-cancer drugs. *Asian Pac. J. Cancer Prev.* **2011**, *12*, 837–851. [PubMed]
30. Kundranda, M.N.; Niu, J. Albumin-bound paclitaxel in solid tumors: Clinical development and future directions. *Drug Des. Dev. Ther.* **2015**, *9*, 3767–3777. [CrossRef]
31. Sofias, A.M.; Dunne, M.; Storm, G.; Allen, C. The battle of "nano" paclitaxel. *Adv. Drug Deliv. Rev.* **2017**, *122*, 20–30. [CrossRef]
32. Xu, X.; Wang, L.; Xu, H.-Q.Q.; Huang, X.-E.E.; Qian, Y.-D.D.; Xiang, J. Clinical comparison between paclitaxel liposome (Lipusu(R)) and paclitaxel for treatment of patients with metastatic gastric cancer. *Asian Pac. J. Cancer Prev.* **2013**, *14*, 2591–2594. [CrossRef]
33. Madaan, A.; Singh, P.; Awasthi, A.; Verma, R.; Singh, A.T.; Jaggi, M.; Mishra, S.K.; Kulkarni, S.; Kulkarni, H. Efficiency and mechanism of intracellular paclitaxel delivery by novel nanopolymer-based tumor-targeted delivery system. NanoxelTM Clin. *Transl. Oncol.* **2013**, *15*, 26–32. [CrossRef]
34. Fernández, M.; Javaid, F.; Chudasama, V. Advances in targeting the folate receptor in the treatment/imaging of cancers. *Chem. Sci.* **2018**, *9*, 790–810. [CrossRef] [PubMed]
35. Chen, C.; Ke, J.; Zhou, X.E.; Yi, W.; Brunzelle, J.S.; Li, J.; Yong, E.L.; Xu, H.E.; Melcher, K. Structural basis for molecular recognition of folic acid by folate receptors. *Nature* **2013**, *500*, 486–489. [CrossRef]
36. Leamon, C.P.; Vlahov, I.R.; Reddy, J.A.; Vetzel, M.; Santhapuram, H.K.; You, F.; Bloomfield, A.; Dorton, R.; Nelson, M.; Kleindl, P.; et al. Folate–vinca alkaloid conjugates for cancer therapy: A structure–activity relationship. *Bioconjug. Chem.* **2014**, *25*, 560–568. [CrossRef]

37. Martínez, A.; Muñiz, E.; Teijón, C.; Iglesias, I.; Teijón, J.M.; Blanco, M.D. Targeting tamoxifen to breast cancer xenograft Tumours: Preclinical efficacy of folate-attached nanoparticles based on alginate-cysteine/disulphide-bond-reduced albumin. *Pharm. Res.* **2014**, *31*, 1264–1274. [CrossRef] [PubMed]
38. Papaioannou, L.; Angelopoulou, A.; Hatziantoniou, S.; Papadimitriou, M.; Apostolou, P.; Papasotiriou, I.; Avgoustakis, K. Folic acid-functionalized gold nanorods for controlled paclitaxel delivery: In vitro evaluation and cell studies. *AAPS PharmSciTech* **2019**, *20*, 13. [CrossRef] [PubMed]
39. Viéville, J.; Tanty, M.; Delsuc, M.A. Polydispersity index of polymers revealed by DOSY NMR. *J. Magn. Reson.* **2011**, *212*, 169–173. [CrossRef]
40. Muller, R.H.; Jacobs, C.; Kayser, O. Nanosuspensions as particulate drug formulations in therapy. Rationale for development and what we can expect for the future. *Adv. Drug Deliv. Rev.* **2001**, *47*, 3–19. [CrossRef]
41. Perez, E.; Martinez, A.; Teijon, C.; Olmo, R.; Teijón, J.M.; Blanco, M.D. Improved antitumor effect of paclitaxel administered in vivo as pH and glutathione-sensitive nanohydrogels. *Int. J. Pharm.* **2015**, *492*, 10–19. [CrossRef]
42. Paul, D.R. Elaborations on the Higuchi model for drug delivery. *Int. J. Pharm.* **2011**, *418*, 13–17. [CrossRef]
43. Machín, R.; Isasi, J.R.; Vélaz, I. Hydrogel matrices containing single and mixed natural cyclodextrins. Mechanisms of drug release. *Eur. Polym. J.* **2013**, *49*, 3912–3920. [CrossRef]
44. Korsmeyer, R.W.; Gurny, R.; Doelker, E.; Buri, P.; Peppas, N.A. Mechanisms of soluble release from porous hydrophilic polymers. *Int. J. Pharmaceut.* **1983**, *15*, 25–35. [CrossRef]
45. Siepmann, J.; Peppas, N.A. Higuchi equation: Derivation, applications, use and misuse. *Int. J. Pharmaceut.* **2011**, *418*, 6–12. [CrossRef]
46. Jhaveri, M.S.; Rait, A.S.; Chung, K.-N.; Trepel, J.B.; Chang, E.H. Antisense oligonucleotides targeted to the human α folate receptor inhibit breast cancer cell growth and sensitize the cells to doxorubicin treatment. *Mol. Cancer Ther.* **2004**, *3*, 1505–1512. [PubMed]
47. Chavez, K.J.; Garimella, S.V.; Lipkowitz, S. Triple negative breast cancer cell lines: One tool in the search for better treatment of triple negative breast cancer. *Breast Dis.* **2010**, *32*, 35–48. [CrossRef] [PubMed]
48. Gadducci, A.; Cosio, S. Neoadjuvant chemotherapy in locally advanced cervical cancer: Review of the literature and perspectives of clinical research. *Anticancer Res.* **2020**, *40*, 4819–4828. [CrossRef] [PubMed]
49. Banu, H.; Sethi, D.K.; Edgar, A.; Sheriff, A.; Rayees, N.; Renuka, N.; Faheem, S.M.; Premkumar, K.; Vasanthakumar, G. Doxorubicin loaded polymeric gold nanoparticles targeted to human folate receptor upon laser photothermal therapy potentiates chemotherapy in breast cancer cell lines. *J. Photochem. Photobiol. B Biol.* **2015**, *149*, 116–128. [CrossRef]
50. Wilhelm, C.; Billotey, C.; Roger, J.; Pons, J.N.; Bacri, J.C.; Gazeau, F. Intracellular uptake of anionic superparamagnetic nanoparticles as a function of their surface coating. *Biomaterials* **2003**, *24*, 1001–1011. [CrossRef]
51. Ke, C.Y.; Mathias, C.J.; Green, M.A. The folate receptor as a molecular target for tumor-selective radionuclide delivery. *Nucl. Med. Biol.* **2003**, *30*, 811–817. [CrossRef]

MDPI
St. Alban-Anlage 66
4052 Basel
Switzerland
Tel. +41 61 683 77 34
Fax +41 61 302 89 18
www.mdpi.com

Polymers Editorial Office
E-mail: polymers@mdpi.com
www.mdpi.com/journal/polymers

www.ingramcontent.com/pod-product-compliance
Lightning Source LLC
LaVergne TN
LVHW070423100526
838202LV00014B/1512